Theoretical Computer Science and
Discrete Mathematics

Theoretical Computer Science and Discrete Mathematics

Editors

Juan Alberto Rodríguez Velázquez
Alejandro Estrada-Moreno

MDPI • Basel • Beijing • Wuhan • Barcelona • Belgrade • Manchester • Tokyo • Cluj • Tianjin

Editors
Juan Alberto Rodríguez
Velázquez
Departament d'Enginyeria
Informàtica i Matemàtiques
Universitat Rovira i Virgili
Tarragona
Spain

Alejandro Estrada-Moreno
Departament d'Enginyeria
Informàtica i Matemàtiques
Universitat Rovira i Virgili
Tarragona
Spain

Editorial Office
MDPI
St. Alban-Anlage 66
4052 Basel, Switzerland

This is a reprint of articles from the Special Issue published online in the open access journal *Symmetry* (ISSN 2073-8994) (available at: www.mdpi.com/journal/symmetry/special_issues/ Computer_Science_Discrete_Mathematics).

For citation purposes, cite each article independently as indicated on the article page online and as indicated below:

LastName, A.A.; LastName, B.B.; LastName, C.C. Article Title. *Journal Name* **Year**, *Volume Number*, Page Range.

ISBN 978-3-0365-3177-9 (Hbk)
ISBN 978-3-0365-3176-2 (PDF)

© 2022 by the authors. Articles in this book are Open Access and distributed under the Creative Commons Attribution (CC BY) license, which allows users to download, copy and build upon published articles, as long as the author and publisher are properly credited, which ensures maximum dissemination and a wider impact of our publications.

The book as a whole is distributed by MDPI under the terms and conditions of the Creative Commons license CC BY-NC-ND.

Contents

About the Editors vii

Preface to "Theoretical Computer Science and Discrete Mathematics" ix

Abel Cabrera Martínez, Alejandro Estrada-Moreno and Juan A. Rodríguez-Velázquez
Secure w-Domination in Graphs
Reprinted from: *Symmetry* **2020**, *12*, 1948, doi:10.3390/sym12121948 .... **1**

Maria Bras-Amorós, Hebert Pérez-Rosés and José Miguel Serradilla-Merinero
Quasi-Ordinarization Transform of a Numerical Semigroup
Reprinted from: *Symmetry* **2021**, *13*, 1084, doi:10.3390/sym13061084 .... **13**

Jorge Martínez Carracedo and Adriana Suárez Corona
Cryptanalysis of a Group Key Establishment Protocol
Reprinted from: *Symmetry* **2021**, *13*, 332, doi:10.3390/sym13020332 .... **31**

Ana Almerich-Chulia, Abel Cabrera Martínez, Frank Angel Hernández Mira and Pedro Martin-Concepcion
From Total Roman Domination in Lexicographic Product Graphs to Strongly Total Roman Domination in Graphs
Reprinted from: *Symmetry* **2021**, *13*, 1282, doi:10.3390/sym13071282 .... **41**

Guangyan Xu, Zailin Guan, Lei Yue, Jabir Mumtaz and Jun Liang
Modeling and Optimization for Multi-Objective Nonidentical Parallel Machining Line Scheduling with a Jumping Process Operation Constraint
Reprinted from: *Symmetry* **2021**, *13*, 1521, doi:10.3390/sym13081521 .... **51**

Abel Cabrera Martínez and Juan A. Rodríguez-Velázquez
Total Domination in Rooted Product Graphs
Reprinted from: *Symmetry* **2020**, *12*, 1929, doi:10.3390/sym12111929 .... **75**

Rasyid Redha Mohd Tahir, Muhammad Asyraf Asbullah, Muhammad Rezal Kamel Ariffin and Zahari Mahad
Determination of a Good Indicator for Estimated Prime Factor and Its Modification in Fermat's Factoring Algorithm
Reprinted from: *Symmetry* **2021**, *13*, 735, doi:10.3390/sym13050735 .... **87**

Armando Maya-López, Fran Casino and Agusti Solanas
Improving Multivariate Microaggregation through Hamiltonian Paths and Optimal Univariate Microaggregation
Reprinted from: *Symmetry* **2021**, *13*, 916, doi:10.3390/sym13060916 .... **109**

Abel Cabrera Martínez, Suitberto Cabrera García, Andrés Carrión García and Angela María Grisales del Rio
On the Outer-Independent Roman Domination in Graphs
Reprinted from: *Symmetry* **2020**, *12*, 1846, doi:10.3390/sym12111846 .... **137**

Anna Bryniarska
The n-Pythagorean Fuzzy Sets
Reprinted from: *Symmetry* **2020**, *12*, 1772, doi:10.3390/sym12111772 .... **149**

Brandon Cortés-Caicedo, Laura Sofía Avellaneda-Gómez, Oscar Danilo Montoya, Lázaro Alvarado-Barrios and César Álvarez-Arroyo
An Improved Crow Search Algorithm Applied to the Phase Swapping Problem in Asymmetric Distribution Systems
Reprinted from: *Symmetry* **2021**, *13*, 1329, doi:10.3390/sym13081329 . . . . **159**

Oscar Danilo Montoya, Andres Arias-Londoño, Luis Fernando Grisales-Noreña, José Ángel Barrios and Harold R. Chamorro
Optimal Demand Reconfiguration in Three-Phase Distribution Grids Using an MI-Convex Model
Reprinted from: *Symmetry* **2021**, *13*, 1124, doi:10.3390/sym13071124 . . . . **179**

Abel Cabrera Martínez, Alejandro Estrada-Moreno and Juan A. Rodríguez-Velázquez
From the Quasi-Total Strong Differential to Quasi-Total Italian Domination in Graphs
Reprinted from: *Symmetry* **2021**, *13*, 1036, doi:10.3390/sym13061036 . . . . **195**

Osmani Tito-Corrioso, Miguel Angel Borges-Trenard, Mijail Borges-Quintana, Omar Rojas and Guillermo Sosa-Gómez
Study of Parameters in the Genetic Algorithm for the Attack on Block Ciphers
Reprinted from: *Symmetry* **2021**, *13*, 806, doi:10.3390/sym13050806 . . . . **209**

Martin Sotola, Pavel Marsalek, David Rybansky, Martin Fusek and Dusan Gabriel
Sensitivity Analysis of Key Formulations of Topology Optimization on an Example of Cantilever Bending Beam
Reprinted from: *Symmetry* **2021**, *13*, 712, doi:10.3390/sym13040712 . . . . **221**

About the Editors

Juan Alberto Rodríguez Velázquez

In 2021 Dr. Rodríguez-Velázquez received a Full Professor Accreditation, in the field of science, from ANECA, the Spanish National Agency for Quality Assessment and Accreditation. Since 2010, he has been an Associated Professor of Applied Mathematics at Rovira i Virgili University. Previously, he has been a lecturer at several Spanish universities, including the Polytechnic University of Catalonia, National Distance Education University, Open University of Catalonia, and Carlos III University of Madrid.

He achieved a Ph.D. Degree in Mathematical Science at the Polytechnic University of Catalonia in 1997. Since then, his main research has focused on graph theory. His current research interests include combinatorics, topology, computational complexity theory, and algorithmics, as well as applications of graph theory to complex network analysis. He has participated in several research projects, has co-authored over 125 journal articles, and has supervised 7 Ph.D. theses.

Alejandro Estrada-Moreno

In 2021, Dr. Estrada-Moreno received an Accreditation of Research granted by the Catalan University Quality Assurance Agency (AQU). Since February 2019, he has been a tenure eligible lecturer (similar to a full-time visiting professor) at Rovira i Virgili University. Previously, he worked as a postdoctoral researcher in the Internet Computing & Systems Optimization (ICSO) research group from the Open University of Catalonia.

He achieved a Ph.D. degree in Mathematical Science at the Universitat Rovira i Virgili in 2016. His current research interests focus on graph theory and combinatorics, computational complexity theory, algorithmics, analytics and operations research, and logistics and transportation. He has participated in several research projects and has co-authored 40 journal articles.

Preface to "Theoretical Computer Science and Discrete Mathematics"

This book includes 15 articles published in the Special Issue "lTheoretical Computer Science and Discrete Mathematics" of *Symmetry* (ISSN 2073-8994). This Special Issue is devoted to original and significant contributions to theoretical computer science and discrete mathematics. The aim was to bring together research papers linking different areas of discrete mathematics and theoretical computer science, as well as applications of discrete mathematics to other areas of science and technology. The Special Issue covers topics in discrete mathematics including (but not limited to) graph theory, cryptography, numerical semigroups, discrete optimization, algorithms, and complexity.

The response to our call for papers had the following statistics:
- 33 submissions;
- 16 publications;
- 17 rejections;
- 15 research articles, 1 correction.

We found the edition and selections of papers for this book very inspiring and rewarding. We thank the editorial staff and reviewers for their great efforts and help during the process.

Juan Alberto Rodríguez Velázquez, Alejandro Estrada-Moreno
Editors

Article
Secure w-Domination in Graphs

Abel Cabrera Martínez, Alejandro Estrada-Moreno * and Juan A. Rodríguez-Velázquez

Departament d'Enginyeria Informàtica i Matemàtiques, Universitat Rovira i Virgili, Av. Països Catalans 26, 43007 Tarragona, Spain; abel.cabrera@urv.cat (A.C.M.); juanalberto.rodriguez@urv.cat (J.A.R.-V.)
* Correspondence: alejandro.estrada@urv.cat

Received: 31 October 2020; Accepted: 23 November 2020; Published: 25 November 2020

Abstract: This paper introduces a general approach to the idea of protection of graphs, which encompasses the known variants of secure domination and introduces new ones. Specifically, we introduce the study of secure w-domination in graphs, where $w = (w_0, w_1, \ldots, w_l)$ is a vector of nonnegative integers such that $w_0 \geq 1$. The secure w-domination number is defined as follows. Let G be a graph and $N(v)$ the open neighborhood of $v \in V(G)$. We say that a function $f : V(G) \longrightarrow \{0, 1, \ldots, l\}$ is a w-dominating function if $f(N(v)) = \sum_{u \in N(v)} f(u) \geq w_i$ for every vertex v with $f(v) = i$. The weight of f is defined to be $\omega(f) = \sum_{v \in V(G)} f(v)$. Given a w-dominating function f and any pair of adjacent vertices $v, u \in V(G)$ with $f(v) = 0$ and $f(u) > 0$, the function $f_{u \to v}$ is defined by $f_{u \to v}(v) = 1$, $f_{u \to v}(u) = f(u) - 1$ and $f_{u \to v}(x) = f(x)$ for every $x \in V(G) \setminus \{u, v\}$. We say that a w-dominating function f is a secure w-dominating function if for every v with $f(v) = 0$, there exists $u \in N(v)$ such that $f(u) > 0$ and $f_{u \to v}$ is a w-dominating function as well. The secure w-domination number of G, denoted by $\gamma_w^s(G)$, is the minimum weight among all secure w-dominating functions. This paper provides fundamental results on $\gamma_w^s(G)$ and raises the challenge of conducting a detailed study of the topic.

Keywords: secure domination; secure Italian domination; weak roman domination; w-domination

1. Introduction

Let $\mathbb{Z}^+ = \{1, 2, 3, \ldots\}$ and $\mathbb{N} = \mathbb{Z}^+ \cup \{0\}$ be the sets of positive and nonnegative integers, respectively. Let G be a graph, $l \in \mathbb{Z}^+$ and $f : V(G) \longrightarrow \{0, \ldots, l\}$ a function. Let $V_i = \{v \in V(G) : f(v) = i\}$ for every $i \in \{0, \ldots, l\}$. We identify f with the subsets $V_0, \ldots, V_l$ associated with it, and thus we use the unified notation $f(V_0, \ldots, V_l)$ for the function and these associated subsets. The weight of f is defined to be

$$\omega(f) = f(V(G)) = \sum_{i=1}^{l} i |V_i|.$$

Let $w = (w_0, \ldots, w_l) \in \mathbb{Z}^+ \times \mathbb{N}^l$ such that $w_0 \geq 1$. As defined in [1], a function $f(V_0, \ldots, V_l)$ is a w-*dominating function* if $f(N(v)) \geq w_i$ for every $v \in V_i$. The w-*domination number* of G, denoted by $\gamma_w(G)$, is the minimum weight among all w-dominating functions. For simplicity, a w-dominating function f of weight $\omega(f) = \gamma_w(G)$ is called a $\gamma_w(G)$-function. For fundamental results on the w-domination number of a graph, we refer the interested readers to the paper by Cabrera et al. [1], where the theory of w-domination in graphs is introduced.

The definition of w-domination number encompasses the definition of several well-known domination parameters and introduces new ones. For instance, we highlight the following particular cases of known domination parameters that we define here in terms of w-domination: the domination number $\gamma(G) = \gamma_{(1,0)}(G) = \gamma_{(1,0,\ldots,0)}(G)$, the total domination number $\gamma_t(G) = \gamma_{(1,1)}(G) = \gamma_{(1,\ldots,1)}(G)$, the k-domination number $\gamma_k(G) = \gamma_{(k,0)}(G)$, the k-tuple domination number $\gamma_{\times k}(G) = \gamma_{(k,k-1)}(G)$, the k-tuple total domination number $\gamma_{\times k,t}(G) = \gamma_{(k,k)}(G)$, the Italian domination number

$\gamma_I(G) = \gamma_{(2,0,0)}(G)$, the total Italian domination number $\gamma_{tI}(G) = \gamma_{(2,1,1)}(G)$, and the $\{k\}$-domination number $\gamma_{\{k\}}(G) = \gamma_{(k,k-1,...,0)}(G)$. In these definitions, the appropriate restrictions on the minimum degree of G are assumed, when needed.

For any function $f(V_0, \ldots, V_l)$ and any pair of adjacent vertices $v \in V_0$ and $u \in V(G) \setminus V_0$, the function $f_{u \to v}$ is defined by $f_{u \to v}(v) = 1$, $f_{u \to v}(u) = f(u) - 1$ and $f_{u \to v}(x) = f(x)$ whenever $x \in V(G) \setminus \{u,v\}$.

We say that a w-dominating function $f(V_0, \ldots, V_l)$ is a *secure w-dominating function* if for every $v \in V_0$ there exists $u \in N(v) \setminus V_0$ such that $f_{u \to v}$ is a w-dominating function as well. The *secure w-domination number* of G, denoted by $\gamma_w^s(G)$, is the minimum weight among all secure w-dominating functions. For simplicity, a secure w-dominating function f of weight $\omega(f) = \gamma_w^s(G)$ is called a $\gamma_w^s(G)$-function. This approach to the theory of secure domination covers the different versions of secure domination known so far. For instance, we emphasize the following cases of known parameters that we define here in terms of secure w-domination.

- The *secure domination number* of G is defined to be $\gamma_s(G) = \gamma_{(1,0)}^s(G)$. In this case, for any secure $(1,0)$-dominating function $f(V_0, V_1)$, the set V_1 is known as a *secure dominating set*. This concept was introduced by Cockayne et al. [2] and studied further in several papers (e.g., [3–9]).
- The *secure total domination number* of a graph G of minimum degree at least one is defined to be $\gamma_{st}(G) = \gamma_{(1,1)}^s(G)$. In this case, for any secure $(1,1)$-dominating function $f(V_0, V_1)$, the set V_1 is known as a *secure total dominating set* of G. This concept was introduced by Benecke et al. [10] and studied further in several papers (e.g., [7,11–14]).
- The *weak Roman domination number* of a graph G is defined to be $\gamma_r(G) = \gamma_{(1,0,0)}^s(G)$. This concept was introduced by Henning and Hedetniemi [15] and studied further in several papers (e.g., [5,6,16,17]).
- The *total weak Roman domination number* of a graph G of minimum degree at least one is defined to be $\gamma_{tr}(G) = \gamma_{(1,1,1)}^s(G)$. This concept was introduced by Cabrera et al. in [12] and studied further in [18].
- The *secure Italian domination number* of G is defined to be $\gamma_I^s(G) = \gamma_{(2,0,0)}^s(G)$. This parameter was introduced by Dettlaff et al. [19].

For the graphs shown in Figure 1, we have the following:

- $\gamma_{(1,1)}^s(G_1) = \gamma_{(2,0)}^s(G_1) = \gamma_{(2,1)}^s(G_1) = \gamma_{(2,0)}(G_1) = \gamma_{(2,1)}G_1) = \gamma_{(1,1,0)}^s(G_1) = \gamma_{(1,1,1)}^s(G_1) = \gamma_{(2,0,0)}^s(G_1) = \gamma_{(2,1,0)}^s(G_1) = \gamma_{(2,0,0)}(G_1) = \gamma_{(2,1,0)}(G_1) = \gamma_{(2,2,0)}(G_1) = \gamma_{(2,2,1)}(G_1) = \gamma_{(2,2,2)}(G_1) = 4$ and $\gamma_{(2,2)}^s(G_1) = \gamma_{(2,2)}(G_1) = \gamma_{(2,2,0)}^s(G_1) = \gamma_{(2,2,1)}^s(G_1) = \gamma_{(2,2,2)}^s(G_1) = \gamma_{(3,0,0)}^s(G_1) = \gamma_{(3,1,0)}^s(G_1) = \gamma_{(3,1,1)}^s(G_1) = \gamma_{(3,2,0)}^s(G_1) = \gamma_{(3,2,1)}^s(G_1) = \gamma_{(3,2,2)}^s(G_1) = \gamma_{(3,0,0)}(G_1) = \gamma_{(3,1,0)}(G_1) = \gamma_{(3,1,1)}(G_1) = \gamma_{(3,2,0)}(G_1) = \gamma_{(3,2,1)}(G_1) = \gamma_{(3,2,2)}(G_1) = 6$.
- $\gamma_{(1,1)}^s(G_2) = \gamma_{(1,1,0)}^s(G_2) = \gamma_{(1,1,1)}^s(G_2) = \gamma_{(2,2,0)}(G_2) = \gamma_{(2,2,1)}(G_2) = \gamma_{(2,2,2)}(G_2) = 3$.
- $\gamma_{(1,1)}^s(G_3) = \gamma_{(1,1,0)}^s(G_3) = \gamma_{(1,1,1)}^s(G_3) = \gamma_{(2,1,0)}(G_3) = \gamma_{(3,0,0)}(G_3) = 3 < 4 = \gamma_{(2,0,0)}^s(G_3) = \gamma_{(2,1,0)}^s(G_3) = \gamma_{(3,1,0)}^s(G_3) = \gamma_{(2,2,0)}(G_3) = \gamma_{(2,2,1)}(G_3) = \gamma_{(2,2,2)}(G_3) = \gamma_{(3,2,0)}(G_3) < 5 = \gamma_{(2,2,0)}^s(G_3) = \gamma_{(3,2,0)}^s(G_3) = \gamma_{(2,2,1)}^s(G_3) = \gamma_{(2,2,2)}^s(G_3) = \gamma_{(3,1,1)}^s(G_3) = \gamma_{(3,2,1)}^s(G_3) = \gamma_{(3,2,1)}(G_3) = \gamma_{(3,2,2)}(G_3) < 6 = \gamma_{(3,2,2)}^s(G_3)$.

This paper is devoted to providing general results on secure w-domination. We assume that the reader is familiar with the basic concepts, notation, and terminology of domination in graph. If this is not the case, we suggest the textbooks [20,21]. For the remainder of the paper, definitions are introduced whenever a concept is needed.

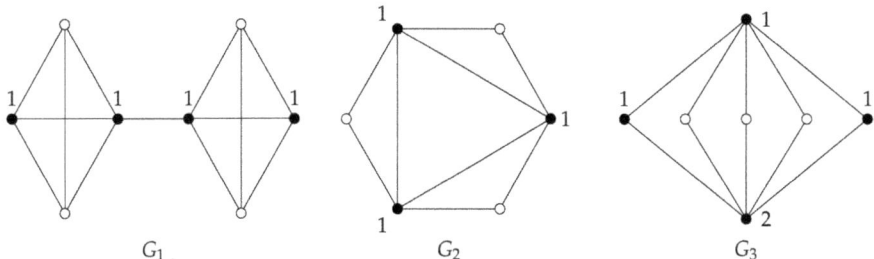

Figure 1. The labels of black-colored vertices describe the positive weights of a $\gamma^s_{(2,1,0)}(G_1)$-function, a $\gamma^s_{(1,1,1)}(G_2)$-function, and a $\gamma^s_{(2,2,2)}(G_3)$-function, respectively.

2. General Results on Secure w-Domination

Given a w-dominating function $f(V_0, \ldots, V_l)$, we introduce the following notation.

- Given $v \in V_0$, we define $M_f(v) = \{u \in V(G) \setminus V_0 : f_{u \to v} \text{ as a } w\text{-dominating function}\}$.
- $M_f(G) = \bigcup_{v \in V_0} M_f(v)$.
- Given $u \in M_f(G)$, we define $D_f(u) = \{v \in V_0 : u \in M_f(v)\}$.
- Given $u \in M_f(G)$, we define $T_f(u) = \{v \in V_0 : u \in M_f(v) \text{ and } f(N(v)) = w_0\}$.

Obviously, if f is a secure w-dominating function, then $M_f(v) \neq \emptyset$ for every $v \in V_0$.

Lemma 1. *Let f be a secure w-dominating function on a graph G, and let $u \in M_f(G)$. If $T_f(u) \neq \emptyset$, then each vertex belonging to $T_f(u)$ is adjacent to every vertex in $D_f(u)$ and, in particular, $G[T_f(u)]$ is a clique.*

Proof. Since $T_f(u) \subseteq D_f(u)$, we only need to suppose the existence of two non-adjacent vertices $v \in T_f(u)$ and $v' \in D_f(u)$ with $v \neq v'$. In such a case, $f_{u \to v'}(N(v)) < w_0$, which is a contradiction. Therefore, the result follows. □

Remark 1 ([1]). *Let G be a graph of minimum degree δ and let $w = (w_0, w_1, \ldots, w_l) \in \mathbb{Z}^+ \times \mathbb{N}^l$. If $w_0 \geq w_1 \geq \cdots \geq w_l$, then there exists a w-dominating function on G if and only if $w_l \leq l\delta$.*

Throughout this section, we repeatedly apply, without explicit mention, the following necessary and sufficient condition for the existence of a secure w-dominating function on G.

Remark 2. *Let G be a graph of minimum degree δ and let $w = (w_0, w_1, \ldots, w_l) \in \mathbb{Z}^+ \times \mathbb{N}^l$. If $w_0 \geq w_1 \geq \cdots \geq w_l$, then there exists a secure w-dominating function on G if and only if $w_l \leq l\delta$.*

Proof. If f is a secure w-dominating function on G, then f is a w-dominating function, and by Remark 1 we conclude that $w_l \leq l\delta$.

Conversely, if $w_l \leq l\delta$, then the function f, defined by $f(v) = l$ for every $v \in V(G)$, is a secure w-dominating function. Therefore, the result follows. □

It was shown by Cabrera et al. [1] that the w-domination numbers satisfy a certain monotonicity. Given two integer vectors $w = (w_0, \ldots, w_l)$ and $w' = (w'_0, \ldots, w'_l)$, we say that $w' \prec w$ if $w'_i \leq w_i$ for every $i \in \{0, \ldots, l\}$. With this notation in mind, we can state the next remark which is a direct consequence of the definition of w-dominating function.

Remark 3. [1] Let G be a graph of minimum degree δ and let $w = (w_0, \ldots, w_l), w' = (w'_0, \ldots, w'_l) \in \mathbb{Z}^+ \times \mathbb{N}^l$ such that $w_i \geq w_{i+1}$ and $w'_i \geq w'_{i+1}$ for every $i \in \{0, \ldots, l-1\}$. If $w' \prec w$ and $w_l \leq l\delta$, then every w-dominating function is a w'-dominating function and, as a consequence,

$$\gamma_{w'}(G) \leq \gamma_w(G).$$

The monotonicity also holds for the case of secure w-domination.

Remark 4. Let G be a graph of minimum degree δ and let $w = (w_0, \ldots, w_l), w' = (w'_0, \ldots, w'_l) \in \mathbb{Z}^+ \times \mathbb{N}^l$ such that $w_i \geq w_{i+1}$ and $w'_i \geq w'_{i+1}$ for every $i \in \{0, \ldots, l-1\}$. If $w' \prec w$ and $w_l \leq l\delta$, then every secure w-dominating function is a secure w'-dominating function and, as a consequence,

$$\gamma^s_{w'}(G) \leq \gamma^s_w(G).$$

Proof. For any $\gamma^s_w(G)$-function f and any $v \in V(G)$ with $f(v) = 0$, there exists $u \in M_f(v)$. Since f and $f_{u \to v}$ are w-dominating functions, by Remark 3, we conclude that, if $w' \prec w$ and $w_l \leq l\delta$, then both f and $f_{u \to v}$ are w'-dominating functions. Therefore, f is a secure w'-dominating function and, as a consequence, $\gamma^s_{w'}(G) \leq \omega(f) = \gamma^s_w(G)$. □

From the following equality chain, we obtain examples of equalities in Remark 4. Graph G_1 is illustrated in Figure 1.

$$\gamma^s_{(3,0,0)}(G_1) = \gamma^s_{(3,1,0)}(G_1) = \gamma^s_{(3,2,0)}(G_1) = \gamma^s_{(3,2,1)}(G_1) = \gamma^s_{(3,2,2)}(G_1).$$

Theorem 1. Let G be a graph of minimum degree δ, and let $w = (w_0, \ldots, w_l) \in \mathbb{Z}^+ \times \mathbb{N}^l$ such that $w_i \geq w_{i+1}$ for every $i \in \{0, \ldots, l-1\}$. If $l\delta \geq w_l$, then the following statements hold.

(i) $\gamma_w(G) \leq \gamma^s_w(G)$.

(ii) If $k \in \mathbb{Z}^+$, then $\gamma_{(k+1,k=w_1,\ldots,w_l)}(G) \leq \gamma^s_{(k,k=w_1,\ldots,w_l)}(G)$.

Proof. Since every secure w-dominating function on G is a w-dominating function on G, (i) follows.

Let $f(V_0, \ldots, V_l)$ be a $\gamma^s_{(k,k=w_1,\ldots,w_l)}(G)$-function. Since f is a $(k, k = w_1, \ldots, w_l)$-dominating function, $f(N(v)) \geq w_i$ for every $v \in V_i$ with $i \in \{1, \ldots, l\}$ and $w_1 = k$. If $V_0 = \emptyset$, then f is a $(k+1, k = w_1, \ldots, w_l)$-dominating function, which implies that $\gamma_{(k+1,k=w_1,\ldots,w_l)}(G) \leq \omega(f) = \gamma^s_{(k,k=w_1,\ldots,w_l)}(G)$. Assume $V_0 \neq \emptyset$. Let $v \in V_0$ and $u \in M_f(v)$. If $f(N(v)) = k$, then $f_{u \to v}(N(v)) = f(N(v)) - 1 = k - 1$, which is a contradiction. Thus, $f(N(v)) \geq k + 1$, which implies that f is a $(k+1, k = w_1, \ldots, w_l)$-dominating function. Therefore, $\gamma_{(k+1,k=w_1,\ldots,w_l)}(G) \leq \omega(f) = \gamma^s_{(k,k=w_1,\ldots,w_l)}(G)$, and (ii) follows. □

The inequalities above are tight. For instance, for any integers $n, n' \geq 4$, we have that $\gamma_{(2,2,2)}(K_n + N_{n'}) = \gamma^s_{(2,2,2)}(K_n + N_{n'}) = 3$ and $\gamma_{(3,2,2)}(K_{2,n}) = \gamma^s_{(2,2,2)}(K_{2,n}) = 5$.

Corollary 1. Let G be a graph of minimum degree δ and order n. Let $w = (w_0, \ldots, w_l) \in \mathbb{Z}^+ \times \mathbb{N}^l$ such that $w_i \geq w_{i+1}$ for every $i \in \{0, \ldots, l-1\}$ and $l\delta \geq w_l$. The following statements hold.

(i) If $n > w_0$, then $\gamma^s_w(G) \geq w_0$.

(ii) If $n > w_0 = w_1$, then $\gamma^s_w(G) \geq w_0 + 1$.

Proof. Assume $n > w_0$. By Theorem 1, we have that $\gamma^s_w(G) \geq \gamma_w(G)$. Now, if $\gamma_w(G) \leq w_0 - 1 < n - 1$, then for any $\gamma_w(G)$-function f there exists at least one vertex $x \in V(G)$ such that $f(x) = 0$ and $f(N(x)) \leq \omega(f) < w_0$, which is a contradiction. Thus, $\gamma^s_w(G) \geq \gamma_w(G) \geq w_0$.

Analogously, if $w_0 = w_1$, then Theorem 1 leads to $\gamma_w^s(G) \geq \gamma_{(w_0+1,w_1,\ldots,w_l)}(G)$. In this case, if $\gamma_{(w_0+1,w_1,\ldots,w_l)}(G) \leq w_0 < n$, then for any $\gamma_{(w_0+1,w_1,\ldots,w_l)}(G)$-function f there exists at least one vertex $x \in V(G)$ such that $f(x) = 0$ and $f(N(x)) \leq \omega(f) < w_0 + 1$, which is a contradiction. Therefore, $\gamma_w^s(G) \geq \gamma_{(w_0+1,w_1,\ldots,w_l)}(G) \geq w_0 + 1$. □

As the following result shows, the bounds above are tight.

Proposition 1. *For any integer n and any $w = (w_0, \ldots, w_l) \in \mathbb{Z}^+ \times \mathbb{N}^l$ such that $w_l \leq \cdots \leq w_0 < n$,*

$$\gamma_w^s(K_n) = \begin{cases} w_0 + 1 & \text{if } w_0 = w_1, \\ w_0 & \text{otherwise.} \end{cases}$$

Proof. Assume $n > w_0$. Let $S \subseteq V(K_n)$ such that $|S| = w_0 + 1$ if $w_0 = w_1$ and $|S| = w_0$ otherwise. In both cases, the function $f(V_0, \ldots, V_l)$, defined by $V_1 = S$, $V_0 = V(G) \setminus V_1$ and $V_j = \varnothing$ whenever $j \notin \{0, 1\}$, is a secure w-dominating function. Hence, $\gamma_w^s(K_n) \leq \omega(f) = |S|$. Therefore, by Corollary 1 the result follows. □

Theorem 2. *Let G be a graph of minimum degree δ, and let $w = (w_0, \ldots, w_l), w' = (w'_0, \ldots, w'_l) \in \mathbb{Z}^+ \times \mathbb{N}^l$ such that $l\delta \geq w_l$, $w_i \geq w_{i+1}$ and $w'_i \geq w'_{i+1}$ for every $i \in \{0, \ldots, l-1\}$. If $w_i \geq w'_{i-1} - 1$ for every $i \in \{1, \ldots, l\}$, and $\max\{w_j - 1, 0\} \geq w'_j$ for every $j \in \{0, \ldots, l\}$, then*

$$\gamma_{w'}^s(G) \leq \gamma_w(G).$$

Proof. Assume that $w_i \geq w'_{i-1} - 1$ for every $i \in \{1, \ldots, l\}$ and $\max\{w_j - 1, 0\} \geq w'_j$ for every $j \in \{0, \ldots, l\}$. Let $f(V_0, \ldots, V_l)$ be a $\gamma_w(G)$-function. We claim that f is a secure w'-dominating function. Since $f(N(x)) \geq w_i \geq w'_i$ for every $x \in V_i$ with $i \in \{0, \ldots, l\}$, we deduce that f is a w'-dominating function. Now, let $v \in V_0$ and $u \in N(v) \cap V_i$ with $i \in \{1, \ldots, l\}$. We differentiate the following cases for $x \in V(G)$.

Case 1. $x = v$. In this case, $f_{u \to v}(v) = 1$ and $f_{u \to v}(N(v)) = f(N(v)) - 1 \geq w_0 - 1 \geq \max\{w_1 - 1, 0\} \geq w'_1$.

Case 2. $x = u$. In this case, $f_{u \to v}(u) = f(u) - 1 = i - 1$ and $f_{u \to v}(N(u)) = f(N(u)) + 1 \geq w_i + 1 \geq w'_{i-1}$.

Case 3. $x \in V(G) \setminus \{u, v\}$. Assume $x \in V_j$. Notice that $f_{u \to v}(x) = f(x) = j$. Now, if $x \notin N(u)$ or $x \in N(u) \cap N(v)$, then $f_{u \to v}(N(x)) = f(N(x)) \geq w_j \geq w'_j$, while if $x \in N(u) \setminus N[v]$, then $f_{u \to v}(N(x)) = f(N(x)) - 1 \geq \max\{w_j - 1, 0\} \geq w'_j$.

According to the three cases above, $f_{u \to v}$ is a w'-dominating function. Therefore, f is a secure w'-dominating function, and so $\gamma_{w'}^s(G) \leq \omega(f) = \gamma_w(G)$. □

The inequality above is tight. For instance, $\gamma_{(1,1,1)}^s(K_{n,n'}) = \gamma_{(2,2,2)}(K_{n,n'}) = 4$ for $n, n' \geq 4$.

From Theorems 1 and 2, we derive the next known inequality chain, where G has minimum degree $\delta \geq 1$, except in the last inequality in which $\delta \geq 2$.

$$\gamma_s(G) \leq \gamma_2(G) \leq \gamma_{\times 2}(G) \leq \gamma_{st}(G) \leq \gamma_{\times 2,t}(G).$$

The following result is a particular case of Theorem 2.

Corollary 2. *Let G be a graph of minimum degree δ, and let $w = (w_0, \ldots, w_l) \in \mathbb{Z}^+ \times \mathbb{N}^l$ and $\mathbf{1} = (1, \ldots, 1)$. If $0 \leq w_{j-1} - w_j \leq 2$ for every $j \in \{1, \ldots, i\}$, where $1 \leq i \leq l$ and $l\delta \geq w_l + 1$, then*

$$\gamma_{(w_0,\ldots,w_i,0,\ldots,0)}^s(G) \leq \gamma_{(w_0+1,\ldots,w_i+1,0,\ldots,0)}(G) \leq \gamma_{w+1}(G).$$

For Graph G_2 illustrated in Figure 1, we have that $\gamma^s_{(1,1)}(G_2) = \gamma^s_{(1,1,0)}(G_2) = \gamma_{(2,2,0)}(G_2) = \gamma^s_{(1,1,1)}(G_2) = \gamma_{(2,2,2)}(G_2) = 3$. Notice that $\gamma^s_w(G_2) = \gamma_{w+1}(G_2)$ for $w = 1 = (1,1,1)$.

Next, we show a class of graphs where $\gamma_w(G) = \gamma_{w+1}(G)$. To this end, we need to introduce some additional notation and terminology. Given the two Graphs G_1 and G_2, the *corona product graph* $G_1 \odot G_2$ is the graph obtained from G_1 and G_2, by taking one copy of G_1 and $|V(G_1)|$ copies of G_2 and joining by an edge every vertex from the ith copy of G_2 with the ith vertex of G_1. For every $x \in V(G_1)$, the copy of G_2 in $G_1 \odot G_2$ associated to x is denoted by $G_{2,x}$.

Theorem 3 ([1]). *Let $G_1 \odot G_2$ be a corona graph where G_1 does not have isolated vertices, and let $w = (w_0, \ldots, w_l) \in \mathbb{Z}^+ \times \mathbb{N}^l$. If $l \geq w_0 \geq \cdots \geq w_l$ and $|V(G_2)| \geq w_0$, then*

$$\gamma_w(G_1 \odot G_2) = w_0|V(G_1)|.$$

From the result above, we deduce that under certain additional restrictions on G_2 and w we can obtain $\gamma^s_w(G_1 \odot G_2) = \gamma_{w+1}(G_1 \odot G_2)$.

Theorem 4. *Let $G_1 \odot G_2$ be a corona graph, where G_1 does not have isolated vertices and G_2 is a triangle-free graph. Let $w = (w_0, \ldots, w_l) \in \mathbb{Z}^+ \times \mathbb{N}^l$ such that $l - 1 \geq w_0 \geq \cdots \geq w_l$. If $|V(G_2)| \geq w_0 + 2$, then*

$$\gamma^s_w(G_1 \odot G_2) = (w_0 + 1)|V(G_1)| = \gamma_{w+1}(G_1 \odot G_2).$$

Proof. Since G_1 does not have isolated vertices, the upper bound $\gamma^s_w(G_1 \odot G_2) \leq (w_0 + 1)|V(G_1)|$ is straightforward, as the function f, defined by $f(x) = w_0 + 1$ for every $x \in V(G_1)$ and $f(x) = 0$ for the remaining vertices of $G_1 \odot G_2$, is a secure w-dominating function.

On the other hand, let $f(V_0, \ldots, V_l)$ be a $\gamma^s_w(G_1 \odot G_2)$-function and suppose that there exists $x \in V(G_1)$ such that $f(V(G_{2,x})) + f(x) \leq w_0$. Since $|V(G_{2,x})| \geq w_0 + 2$, there exist at least two different vertices $u, v \in V(G_{2,x}) \cap V_0$. Hence, $f(N(u)) = f(N(v)) = w_0$, which implies that u and v are adjacent and, since G_2 is a triangle-free graph, $f(x) = w_0$ and $f(y) = 0$ for every $y \in V(G_{2,x})$. Thus, by Lemma 1, we conclude that $G_{2,x}$ is a clique, which is a contradiction as $|V(G_2)| \geq 3$ and G_2 is a triangle-free graph. This implies that $f(V(G_{2,x})) + f(x) \geq w_0 + 1$ for every $x \in V(G_1)$, and so $\gamma^s_w(G_1 \odot G_2) = \omega(f) \geq (w_0 + 1)|V(G_1)|$.

Therefore, $\gamma^s_w(G_1 \odot G_2) = (w_0 + 1)|V(G_1)|$, and by Theorem 3 we conclude that $\gamma_{w+1}(G_1 \odot G_2) = (w_0 + 1)|V(G_1)|$, which completes the proof. □

Theorem 5. *Let G be a graph of minimum degree δ and $l \geq 2$ an integer. For any $(w_0, \ldots, w_{l-1}) \in \mathbb{Z}^+ \times \mathbb{N}^{l-1}$ with $w_0 \geq \cdots \geq w_{l-1}$ and $l\delta \geq w_{l-1}$,*

$$\gamma^s_{(w_0,\ldots,w_{l-1},w_l=w_{l-1})}(G) \leq \gamma_{(w_0,\ldots,w_{l-1})}(G) + \gamma(G).$$

Proof. Let $f(V_0, \ldots, V_{l-1})$ be a $\gamma_{(w_0,\ldots,w_{l-1})}(G)$-function and S a $\gamma(G)$-set. We define a function $g(W_0, \ldots, W_l)$ as follows. Let $W_l = V_{l-1} \cap S$, $W_0 = V_0 \setminus S$, and $W_i = (V_{i-1} \cap S) \cup (V_i \setminus S)$ for every $i \in \{1, \ldots, l-1\}$.

We claim that g is a secure $(w_0, \ldots, w_{l-1}, w_l = w_{l-1})$-dominating function. First, we observe that, if $x \in W_i \cap S$ with $i \in \{1, \ldots, l\}$, then $x \in V_{i-1}$ and $g(N(x)) \geq f(N(x)) \geq w_{i-1} \geq w_i$. Moreover, if $x \in W_i \setminus S$ with $i \in \{0, \ldots, l-1\}$, then $x \in V_i$ and $g(N(x)) \geq f(N(x)) \geq w_i$. Hence, g is a $(w_0, \ldots, w_{l-1}, w_l = w_{l-1})$-dominating function.

Now, let $v \in W_0 = V_0 \setminus S$. Notice that there exists a vertex $u \in N(v) \cap V_{i-1} \cap S$ with $i \in \{1, \ldots, l\}$. Hence, $u \in N(v) \cap W_i$. We differentiate the following cases for $x \in V(G)$.

Case 1. $x = v$. In this case, $g_{u \to v}(v) = 1$ and, as $N(v) \cap S \neq \emptyset$, we obtain that $g_{u \to v}(N(v)) = g(N(v)) - 1 \geq f(N(v)) \geq w_0 \geq w_1$.

Case 2. $x = u$. In this case, $g_{u \to v}(u) = g(u) - 1 = i - 1$ and $g_{u \to v}(N(u)) = g(N(u)) + 1 \geq f(N(u)) + 1 \geq w_{i-1} + 1 > w_{i-1}$.

Case 3. $x \in V(G) \setminus \{u, v\}$. Assume $x \in W_j$. Notice that $g_{u \to v}(x) = g(x) = j$. If $x \notin N(u)$ or $x \in N(u) \cap N(v)$, then $g_{u \to v}(N(x)) = g(N(x)) \geq f(N(x)) \geq w_j$.

Moreover, if $x \in (N(u) \setminus N[v]) \cap S$, then $x \in V_{j-1}$ and so $g_{u \to v}(N(x)) = g(N(x)) - 1 \geq f(N(x)) \geq w_{j-1} \geq w_j$. Finally, if $x \in (N(u) \setminus N[v]) \setminus S$, then $x \in V_j$ and therefore $g_{u \to v}(N(x)) = g(N(x)) - 1 \geq f(N(x)) \geq w_j$.

According to the three cases above, $g_{u \to v}$ is a $(w_0, \ldots, w_{l-1}, w_l = w_{l-1})$-dominating function. Therefore, f is a secure $(w_0, \ldots, w_{l-1}, w_l = w_{l-1})$-dominating function, and so $\gamma^s_{(w_0, \ldots, w_{l-1}, w_l = w_{l-1})}(G) \leq w(g) \leq w(f) + |S| = \gamma^s_{(w_0, \ldots, w_{l-1})}(G) + \gamma(G)$. □

From Theorem 5, we derive the next known inequalities, which are tight.

Corollary 3. *For a graph G, the following statements hold.*

- *Ref. [15] $\gamma_r(G) \leq 2\gamma(G)$.*

- *Ref. [12] $\gamma_{tr}(G) \leq \gamma_t(G) + \gamma(G)$, where G has minimum degree at least one.*

- *Ref. [19] $\gamma^s_1(G) \leq \gamma_2(G) + \gamma(G)$.*

To establish the following result, we need to define the following parameter.

$$v^s_{(w_0, \ldots, w_l)}(G) = \max\{|V_0| : f(V_0, \ldots, V_l) \text{ is a } \gamma^s_{(w_0, \ldots, w_l)}(G)\text{-function.}\}$$

In particular, for $l = 1$ and a graph G of order n, we have that $v^s_{(w_0, w_1)}(G) = n - \gamma^s_{(w_0, w_1)}(G)$.

Theorem 6. *Let G be a graph of minimum degree δ and order n. The following statements hold for any $(w_0, \ldots, w_l) \in \mathbb{Z}^+ \times \mathbb{N}^l$ with $w_0 \geq \cdots \geq w_l$.*

(i) *If there exists $i \in \{1, \ldots, l-1\}$ such that $i\delta \geq w_i$, then $\gamma^s_{(w_0, \ldots, w_l)}(G) \leq \gamma^s_{(w_0, \ldots, w_i)}(G)$.*

(ii) *If $l \geq i+1 > w_0$, then $\gamma^s_{(w_0, \ldots, w_i, 0, \ldots, 0)}(G) \leq (i+1)\gamma(G)$.*

(iii) *Let $k, i \in \mathbb{Z}^+$ such that $l \geq ki$, and let $(w'_0, w'_1, \ldots, w'_i) \in \mathbb{Z}^+ \times \mathbb{N}^i$. If $i\delta \geq w'_i$ and $w_{kj} = kw'_j$ for every $j \in \{0, 1, \ldots, i\}$, then $\gamma^s_{(w_0, \ldots, w_l)}(G) \leq k\gamma^s_{(w'_0, \ldots, w'_i)}(G)$.*

(iv) *Let $k \in \mathbb{Z}^+$ and $\beta_1, \ldots, \beta_k \in \mathbb{Z}^+$. If $l\delta \geq k + w_l > k$ and $w_0 + k \geq \beta_1 \geq \cdots \geq \beta_k \geq w_l + k$, then $\gamma^s_{(w_0+k, \beta_1, \ldots, \beta_k, w_l+k, \ldots, w_l+k)}(G) \leq \gamma^s_{(w_0, \ldots, w_l)}(G) + k(n - v^s_{(w_0, \ldots, w_l)}(G))$.*

(v) *If $l\delta \geq w_l \geq l \geq 2$, then $\gamma^s_{(w_0, \ldots, w_l)}(G) \leq l\gamma^s_{(w_0 - l + 1, w_l - l + 1)}(G)$.*

Proof. If there exists $i \in \{1, \ldots, l-1\}$ such that $i\delta \geq w_i$, then for any $\gamma^s_{(w_0, \ldots, w_i)}(G)$-function $f(V_0, \ldots, V_i)$ we define a secure $(w_0, \ldots, w_l)$-dominating function $g(W_0, \ldots, W_l)$ by $W_j = V_j$ for every $j \in \{0, \ldots, i\}$ and $W_j = \emptyset$ for every $j \in \{i+1, \ldots, l\}$. Hence, $\gamma^s_{(w_0, \ldots, w_l)}(G) \leq w(g) = w(f) = \gamma^s_{(w_0, \ldots, w_i)}(G)$. Therefore, (i) follows.

Now, assume $l \geq i + 1 > w_0$. Let S be a $\gamma(G)$-set. Let f be the function defined by $f(v) = i + 1$ for every $v \in S$ and $f(v) = 0$ for the remaining vertices. Since $i + 1 > w_0$, we can conclude that f is a secure $(w_0, \ldots, w_i, 0 \ldots, 0)$-dominating function. Therefore, $\gamma^s_{(w_0, \ldots, w_i, 0 \ldots, 0)}(G) \leq w(f) = (i+1)|S| = (i+1)\gamma(G)$, which implies that (ii) follows.

To prove (iii), assume that $l \geq ki$, $i\delta \geq w_i'$ and $w_{kj} = kw_j'$ for every $j \in \{0,\ldots,i\}$. Let $f'(V_0',\ldots,V_i')$ be a $\gamma^s_{(w_0',\ldots,w_i')}(G)$-function. We construct a function $f(V_0,\ldots,V_l)$ as $f(v) = kf'(v)$ for every $v \in V(G)$. Hence, $V_{kj} = V_j'$ for every $j \in \{0,\ldots,i\}$, while $V_j = \emptyset$ for the remaining cases. Thus, for every $v \in V_{kj}$ with $j \in \{0,\ldots,i\}$ we have that $f(N(v)) = kf'(N(v)) \geq kw_j' = w_{kj}$, which implies that f is a $(w_0,\ldots,w_l)$-dominating function. Now, for every $x \in V_0$, there exists $y \in M_{f'}(x)$. Hence, for every $v \in V_{kj}$ with $j \in \{0,\ldots,i\}$, we have that $f_{y \to x}(N(v)) = kf'_{y \to x}(N(v)) \geq kw_j' = w_{kj}$, which implies that $f_{y \to x}$ is a $(w_0,\ldots,w_l)$-dominating function. Therefore, f is a secure $(w_0,\ldots,w_l)$-dominating function, and so $\gamma^s_{(w_0,\ldots,w_l)}(G) \leq \omega(f) = k\omega(f') = k\gamma^s_{(w_0',\ldots,w_i')}(G)$. Therefore, (iii) follows.

Now, assume that $l\delta \geq k + w_l > k$ and $w_0 + k \geq \beta_1 \geq \cdots \geq \beta_k \geq w_l + k$. Let $g(W_0,\ldots,W_l)$ be a $\gamma^s_{(w_0,\ldots,w_l)}(G)$-function. We construct a function $f(V_0,\ldots,V_{l+k})$ as $f(v) = g(v) + k$ for every $v \in V(G) \setminus W_0$ and $f(v) = 0$ for every $v \in W_0$. Hence, $V_{j+k} = W_j$ for every $j \in \{1,\ldots,l\}$, $V_0 = W_0$ and $V_j = \emptyset$ for the remaining cases. Thus, if $v \in V_{j+k}$ and $j \in \{1,\ldots,l\}$, then $f(N(v)) \geq g(N(v)) + k \geq w_j + k$, and if $v \in V_0$, then $f(N(v)) \geq g(N(v)) + k \geq w_0 + k$. This implies that f is a $(w_0 + k, \beta_1,\ldots,\beta_k, w_1 + k,\ldots,w_l + k)$-dominating function. Now, for every $x \in V_0 = W_0$, there exists $y \in M_g(x)$. Hence, if $v \in V_{j+k}$ and $j \in \{1,\ldots,l\}$, then $f_{y \to x}(N(v)) \geq g_{y \to x}(N(v)) + k \geq w_j + k$, and if $v \in V_0$, then $f_{y \to x}(N(v)) \geq g_{y \to x}(N(v)) + k \geq w_0 + k$. This implies that $f_{y \to x}$ is a $(w_0 + k, \beta_1,\ldots,\beta_k, w_1 + k,\ldots,w_l + k)$-dominating function, and so f is a secure $(w_0 + k, \beta_1,\ldots,\beta_k, w_1 + k,\ldots,w_l + k)$-dominating function. Therefore, $\gamma^s_{(w_0+k,\beta_1,\ldots,\beta_k,w_1+k,\ldots,w_l+k)}(G) \leq \omega(f) = \omega(g) + k\sum_{j=1}^{l}|W_j| = \gamma^s_{(w_0,\ldots,w_l)}(G) + k(n - |W_0|) \leq \gamma^s_{(w_0,\ldots,w_l)}(G) + k(n - v^s_{(w_0,\ldots,w_l)}(G))$, concluding that (iv) follows.

Furthermore, if $l\delta \geq w_l \geq l \geq 2$, then, by applying (iv) for $k = l - 1$, we deduce that

$$\gamma^s_{(w_0,\ldots,w_l)}(G) \leq \gamma^s_{(w_0-l+1,w_l-l+1)}(G) + (l-1)(n - v^s_{(w_0-l+1,w_l-l+1)}(G)) = l\gamma^s_{(w_0-l+1,w_l-l+1)}(G).$$

Therefore, (v) follows. □

In the next subsections, we consider several applications of Theorem 6 where we show that the bounds are tight. For instance, the following particular cases is of interest.

Corollary 4. *Let G be a graph of minimum degree δ, and let $k, l, w_2, \ldots, w_l \in \mathbb{Z}^+$ with $k \geq w_2 \geq \cdots \geq w_l$.*

(i') *If $\delta \geq k$, then $\gamma^s_{(k+1,k,w_2,\ldots,w_l)}(G) \leq \gamma^s_{(k+1,k)}(G)$.*

(ii') *If $\delta \geq k$, then $\gamma^s_{(k,k,w_2,\ldots,w_l)}(G) \leq \gamma^s_{(k,k)}(G)$.*

(iii') *If $l\delta \geq k \geq l \geq 2$, then $\gamma^s_{\underbrace{(k,k,\ldots,k)}_{l+1}}(G) \leq l\gamma^s_{(k-l+1,k-l+1)}(G).$*

(iv') *Let $i \in \mathbb{Z}^+$. If $l \geq ki$ and $\delta \geq 1$, then $\gamma^s_{\underbrace{(k,\ldots,k)}_{l+1}}(G) \leq k\gamma^s_{\underbrace{(1,\ldots,1)}_{i+1}}(G).$*

Proof. If $\delta \geq k$, then by Theorem 6 (i) we conclude that (i') and (ii') follow. If $l\delta \geq k \geq l \geq 2$, then by Theorem 6 (v) we deduce (iii'). Finally, if $l \geq k$ and $\delta \geq 1$, then by Theorem 6 (iii) we deduce that (iv') follows. □

To show that the inequalities above are tight, we consider the following examples. For (i'), we have $\gamma^s_{(2,1,1)}(K_1 + (K_2 \cup K_2)) = \gamma^s_{(2,1)}(K_1 + (K_2 \cup K_2)) = 3$. For (ii') we have $\gamma^s_{(k,k,w_2,\ldots,w_l)}(G) = \gamma^s_{(k,k)}(G) = k + 1$ for every graph G with $k + 1$ universal vertices. Finally, for (iii') and (iv'), we take $l = k = 2$ and $\gamma^s_{(2,2,2)}(K_2 + N_n) = 2\gamma^s_{(1,1)}(K_2 + N_n) = 4$ for every $n \geq 2$.

We already know that $\gamma_t(G) = \gamma_{(1,1)}(G) = \gamma_{(1,1,w_2,\ldots,w_l)}(G)$, for every $w_2,\ldots,w_l \in \{0,1\}$. In contrast, the picture is quite different for the case of secure $(1,1)$-domination, as there are graphs where the gap $\gamma^s_{(1,1)}(G) - \gamma^s_{(1,\ldots,1)}(G)$ is arbitrarily large. For instance, $\lim_{n\to\infty} \gamma^s_{(1,1)}(K_{1,n-1}) = +\infty$, while, if $l \geq 2$, then $\lim_{n\to+\infty} \gamma^s_{\underbrace{(1,\ldots,1)}_{l+1}}(K_{1,n-1}) = 3$.

Proposition 2. *Let G be a graph of order n. Let $w = (w_0, \ldots, w_l) \in \mathbb{Z}^+ \times \mathbb{N}^l$ such that $w_0 \geq \cdots \geq w_l$. If G' is a spanning subgraph of G with minimum degree $\delta' \geq \frac{w_1}{l}$, then*

$$\gamma^s_w(G) \leq \gamma^s_w(G').$$

Proof. Let $E^- = \{e_1, \ldots, e_k\}$ be the set of all edges of G not belonging to the edge set of G'. Let $G'_0 = G$ and, for every $i \in \{1, \ldots, k\}$, let $X_i = \{e_1, \ldots, e_i\}$ and $G'_i = G - X_i$, the edge-deletion subgraph of G induced by $E(G) \setminus X_i$.

For any $\gamma^s_w(G'_i)$-function f and any $v \in V(G'_i) = V(G)$ with $f(v) = 0$, there exists $u \in M_f(v)$. Since f and $f_{u\to v}$ are w-dominating functions on G'_i, both are w-dominating functions on G'_{i-1}, and so we can conclude that f is a secure w-dominating function on G'_{i-1}, which implies that $\gamma^s_w(G'_{i-1}) \leq \gamma^s_w(G'_i)$. Hence, $\gamma^s_w(G) = \gamma^s_w(G'_0) \leq \gamma^s_w(G'_1) \leq \cdots \leq \gamma^s_w(G'_k) = \gamma^s_w(G')$. □

As a simple example of equality in Proposition 2 we can take any graph G of order n, having $n' + 1 \geq 2$ universal vertices. In such a case, for $n' = w_1 \geq \cdots \geq w_l$ we have that

$$\gamma^s_{(n', n'=w_1, \ldots, w_l)}(K_n) = \gamma^s_{(n', n'=w_1, \ldots, w_l)}(G) = \gamma^s_{(n', n')}(K_n) = \gamma^s_{(n', n')}(G) = n' + 1.$$

From Proposition 2, we obtain the following result.

Corollary 5. *Let G be a graph of order n and $w = (w_0, \ldots, w_l) \in \mathbb{Z}^+ \times \mathbb{N}^l$ such that $w_0 \geq \cdots \geq w_l$.*

- *If G is a Hamiltonian graph and $w_l \leq 2l$, then $\gamma^s_w(G) \leq \gamma^s_w(C_n)$.*

- *If G has a Hamiltonian path and $w_l \leq l$, then $\gamma^s_w(G) \leq \gamma^s_w(P_n)$.*

To derive some lower bounds on $\gamma^s_w(G)$, we need to establish the following lemma.

Lemma 2 ([1]). *Let G be a graph with no isolated vertex, maximum degree Δ and order n. For any w-dominating function $f(V_0, \ldots, V_l)$ on G such that $w_0 \geq \cdots \geq w_l$,*

$$\Delta \omega(f) \geq w_0 n + \sum_{i=1}^{l}(w_i - w_0)|V_i|.$$

Theorem 7. *Let G be a graph with no isolated vertex, maximum degree Δ and order n. Let $w = (w_0, \ldots, w_l) \in \mathbb{Z}^+ \times \mathbb{N}^l$ such that $w_0 \geq \cdots \geq w_l$ and $l\delta \geq w_l$. The following statements hold.*

- *If $w_0 = w_1$ and $w_0 - w_i \leq i$ for every $i \in \{2, \ldots, l\}$, then $\gamma^s_w(G) \geq \left\lceil \frac{(w_0+1)n}{\Delta+1} \right\rceil$.*

- *If $w_0 = w_1$, then $\gamma^s_w(G) \geq \left\lceil \frac{(w_0+1)n}{\Delta+w_0} \right\rceil$.*

- *If $w_0 = w_1 + 1$ and $w_0 - w_i \leq i$ for every $i \in \{2, \ldots, l\}$, then $\gamma^s_w(G) \geq \left\lceil \frac{w_0 n}{\Delta+1} \right\rceil$.*

- $\gamma^s_w(G) \geq \left\lceil \frac{w_0 n}{\Delta+w_0} \right\rceil$.

Proof. Let $w_0 = w_1$ and $w_0 - w_i \leq i$ for every $i \in \{2, ..., l\}$. Let $f(V_0, \ldots, V_l)$ be a $\gamma_{(w_0+1, w_1, \ldots, w_l)}(G)$-function. By Lemma 2, we deduce the following.

$$\Delta\omega(f) \geq (w_0+1)n + \sum_{i=1}^{l}(w_i - w_0)|V_i|$$

$$\geq (w_0+1)n - \sum_{i=1}^{l} i|V_i|$$

$$= (w_0+1)n - \omega(f).$$

Therefore, Theorem 1 (ii) leads to $\gamma_w^s(G) \geq \omega(f) \geq \left\lceil \frac{(w_0+1)n}{\Delta+1} \right\rceil$.

The proof of the remaining items is completely analogous. In the last two cases, we consider that $f(V_0, \ldots, V_l)$ is a $\gamma_w(G)$-function, and we apply Theorem 1 (i) instead of (ii). □

The bounds above are sharp. For instance, $\gamma_{(1,1,0)}^s(G) \geq \left\lceil \frac{2n}{\Delta+1} \right\rceil$ is achieved by Graph G_2 shown in Figure 1, the bound $\gamma_{(k,k,0)}^s(G) \geq \left\lceil \frac{(k+1)n}{\Delta+k} \right\rceil$ is achieved by $G \cong K_n$ for every $n > k(k-1) > 0$, the bound $\gamma_{(2,1,1)}^s(G) \geq \left\lceil \frac{2n}{\Delta+1} \right\rceil$ is achieved by the corona graph $K_2 \odot K_{n'}$ with $n' \geq 4$, while $\gamma_{(2,0,0)}^s(G) \geq \left\lceil \frac{2n}{\Delta+2} \right\rceil$ is achieved by $G \cong C_5$, $G \cong K_n$ and $G \cong K_{n'} \cup K_{n'}$ with $n \geq 2$ and $n' \geq 4$.

To conclude the paper, we consider the problem of characterizing the graphs G and the vectors w for which $\gamma_w^s(G)$ takes small values. It is readily seen that $\gamma_{(w_0,\ldots,w_l)}^s(G) = 1$ if and only if $w_0 = 1$, $w_1 = 0$ and $G \cong K_n$. Next, we consider the case $\gamma_w^s(G) = 2$.

Theorem 8. *Let $w = (w_0, \ldots, w_l) \in \mathbb{Z}^+ \times \mathbb{N}^l$ such that $w_0 \geq \cdots \geq w_l$. For a graph G of order at least three, $\gamma_{(w_0,\ldots,w_l)}^s(G) = 2$ if and only if one of the following conditions holds.*

(i) $w_2 = 0$, $\gamma(G) = 1$ *and one of the following conditions holds.*

- $w_0 = w_1 = 1$.
- $w_0 = 1$, $w_1 = 0$, *and* $G \not\cong K_n$.
- $w_0 = 2$, $w_1 \in \{0, 1\}$ *and* $G \cong K_n$.

(ii) $w_0 = 1$, $w_1 = 0$, *and* $\gamma_{(1,0)}^s(G) = 2$.

(iii) $w_0 = w_1 = 1$ *and* $\gamma_{(1,1)}^s(G) = 2$.

(iv) $w_0 = 2$, $w_1 \in \{0, 1\}$, *and* $G \cong K_n$.

Proof. Assume first that $\gamma_{(w_0,\ldots,w_l)}^s(G) = 2$ and let $f(V_0, \ldots, V_l)$ be a $\gamma_{(w_0,\ldots,w_l)}^s(G)$-function. Notice that $(w_0, w_1) \in \{(1,0), (1,1), (2,0), (2,1)\}$ and $|V_2| \in \{0, 1\}$.

Firstly, we consider that $|V_2| = 1$, i.e., $V_2 = \{u\}$ for some universal vertex $u \in V(G)$. In this case, $w_2 = 0$, $\gamma(G) = 1$, and $V_i = \emptyset$ for every $i \neq 0, 2$. By Lemma 1, if $w_0 = 2$, then $G[T_f(u)] = G[V(G) \setminus \{u\}]$ is a clique, which implies that $G \cong K_n$. Obviously, in such a case, $w_1 < 2$. Finally, the case, $w_0 = 1$ and $w_1 = 0$ leads to $G \not\cong K_n$, as $\gamma_{(1,0...,0)}^s(K_n) = 1$. Therefore, (i) follows.

From now on, assume that $V_2 = \emptyset$. Hence, $V_i = \emptyset$ for every $i \neq 0, 1$. If $w_0 = 1$ and $w_1 = 0$, then $G \not\cong K_n$ and V_1 is a secure dominating set. Therefore, (ii) follows. If $w_0 = w_1 = 1$, then V_1 is a secure total dominating set of cardinality two, and so $\gamma_{(1,1)}^s(G) = 2$. Therefore, (iii) follows. Finally, assume $w_0 = 2$. In this case, V_1 is a double dominating set of cardinality two, and by Lemma 1 we know that $G[T_f(x)] = G[V(G) \setminus V_1]$ is a clique for any $x \in V_1$. Hence, $G \cong K_n$ and, in such a case, $w_1 < 2$. Therefore, (iv) follows.

Conversely, if one of the four conditions holds, then it is easy to check that $\gamma_{(w_0,\ldots,w_l)}^s(G) = 2$, which completes the proof. □

Author Contributions: All authors contributed equally to this work. Investigation, A.C.M., A.E.-M., and J.A.R.-V.; and Writing—review and editing, A.C.M., A.E.-M., and J.A.R.-V. All authors have read and agreed to the published version of the manuscript.

Funding: This research received no external funding.

Conflicts of Interest: The authors declare no conflict of interest.

References

1. Cabrera Martínez, A.; Estrada-Moreno, A.; Rodríguez-Velázquez, J.A. From Italian domination in lexicographic product graphs to w-domination in graphs. *arXiv* **2020**, arXiv:2011.05371.
2. Cockayne, E.J.; Grobler, P.J.P.; Gründlingh, W.R.; Munganga, J.; van Vuuren, J.H. Protection of a graph. *Util. Math.* **2005**, *67*, 19–32.
3. Boumediene Merouane, H.; Chellali, M. On secure domination in graphs. *Inform. Process. Lett.* **2015**, *115*, 786–790.
4. Burger, A.P.; Henning, M.A.; van Vuuren, J.H. Vertex covers and secure domination in graphs. *Quaest. Math.* **2008**, *31*, 163–171.
5. Chellali, M.; Haynes, T.W.; Hedetniemi, S.T. Bounds on weak Roman and 2-rainbow domination numbers. *Discret. Appl. Math.* **2014**, *178*, 27–32.
6. Cockayne, E.J.; Favaron, O.; Mynhardt, C.M. Secure domination, weak Roman domination and forbidden subgraphs. *Bull. Inst. Combin. Appl.* **2003**, *39*, 87–100.
7. Klostermeyer, W.F.; Mynhardt, C.M. Secure domination and secure total domination in graphs. *Discuss. Math. Graph Theory* **2008**, *28*, 267–284.
8. Valveny, M.; Rodríguez-Velázquez, J.A. Protection of graphs with emphasis on Cartesian product graphs. *Filomat* **2019**, *33*, 319–333.
9. Klein, D.J.; Rodríguez-Velázquez, J.A. Protection of lexicographic product graphs. *Discuss. Math. Graph Theory* **2019**, in press.
10. Benecke, S.; Cockayne, E.J.; Mynhardt, C.M. Secure total domination in graphs. *Util. Math.* **2007**, *74*, 247–259.
11. Cabrera Martínez, A.; Montejano, L.P.; Rodríguez-Velázquez, J.A. On the secure total domination number of graphs. *Symmetry* **2019**, *11*, 1165.
12. Cabrera Martínez, A.; Montejano, L.P.; Rodríguez-Velázquez, J.A. Total weak Roman domination in graphs. *Symmetry* **2019**, *11*, 831.
13. Duginov, O. Secure total domination in graphs: bounds and complexity. *Discret. Appl. Math.* **2017**, *222*, 97–108.
14. Kulli, V.R.; Chaluvaraju, B.; Kumara, M. Graphs with equal secure total domination and inverse secure total domination numbers. *J. Inf. Optim. Sci.* **2018**, *39*, 467–473.
15. Henning, M.A.; Hedetniemi, S.T. Defending the Roman Empire—A new strategy. *Discret. Math.* **2003**, *266*, 239–251.
16. Cabrera Martínez, A.; Yero, I.G. Constructive characterizations concerning weak Roman domination in trees. *Discret. Appl. Math.* **2020**, *284*, 384–390.
17. Valveny, M.; Pérez-Rosés, H.; Rodríguez-Velázquez, J.A. On the weak Roman domination number of lexicographic product graphs. *Discret. Appl. Math.* **2019**, *263*, 257–270.
18. Cabrera Martínez, A.; Rodríguez-Velázquez, J.A. Total protection of lexicographic product graphs. *Discuss. Math. Graph Theory.* **2020**, in press.
19. Dettlaff, M.; Lemańska, M.; Rodríguez-Velázquez, J.A. Secure Italian domination in graphs. *J. Comb. Optim.*. **2020**, in press.

20. Haynes, T.W.; Hedetniemi, S.T.; Slater, P.J. *Domination in Graphs: Advanced Topics*; Chapman and Hall/CRC Pure and Applied Mathematics Series; Marcel Dekker, Inc.: New York, NY, USA, 1998.
21. Haynes, T.W.; Hedetniemi, S.T.; Slater, P.J. *Fundamentals of Domination in Graphs*; Chapman and Hall/CRC Pure and Applied Mathematics Series; Marcel Dekker, Inc.: New York, NY, USA, 1998.

Publisher's Note: MDPI stays neutral with regard to jurisdictional claims in published maps and institutional affiliations.

© 2020 by the authors. Licensee MDPI, Basel, Switzerland. This article is an open access article distributed under the terms and conditions of the Creative Commons Attribution (CC BY) license (http://creativecommons.org/licenses/by/4.0/).

Article

Quasi-Ordinarization Transform of a Numerical Semigroup

Maria Bras-Amorós *, Hebert Pérez-Rosés and José Miguel Serradilla-Merinero

Departament d'Enginyeria Informàtica i Matemàtiques, Universitat Rovira i Virgili, 43007 Tarragona, Spain; hebert.perez@urv.cat (H.P.-R.); josemiguel.serradilla@estudiants.urv.cat (J.M.S.-M.)
* Correspondence: maria.bras@urv.cat

Abstract: In this study, we present the notion of the quasi-ordinarization transform of a numerical semigroup. The set of all semigroups of a fixed genus can be organized in a forest whose roots are all the quasi-ordinary semigroups of the same genus. This way, we approach the conjecture on the increasingness of the cardinalities of the sets of numerical semigroups of each given genus. We analyze the number of nodes at each depth in the forest and propose new conjectures. Some properties of the quasi-ordinarization transform are presented, as well as some relations between the ordinarization and quasi-ordinarization transforms.

Keywords: numerical semigroup; forest; ordinarization transform; quasi-ordinarization transform

Citation: Bras-Amorós, M.;
Pérez-Rosés, H.; Serradilla-Merinero,
J.M. Quasi-Ordinarization Transform
of a Numerical Semigroup. *Symmetry*
2021, *13*, 1084. https://doi.org/
10.3390/sym13061084

Academic Editor: Michel Planat

Received: 9 February 2021
Accepted: 4 June 2021
Published: 17 June 2021

Publisher's Note: MDPI stays neutral with regard to jurisdictional claims in published maps and institutional affiliations.

Copyright: © 2021 by the authors. Licensee MDPI, Basel, Switzerland. This article is an open access article distributed under the terms and conditions of the Creative Commons Attribution (CC BY) license (https://creativecommons.org/licenses/by/4.0/).

1. Introduction

A numerical semigroup is a cofinite submonoid of $\mathbb{N}_0$ under addition, where $\mathbb{N}_0$ is the set of nonnegative integers.

While the symmetry of structures has traditionally been studied with the aid of groups, it is also possible to relax the definition of symmetry, so as to describe some forms of symmetry that arise in quasicrystals, fractals, and other natural phenomena, with the aid of semigroups or monoids, rather than groups. For example, Rosenfeld and Nordahl [1] lay the groundwork for such a theory of symmetry based on semigroups and monoids, and they cite some applications in chemistry.

Suppose that Λ is a numerical semigroup. The elements in the complement $\mathbb{N}_0 \setminus \Lambda$ are called the *gaps* of the semigroup and the number of gaps is its *genus*. The *Frobenius number* is the largest gap and the *conductor* is the non-gap that equals the Frobenius number plus one. The first non-zero non-gap of a numerical semigroup (usually denoted by m) is called its *multiplicity*. An *ordinary* semigroup is a numerical semigroup different from $\mathbb{N}_0$ in which all gaps are in a row. The non-zero non-gaps of a numerical semigroup that are not the result of the sum of two smaller non-gaps are called the *generators* of the numerical semigroup. It is easy to deduce that the set of generators of a numerical semigroup must be co-prime. One general reference for numerical semigroups is [2].

To illustrate all these definitions, consider the well-tempered harmonic semigroup $H = \{0, 12, 19, 24, 28, 31, 34, 36, 38, 40, 42, 43, 45, 46, 47, 48, \dots\}$, where we use "$\dots$" to indicate that the semigroup consecutively contains all the integers from the number that precedes the ellipsis. The semigroup H arises in the mathematical theory of music [3]. It is obviously cofinite and it contains zero. One can also check that it is closed under addition. Hence, it is a numerical semigroup. Its Frobenius number is 44, its conductor is 45, its genus is 33, and its multiplicity is 12. Its generators are $\{12, 19, 28, 34, 42, 45, 49, 51\}$.

The number of numerical semigroups of genus g is denoted n_g. It was conjectured in [4] that the sequence n_g asymptotically behaves as the Fibonacci numbers. In particular, it was conjectured that each term in the sequence is larger than the sum of the two previous terms, that is, $n_g \geqslant n_{g-1} + n_{g-2}$ for $g \geqslant 2$, with each term being increasingly similar to the sum of the two previous terms as g approaches infinity, more precisely $\lim_{g \to \infty} \frac{n_g}{n_{g-1} + n_{g-2}} = 1$ and, equivalently, $\lim_{g \to \infty} \frac{n_g}{n_{g-1}} = \phi = \frac{1+\sqrt{5}}{2}$. A number of papers deal with the sequence

n_g [5–20]. Alex Zhai proved the asymptotic Fibonacci-like behavior of n_g [21]. However, it remains unproven that n_g is increasing. This was already conjectured by Bras-Amorós in [22]. More information on n_g, as well as the list of the first 73 terms can be found in entry A007323 of The On-Line Encyclopedia of Integer Sequences [23].

It is well known that all numerical semigroups can be organized in an infinite tree $\mathcal{T}$ whose root is the semigroup $\mathbb{N}_0$ and in which the parent of a numerical semigroup Λ is the numerical semigroup Λ' obtained by adjoining to Λ its Frobenius number. For instance, the parent of the semigroup $H = \{0, 12, 19, 24, 28, 31, 34, 36, 38, 40, 42, 43, 45, 46, 47, 48, \dots\}$ is the semigroup $H' = \{0, 12, 19, 24, 28, 31, 34, 36, 38, 40, 42, 43, \mathbf{44}, 45, 46, 47, 48, \dots\}$. In turn, the children of a numerical semigroup are the semigroups we obtain by taking away the generators one by one that are larger than or equal to the conductor of the semigroup. The parent of a numerical semigroup of genus g has genus $g - 1$ and all numerical semigroups are in $\mathcal{T}$, at a depth equal to its genus. In particular, n_g is the number of nodes of $\mathcal{T}$ at depth g. This construction was already considered in [24]. Figure 1 shows the tree up to depth 7.

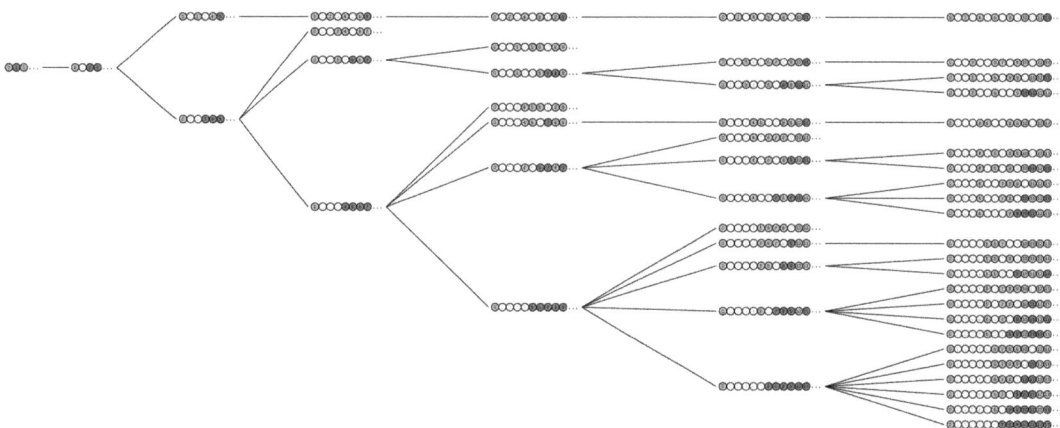

Figure 1. The tree $\mathcal{T}$ up to depth 7. White dots refer to the gaps, dark gray dots to the generators and the light gray ones to the elements of the semigroups that are not generators.

In [9], a new tree construction is introduced as follows. The *ordinarization transform* of a non-ordinary semigroup Λ with Frobenius number F and multiplicity m is the set $\Lambda' = \Lambda \setminus \{m\} \cup \{F\}$. For instance, the ordinarization transform of the semigroup $H = \{0, \mathbf{12}, 19, 24, 28, 31, 34, 36, 38, 40, 42, 43, 45, 46, 47, 48, \dots\}$ is the semigroup $H' = \{0, 19, 24, 28, 31, 34, 36, 38, 40, 42, 43, \mathbf{44}, 45, 46, 47, 48, \dots\}$ The ordinarization transform of an ordinary semigroup is then defined to be itself. Note that the genus of the ordinarization transform of a semigroup is the genus of the semigroup.

The definition of the ordinarization transform of a numerical semigroup allows the construction of a tree $\mathcal{T}_g$ on the set of all numerical semigroups of a given genus rooted at the unique ordinary semigroup of this genus, where the parent of a semigroup is its ordinarization transform and the children of a semigroup are the semigroups obtained by taking away the generators one by one that are larger than the Frobenius number and adding a new non-gap smaller than the multiplicity in a licit place. To illustrate this construction with an example in Figure 2, we depicted $\mathcal{T}_7$.

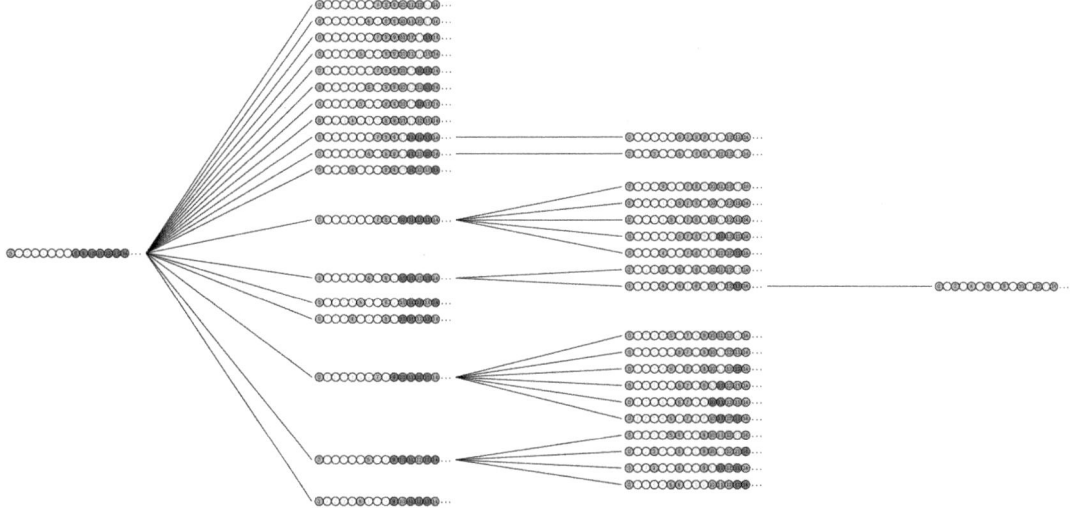

Figure 2. The whole tree $\mathcal{T}_7$.

One significant difference between $\mathcal{T}_g$ and $\mathcal{T}$ is that the first one has only a finite number of nodes. In fact, it has n_g nodes, while $\mathcal{T}$ is an infinite tree. It was conjectured in [9] that the number of numerical semigroups in $\mathcal{T}_g$ at a given depth is at most the number of numerical semigroups in $\mathcal{T}_{g+1}$ at the same depth. This was proved in the same reference for the lowest and largest depths. This conjecture would prove that $n_{g+1} \geqslant n_g$.

In Section 2, we will construct the quasi-ordinarization transform of a general semigroup, paralleling the ordinarization transform. If the quasi-ordinarization transform is applied repeatedly to a numerical semigroup, it ends up in a quasi-ordinary semigroup. In Section 3, we define the quasi-ordinarization number of a semigroup as the number of successive quasi-ordinarization transforms of the semigroup that give a quasi-ordinary semigroup. Section 4 analyzes the number of numerical semigroups of a given genus and a given quasi-ordinarization number in terms of the given parameters. We present the conjecture that the number of numerical semigroups of a given genus and a fixed quasi-ordinarization number increases with the genus and we prove it for the largest quasi-ordinarization numbers. In Section 5, we present the forest of semigroups of a given genus that is obtained when connecting each semigroup to its quasi-ordinarization transform. The forest corresponding to genus g is denoted $\mathcal{F}_g$. Section 6 analyzes the relationships between $\mathcal{T}$, $\mathcal{T}_g$, and $\mathcal{F}_g$.

From the perspective of the forests of numerical semigroups here presented, the conjecture in Section 4 translates to the conjecture that the number of numerical semigroups in $\mathcal{F}_g$ at a given depth is at most the number of numerical semigroups in $\mathcal{F}_{g+1}$ at the same depth. The results in Section 4 provide a proof of the conjecture for the largest depths. Proving this conjecture for all depths, would prove that $n_{g+1} \geqslant n_g$. Hence, we expect our work to contribute to the proof of the conjectured increasingness of the sequence n_g (A007323).

2. Quasi-Ordinary Semigroups and Quasi-Ordinarization Transform

Quasi-ordinary semigroups are those semigroups for which $m = g$ and so, there is a unique gap larger than m. The *sub-Frobenius number* of a non-ordinary semigroup Λ with Frobenius number F is the Frobenius number of $\Lambda \cup \{F\}$.

The *subconductor* of a semigroup with Frobenius number F is the smallest nongap in the interval of nongaps immediatelly previous to F. For instance, the subconductor of the above example, $H = \{0, 12, 19, 24, 28, 31, 34, 36, 38, 40, 42, 43, 45, 46, 47, 48, \dots\}$, is 42.

Lemma 1. *Let Λ be a non-ordinary and non quasi-ordinary semigroup, with multiplicity m, genus g, and sub-Frobenius number f. Then, $\Lambda \cup \{f\} \setminus \{m\}$ is another numerical semigroup of the same genus g.*

Proof. Since Λ is already a numerical semigroup, it is enough to see that $F - f$ is not in $\Lambda \cup \{f\} \setminus \{m\}$, where F is the Frobenius number of Λ. Notice that for a non-ordinary numerical semigroup, the difference between its Frobenius number and its sub-Frobenius number needs to be less than the multiplicity of the semigroup; hence, $F - f \notin \Lambda$. So, the only option for $F - f$ to be in $\Lambda \cup \{f\} \setminus \{m\}$ is that $F - f = f$. In this case, any integer between 1 and $f - 1$ must be a gap, since the integers between $F - 1$ and $F - f + 1$ are nongaps. In this case, Λ would be quasi-ordinary, contradicting the hypotheses. □

Definition 1. *The* quasi-ordinarization transform *of a non-ordinary and non quasi-ordinary numerical semigroup Λ, with multiplicity m, genus g and sub-Frobenius number f, is the numerical semigroup $\Lambda \cup \{f\} \setminus \{m\}$.*

The quasi-ordinarization *of either an ordinary or quasi-ordinary semigroup is defined to be itself.*

As an example, the quasi-ordinarization of the well-tempered harmonic semigroup $H = \{0, 12, 19, 24, 28, 31, 34, 36, 38, 40, 42, 43, 45, 46, 47, 48, \ldots\}$ used in the previous examples is $H' = \{0, 19, 24, 28, 31, 34, 36, 38, 40, 41, 42, 43, 45, 46, 47, 48, \ldots\}$.

Remark 1. *In the ordinarization and quasi-ordinarization transform process, we replace the multiplicity by the largest and second largest gap, respectively, and we obtain numerical semigroups. In general, if we replace the multiplicity by the third largest gap, we do not obtain a numerical semigroup.*

See for instance $\{0, 2, 4, 6, 8, 10, 11, \ldots\}$. Replacing 2 by 5, we obtain $\{0, 4, 5, 6, 8, 10, 11, \ldots\}$, which is not a numerical semigroup since $9 = 4 + 5$ is not in the set.

3. Quasi-Ordinarization Number

Next, lemma explicits that there is only one quasi-ordinary semigroup with genus g and conductor c where $c \leqslant 2g$.

Lemma 2. *For each of the positive integers g and c with $c \leqslant 2g$, the semigroup $\{0, g, g + 1, \ldots, c - 2, c, c + 1 \ldots\}$ is the unique quasi-ordinary semigroup of genus g and conductor c.*

The quasi-ordinarization transform of a non-ordinary semigroup of genus g and conductor c can be applied subsequently and at some step, we will attain the quasi-ordinary semigroup of that genus and conductor, that is, the numerical semigroup $\{0, g, g + 1, \ldots, c - 2, c, c + 1, \ldots\}$. The number of such steps is defined to be the *quasi-ordinarization number* of Λ.

We denote by $\varrho_{g,q}$, the number of numerical semigroups of genus g and quasi-ordinarization number q. In Table 1, one can see the values of $\varrho_{g,q}$ for genus up to 45. It has been computed by an exhaustive exploration of the semigroup tree using the RGD algorithm [12].

Lemma 3. *The quasi-ordinarization number of a non-ordinary numerical semigroup of genus g coincides with the number of non-zero non-gaps of the semigroup that are smaller than or equal to $g - 1$.*

Proof. A non-ordinary numerical semigroup of genus g is non-quasi-ordinary if and only if its multiplicity is at most $g - 1$. Consequently, we can repeatedly apply the quasi-ordinarization transform to a numerical semigroup while its multiplicity is at most $g - 1$. Furthermore, the number of consecutive transforms that we can apply before obtaining the quasi-ordinary semigroup is hence the number of its non-zero non-gaps that are at most the genus minus one. □

Table 1. Number of semigroups of each genus and quasi-ordinarization number.

$\varrho_{g,q}$	$g=1$	$g=2$	$g=3$	$g=4$	$g=5$	$g=6$	$g=7$	$g=8$	$g=9$	$g=10$
$q=0$	1	2	3	4	5	6	7	8	9	10
$q=1$			1	3	6	15	24	42	61	93
$q=2$					1	2	7	16	43	89
$q=3$							1	1	4	11
$q=4$									1	1
sum=	1	2	4	7	12	23	39	67	118	204
$\varrho_{g,q}$	$g=11$	$g=12$	$g=13$	$g=14$	$g=15$	$g=16$	$g=17$	$g=18$	$g=19$	$g=20$
$q=0$	11	12	13	14	15	16	17	18	19	20
$q=1$	123	174	219	291	355	453	537	666	774	936
$q=2$	176	327	538	903	1379	2127	3022	4441	5979	8417
$q=3$	30	75	209	448	990	1894	3575	6367	10,796	17,960
$q=4$	2	3	19	34	106	295	829	1847	4447	9019
$q=5$	1	1	2	2	9	18	55	116	403	986
$q=6$			1	1	2	2	7	9	36	48
$q=7$					1	1	2	2	7	7
$q=8$							1	1	2	2
$q=9$									1	1
sum=	343	592	1001	1693	2857	4806	8045	13,467	22,464	37,396
$\varrho_{g,q}$	$g=21$	$g=22$	$g=23$	$g=24$	$g=25$	$g=26$	$g=27$	$g=28$	$g=29$	$g=30$
$q=0$	21	22	23	24	25	26	27	28	29	30
$q=1$	1072	1272	1437	1680	1878	2166	2401	2739	3012	3405
$q=2$	10,966	14,826	18,774	24,770	30,539	39,321	47,697	60,083	71,711	88,938
$q=3$	28,265	44,272	66,046	99,525	140,960	204,611	281,077	394,617	525,838	720,977
$q=4$	18,673	35,178	65,533	115,252	197,836	329,568	533,479	848,091	1,304,275	2,001,344
$q=5$	2981	7165	17,640	37,770	84,075	166,465	331,872	615,860	1,135,074	1,989,842
$q=6$	181	464	1383	3603	11,141	26,864	67,991	153,882	352,322	727,680
$q=7$	25	37	94	170	652	1679	5300	14,899	42738	107,050
$q=8$	7	7	23	24	85	99	321	715	2506	7073
$q=9$	2	2	7	7	23	23	69	83	233	331
$q=10$	1	1	2	2	7	7	23	23	68	70
$q=11$			1	1	2	2	7	7	23	23
$q=12$					1	1	2	2	7	7
$q=13$							1	1	2	2
$q=14$									1	1
sum=	62,194	103,246	170,963	282,828	467,224	770,832	1,270,267	2,091,030	3,437,839	5,646,773

For a numerical semigroup Λ, we will consider its enumeration λ, that is, the unique increasing bijective map between $\mathbb{N}_0$ and Λ. The element $\lambda(i)$ is then denoted λ_i. As a consequence of the previous lemma, for a numerical semigroup Λ with quasi-ordinarization number equal to q, the non-gaps that are at most $g-1$ are exactly $\lambda_0 = 0, \lambda_1, \ldots, \lambda_q$.

Lemma 4. *The maximum quasi-ordinarization number of a non-ordinary semigroup of genus g is $\lfloor \frac{g-1}{2} \rfloor$.*

Proof. Let Λ be a numerical semigroup with quasi-ordinarization number equal to q. Since the Frobenius number F is at most $2g - 1$, the total number of gaps from 1 to $2g - 1$ is g, and so the number of non-gaps from 1 to $2g - 1$ is $g - 1$. The number of those non-gaps that are larger than $g - 1$ is $g - 1 - q$. On the other hand, $\lambda_q + \lambda_1, \lambda_q + \lambda_2, \ldots, 2\lambda_q$ are different non-gaps between g and $2g - 1$. So, the number of non-gaps between g and $2g - 1$ is at least q. All these results imply that $g - 1 - q \geqslant q$ and so, $q \leqslant \frac{g-1}{2}$.

On the other hand, the bound stated in the lemma is attained by the hyperelliptic numerical semigroup

$$\{0, 2, 4, \ldots, 2\left\lfloor\frac{g-1}{2}\right\rfloor, 2\left(\left\lfloor\frac{g-1}{2}\right\rfloor + 1\right), \ldots, 2g, 2g + 1, 2g + 2, \ldots\}. \tag{1}$$

□

We will next see that the maximum ordinarization number stated in the previous lemma is attained uniquely by the numerical semigroup in (1). To prove this result, we will need the next lemma. Let us recall that $A + B = \{a + b : a \in A, b \in B\}$ and that $\#A$ denotes the cardinality of A.

Lemma 5. Consider a finite subset $A = \{a_1 < \cdots < a_n\} \subseteq \mathbb{N}_0$.

1. The set $A + A$ contains at least $2n - 1$ elements
2. If $n \geqslant 1$, the set $A + A$ contains exactly $2n - 1$ elements if and only if there exists a positive integer α such that $a_i = a_1 + (i - 1)\alpha$ for all $i \leqslant n$.
3. If $n \geqslant 4$, the set $A + A$ contains exactly $2n$ elements if and only if either

   - there exists a positive integer α such that $a_i = a_1 + \alpha(i - 1)$ for all i with $1 \leqslant i < n$ and $a_n = a_1 + n\alpha$,
   - there exists a positive integer α such that $a_i = a_1 + i\alpha$ for all i with $2 \leqslant i \leqslant n$.

Proof. The first item stems from the fact that if $A = \{a_1, \ldots, a_n\}$, then $A + A$ must contain at least $2a_1, a_1 + a_2, a_1 + a_3, \ldots, a_1 + a_n, a_2 + a_n, a_3 + a_n, \ldots, a_{n-1} + a_n, 2a_n$, which are all different.

The second item easily follows from the fact that if $A + A$ has $2n - 1$ elements, then $A + A$ must be exactly the set $2a_1, a_1 + a_2, a_1 + a_3, \ldots, a_1 + a_n, a_2 + a_n, a_3 + a_n, \ldots, a_{n-1} + a_n, 2a_n$. Indeed, in this case, the increasing set $\{a_1 + a_3, \ldots, a_1 + a_n, a_2 + a_n, a_3 + a_n, \ldots, a_{n-1} + a_n, 2a_n\}$ must coincide with the increasing set $\{2a_2, a_2 + a_3, a_2 + a_4, \ldots, a_2 + a_n, a_3 + a_n, \ldots, a_{n-1} + a_n, 2a_n\}$, having as a consequence that $2a_2 = a_1 + a_3$ and so, $a_2 = \frac{a_1 + a_3}{2} = a_1 + \frac{a_3 - a_1}{2}$, and $a_3 = 2a_2 - a_1 = a_1 + 2\frac{a_3 - a_1}{2}$. Hence,

$$a_2 = a_1 + \frac{a_3 - a_1}{2}$$
$$a_3 = a_1 + 2\frac{a_3 - a_1}{2}$$

Similarly, one can show that $2a_3 = a_2 + a_4$ and, so, $a_4 = 2a_3 - a_2 = 2a_1 + 4\frac{a_3 - a_1}{2} - a_1 - \frac{a_3 - a_1}{2} = a_1 + 3\frac{a_3 - a_1}{2}$. It equally follows that

$$a_4 = a_1 + 3\frac{a_3 - a_1}{2}$$
$$a_5 = a_1 + 4\frac{a_3 - a_1}{2}$$
$$\vdots$$

For the third item, one direction of the proof is obvious, so we just need to prove the other one, that is, if the sum contains $2n$ elements, then $a_1, \ldots, a_n$ must be as stated.

We will proceed by induction. Suppose that $n = 4$ and that the set $A + A$ contains exactly 8 elements. Since the ordered sequence

$$2a_1 < a_1 + a_2 < 2a_2 < a_2 + a_3 < 2a_3 < a_3 + a_4 < 2a_4 \qquad (2)$$

already contains 7 elements, then necessarily two of the elements $a_1 + a_3, a_1 + a_4, a_2 + a_4$ coincide with one element in (2) and the third one is not in (2). So, at least one of $a_1 + a_3$ and $a_2 + a_4$ must be in (2).

Suppose first that $a_1 + a_3$ is in (2). Then, necessarily $a_1 + a_3 = 2a_2$, which means that $a_2 - a_1 = a_3 - a_2$. Hence, there exists α (in fact, $\alpha = a_2 - a_1$) such that $a_2 = a_1 + \alpha$ and $a_3 = a_1 + 2\alpha$. Now, the elements

$$2a_1 < a_1 + a_2 < 2a_2 < a_2 + a_3 < 2a_3 \qquad (3)$$

are equally separated by the same separation α. That is,

$$\begin{aligned}
(a_1 + a_2) - (2a_1) &= \alpha \\
(2a_2) - (a_1 + a_2) &= \alpha \\
(a_2 + a_3) - (2a_2) &= \alpha \\
(2a_3) - (a_2 + a_3) &= \alpha.
\end{aligned}$$

Additionally, the elements

$$a_4 + a_1 < a_4 + a_2 < a_4 + a_3 \qquad (4)$$

are equally separated by the same separation α. That is,

$$\begin{aligned}
(a_4 + a_3) - (a_4 + a_2) &= \alpha \\
(a_4 + a_2) - (a_4 + a_1) &= \alpha.
\end{aligned}$$

Furthermore, $A + A$ must contain all the elements in (3) and (4) as well as the element $2a_4$, which is not in (3), nor in (4). Since $\#(A + A) = 8$, this means that there must be exatly one element that is both in (3) and (4). The only way for this to happen is that $2a_3 = a_4 + a_1$. Consequently, $a_4 + a_1 = 2a_1 + 4\alpha$, and so, $a_4 = a_1 + 4\alpha$. This proves the result in the first case.

For the case in which $a_2 + a_4$ is in (2), it is necessary that $a_2 + a_4 = 2a_3$, which means that $a_3 - a_2 = a_4 - a_3$. Hence, there exists β (in fact, $\beta = a_3 - a_2$) such that $a_3 = a_2 + \beta$ and $a_4 = a_2 + 2\beta$. Now, the elements

$$2a_2 < a_2 + a_3 < 2a_3 < a_3 + a_4 < 2a_4 \qquad (5)$$

are equally separated by the same separation β. That is,

$$\begin{aligned}
(a_2 + a_3) - (2a_2) &= \beta \\
(2a_3) - (a_2 + a_3) &= \beta \\
(a_3 + a_4) - (2a_3) &= \beta \\
(2a_4) - (a_3 + a_4) &= \beta.
\end{aligned}$$

Additionally, the elements

$$a_1 + a_2 < a_1 + a_3 < a_1 + a_4 \qquad (6)$$

are equally separated by the same separation β. That is,

$$(a_1 + a_3) - (a_1 + a_2) = \beta$$
$$(a_1 + a_4) - (a_1 + a_3) = \beta.$$

Now, $A + A$ must contain all the elements in (5) and (6), as well as the element $2a_1$, which is not in (5), nor in (6). Since $\#(A + A) = 8$, this means that there must be exactly one element that is both in (5) and in (6). The only way for this to happen is that $a_1 + a_4 = 2a_2$. Consequently, $a_1 + a_4 = 2a_1 + 4\alpha$, and so, $a_4 = a_1 + 4\alpha$. Hence, $a_2 = a_1 + 2\beta$, $a_3 = a_1 + 3\beta$, $a_4 = a_1 + 4\beta$. This proves the result in the second case and concludes the proof for $n = 4$.

Now, let us prove the result for a general $n > 4$. We will denote A_n the set $\{a_1, \ldots, a_n\}$. Notice that $A_1 + A_1 = \{2a_1\}$ while, if $i > 1$, then $\{a_{i-1} + a_i, 2a_i\} \subseteq (A_i + A_i) \setminus (A_{i-1} + A_{i-1})$, hence, $\#((A_i + A_i) \setminus (A_{i-1} + A_{i-1})) \geq 2$. Consequently, if $\#(A_n + A_n) = 2n$, we can affirm that there exists exactly one integer s such that $\#(A_r + A_r) = 2r - 1$, for all $r < s$ and $\#(A_r + A_r) = 2r$ for all $r \geq s$.

If $s = n$, then we already have, by the second item of the lemma, that $a_i = a_1 + (i-1)\gamma$ for a given positive integer γ for all $i < n$.

On one hand,

$$A_{n-1} + A_{n-1} = \{2a_1, 2a_1 + \gamma, 2a_1 + 2\gamma, 2a_1 + 3\gamma, \ldots, 2a_1 + (2n-4)\gamma\}, \qquad (7)$$

which has $2n - 3$ elements. On the other hand,

$$A_{n-1} + a_n = \{(a_1 + a_n), (a_1 + a_n) + \gamma, (a_1 + a_n) + 2\gamma, (a_1 + a_n) + 3\gamma, \ldots, (a_1 + a_n) + (n-2)\gamma\}, \qquad (8)$$

has $n - 1$ elements.

Now, $A + A = (A_{n-1} + A_{n-1}) \cup (A_{n-1} + a_n) \cup (2a_n)$. By the inclusion–exclusion principle, and since $2a_n$ is not in $(A_{n-1} + A_{n-1}) \cup (A_{n-1} + a_n)$,

$$\#((A_{n-1} + A_{n-1}) \cap (A_{n-1} + a_n)) = \#(A_{n-1} + A_{n-1}) + \#(A_{n-1} + a_n) + 1 - \#(A + A)$$
$$= (2n - 3) + (n - 1) + 1 - 2n$$
$$= n - 3$$

By (7) and (8), we conclude that $(a_1 + a_n) + (n - 4)\gamma = 2a_1 + (2n - 4)\gamma$, that is, $a_n = a_1 + n\gamma$. Hence, the result follows with $\alpha = \gamma$.

On the contrary, if $s < n$, then, since $\#(A_{n-1} + A_{n-1}) = 2(n-1)$, we can apply the induction hypothesis and affirm that either one of the following cases, (a) or (b), holds.

(a) There exists a positive integer α_{n-1} such that $a_i = a_1 + \alpha_{n-1}(i-1)$ for all i with $1 \leq i < n - 1$ and $a_{n-1} = a_1 + (n-1)\alpha$;
(b) There exists a positive integer α such that $a_i = a_1 + i\alpha_{n-1}$ for all i with $2 \leq i \leq n - 1$.

In case (a), we will have

$$A_{n-1} + A_{n-1} = \{2a_1, 2a_1 + \alpha_{n-1}, 2a_1 + 2\alpha_{n-1}, \ldots$$
$$\ldots, 2a_1 + (2n-4)\alpha_{n-1}, 2a_1 + (2n-2)\alpha_{n-1}\},$$

and

$$A_{n-1} + a_n = \{(a_1 + a_n), (a_1 + a_n) + \alpha_{n-1}, (a_1 + a_n) + 2\alpha_{n-1}, \ldots$$
$$\ldots, (a_1 + a_n) + (n-3)\alpha_{n-1}, (a_1 + a_n) + (n-1)\alpha_{n-1}\},$$

In case (b), we will have

$$A_{n-1} + A_{n-1} = \{2a_1, 2a_1 + 2\alpha_{n-1}, 2a_1 + 3\alpha_{n-1}, \ldots$$
$$\ldots, 2a_1 + (2n-3)\alpha_{n-1}, 2a_1 + (2n-2)\alpha_{n-1}\},$$

and

$$A_{n-1} + a_n = \{(a_1 + a_n), (a_1 + a_n) + 2\alpha_{n-1}, (a_1 + a_n) + 3\alpha_{n-1}, \ldots$$
$$\ldots, (a_1 + a_n) + (n-1)\alpha_{n-1}\},$$

Now,

$$
\begin{aligned}
\#((A_{n-1}+A_{n-1})\cap(A_{n-1}+a_n)) &= \#(A_{n-1}+A_{n-1})+\#(A_{n-1}+a_n)+1-\#(A+A)\\
&= \#(A_{n-1}+A_{n-1})-n\\
&= n-2.
\end{aligned}
$$

This is only possible in case (b) with

$$(A_{n-1}+A_{n-1})\cap(A_{n-1}+a_n) = \{(a_1+a_2),(a_1+a_n)+2\alpha_{n-1},(a_1+a_n)+3\alpha_{n-1},\ldots\\
\ldots,(a_1+a_n)+(n-2)\alpha_{n-1}\},$$

and, hence, with $(a_1+a_n)+(n-2)\alpha_{n-1} = 2a_1+(2n-2)\alpha_{n-1}$, that is, $a_n = a_1+n\alpha_{n-1}$, hence yielding the result with $\alpha = \alpha_{n-1}$. □

Lemma 6. *Let $g > 2$ and $g \neq 4, g \neq 6$. The unique non-quasi-ordinary numerical semigroup of genus g and quasi-ordinarization number $\lfloor \frac{g-1}{2} \rfloor$ is $\{0,2,4,\ldots,2g,2g+2,2g+3\ldots\}$.*

Proof. If $g = 3$, there is only one numerical semigroup non-ordinary and non-quasi-ordinary as we can observe in Figure 1, and it is exactly $\{0,2,4,6,\ldots\}$, which indeed, has a quasi-ordinarization number $\lfloor \frac{g-1}{2} \rfloor$ and it is of the form $\{0,2,4,\ldots,2g,2g+1,2g+2,\ldots\}$. The case $g = 4$ and $g = 6$ are excluded from the statement (and analyzed in Remark 2). So, we can assume that either $g = 5$ or $g > 6$.

Suppose that the quasi-ordinarization number of Λ is $\lfloor \frac{g-1}{2} \rfloor$. Since $\lambda_{\lfloor \frac{g-1}{2} \rfloor} \leq g-1$, we know that the set of all non-gaps between 0 and $2g-2$ must contain all the sums

$$\Sigma = \{\lambda_i + \lambda_j : 0 \leq i,j \leq \lfloor \frac{g-1}{2} \rfloor\}.$$

However, the number of non-gaps between 0 and $2g-2$ is either $g-1$ or g depending on whether $2g-1$ is a gap or not. So, $\#\Sigma \leq g$. On the other hand, by Lemma 5, $\#\Sigma \geq 2\lfloor \frac{g-1}{2} \rfloor + 1$.

If g is odd, we get that $2\lfloor \frac{g-1}{2} \rfloor + 1 = g$ and so, $\#\Sigma = g$. Then, by the second item in Lemma 5, we get that $\lambda_i = i\lambda_1$ for $i \leq \frac{g-1}{2}$. Now, $\lambda_{\frac{g-1}{2}} = \frac{g-1}{2}\lambda_1$ and, since $\lambda_{\frac{g-1}{2}} \leq g-1$, one can deduce that $\lambda_1 \leq 2$. If $\lambda_1 = 1$ this contradicts $g > 1$. So, $\lambda_i = 2i$ for $0 \leq i \leq \frac{g-1}{2}$ and the remaining non-gaps between g and $2g$ are necessarily $\lambda_i = 2i$ for $i = \frac{g-1}{2}+1$ to $i = g$.

If g is even, then $g-1 \leq \#\Sigma \leq g$. If $\#\Sigma = g$, then, since the number of summands in the sum Σ is at least 4 (because we excluded the even genera 4 and 6), we can apply the third item in Lemma 5. Then, we obtain $\lambda_{\frac{g}{2}-1} = \frac{g}{2}\lambda_1$. This, together with $\lambda_{\frac{g}{2}-1} \leq g-1$ implies that $\lambda_1 \leq 2\frac{g-1}{g} < 2$. So, $\lambda_1 = 1$, contradicting $g > 1$. Hence, it must be $\Sigma = g-1$. If $\Sigma = g-1$, then, by the second item in Lemma 5, we obtain $\lambda_i = i\lambda_1$ for all $i \leq \frac{g}{2}-1$. Now, $\lambda_{\frac{g}{2}-1} = (\frac{g}{2}-1)\lambda_1$ and, since $\lambda_{\frac{g}{2}-1} \leq g-1$, one can deduce that $\lambda_1 \leq 2\frac{g-1}{g-2}$. However, $2\frac{g-1}{g-2} < 3$ if $g \geq 5$. So, λ_1 con only be 1 or 2. If $\lambda_1 = 1$ this contradicts $g > 1$. So, $\lambda_i = 2i$ for $0 \leq i \leq \frac{g}{2}-1$ and the remaining non-gaps between g and $2g$ are necessarily $\lambda_i = 2i$ for $i = \frac{g}{2}$ to $i = g$. □

Remark 2. *For $g = 4$, the maximum quasi-ordinarization number $\lfloor \frac{g-1}{2} \rfloor = 1$ is, in fact, attained by three of the 7 semigroups of genus 4. The semigroups whose quasi-ordinarization number is maximum are $\{0,3,6,\ldots\}, \{0,2,4,6,8,\ldots\}, \{0,3,5,6,8,\ldots\}$.*

For $g = 6$, the maximum quasi-ordinarization number $\lfloor \frac{g-1}{2} \rfloor = 2$ is, in fact, attained by two of the 23 semigroups of genus 6. The semigroups whose quasi-ordinarization number is maximum are $\{0,2,4,6,8,10,12,\ldots\}, \{0,4,5,8,9,10,12,\ldots\}$.

Hence, $g = 4$ and $g = 6$ are exceptional cases.

4. Analysis of $\varrho_{g,q}$

Let us denote $o_{g,r}$, the number of numerical semigroups of genus g and ordinarization number r and $\varrho_{g,q}$, the number of numerical semigroups of genus g and quasi-ordinarization number r.

We can observe a behavior of $\varrho_{g,q}$ very similar to the behavior of $o_{g,r}$ introduced in [9]. Indeed, for odd g and large r, it holds $\varrho_{g,q} = o_{g,r}$ and for even g and large q, it holds $\varrho_{g,q} = o_{g,r+1}$. We will give a partial proof of these equalities at the end of this section.

It is conjcetured in [9] that, for each genus $g \in \mathbb{N}_0$ and each ordinarization number $r \in \mathbb{N}_0$,

$$o_{g,r} \leqslant o_{g+1,r}.$$

We can write the new conjecture below paralleling this.

Conjecture 1. *For each genus $g \in \mathbb{N}_0$ and each quasi-ordinarization number $q \in \mathbb{N}_0$,*

$$\varrho_{g,q} \leqslant \varrho_{g+1,q}.$$

Now, we will provide some results on $\varrho_{g,q}$ for high quasi-ordinarization numbers. First, we will need Freĭman's Theorem [25,26] as formulated in [27].

Theorem 2 (Freĭman). *Let A be a set of integers such that $\#A = k \geqslant 3$. If $\#(A + A) \leqslant 3k - 4$, then A is a subset of an arithmetic progression of length $\#(A + A) - k + 1 \leqslant 2k - 3$.*

The next lemma is a consequence of Freĭman's Theorem. The lemma shows that the first non-gaps of numerical semigroups of large quasi-ordinarization number must be even.

Lemma 7. *If a semigroup Λ of genus g has quasi-ordinarization number q with $\frac{g+1}{3} \leqslant q \leqslant \lfloor \frac{g-1}{2} \rfloor$ then all its non-gaps which are less than or equal to $g - 1$ are even.*

Proof. Suppose that Λ is a semigroup of genus g and quasi-ordinarization number $q \geqslant \frac{g+1}{3}$.

Since the quasi-ordinarization is q, this means that $\lambda_0 = 0, \lambda_1, \ldots, \lambda_q \leqslant g - 1$ and $\lambda_{q+1} \geqslant g$, hence $\Lambda \cap [0, g-1] = \{\lambda_0, \lambda_1, \ldots, \lambda_q\}$. Let $A = \Lambda \cap [0, g-1]$. By the previous equality, $\#A = q + 1$. We have that the elements in $A + A$ are upper bounded by $2g - 2$ and so $A + A \subseteq \Lambda \cap [0, 2g - 2]$. Then, $\#(A + A) \leqslant \#(\Lambda \cap [0, 2g - 2]) < \#(\Lambda \cap [0, 2g])$. Since the Frobenius number of Λ is at most $2g - 1$, $\#(\Lambda \cap [0, 2g]) = g + 1$ and, so, $\#(A + A) \leqslant g$. Now, since $q \geqslant \frac{g+1}{3}$, we have $g \leqslant 3q - 1 = 3(q + 1) - 4 = 3(\#A) - 4$ and we can apply Theorem 2 with $k = q + 1$. Thus, we have that A is a subset of an arithmetic progression of length at most $g - k + 1 = g - q$.

Let $d(A)$ be the difference between consecutive terms of this arithmetic progression. The number $d(A)$ can not be larger than or equal to three since otherwise $\lambda_q \geqslant q \cdot d(A) \geqslant 3q \geqslant 3\frac{g+1}{3} > g$, a contradiction with q being the quasi-ordinarization number.

If $d(A) = 1$, then $A \subseteq [0, g - q - 1]$ and so $\Lambda \cap [g - q, g - 1] = \emptyset$. We claim that in this case $A \subseteq \{0\} \cup [\lceil \frac{g}{2} \rceil, g - q - 1]$. Indeed, suppose that $x \in A$. Then, $2x$ satisfies either $2x \leqslant g - q - 1$ or $2x \geqslant g$. If the second inequality is satisfied, then it is obvious that $x \in \{0\} \cup [\lceil \frac{g}{2} \rceil, g - q - 1]$. If the first inequality is satisfied, then we will prove that $mx \leqslant g - q - 1$ for all $m \geqslant 2$ by induction on m and this leads to $x = 0$. Indeed, if $mx \leqslant g - q - 1$, then $x \leqslant \frac{g-q-1}{m} \leqslant \frac{g - \frac{g+1}{3} - 1}{m} = \frac{2g-4}{3m} < \frac{2g}{3m}$. Now, $(m + 1)x < \frac{2g(m+1)}{3m} = \frac{(2m+2)g}{3m}$ and since $m \geqslant 2$, we have $(m + 1)x < \frac{(2m+m)g}{3m} = g$ and so $(m + 1)x \leqslant g - 1$. Since $(m + 1)x$ is in $\Lambda \cap [0, g - 1] = A \subseteq [0, g - q - 1]$, this means that $(m + 1)x \leqslant g - q - 1$ and this proves the claim.

Now, $A \subseteq \{0\} \cup [\lceil \frac{g}{2} \rceil, g - q - 1]$ together with $\#A = q + 1$ implies that $q \leqslant g - q - \lceil \frac{g}{2} \rceil = \lfloor \frac{g}{2} \rfloor - q \leqslant \frac{g}{2} - \frac{g+1}{3} = \frac{g-2}{6} < q$, a contradiction.

So, we deduce that $d(A) = 2$, leading to the proof of the lemma. □

The next lemma was proved in [9].

Lemma 8. *Suppose that a numerical semigroup Λ has ω gaps between 1 and $n - 1$ and $n \geqslant 2\omega + 2$, then*
1. *$n \in \Lambda$,*
2. *the Frobenius number of Λ is smaller than n,*
3. *the genus of Λ is ω.*

Let Λ be a numerical semigroup. As in [9], let us say that a set $B \subset \mathbb{N}_0$ is Λ-closed if for any $b \in B$ and any λ in Λ, the sum $b + \lambda$ is either in B or it is larger than $\max(B)$. If B is Λ-closed, so is $B - \min(B)$. Indeed, $b - \min(B) + \lambda = (b + \lambda) - \min(B)$ is either in $B - \min(B)$ or it is larger than $\max(B) - \min(B) = \max(B - \min(B))$. The new Λ-closed set contains 0. We will denote by $C(\Lambda, i)$, the Λ-closed sets of size i that contain 0.

Let $\mathcal{S}_\gamma$ be the set of numerical semigroups of genus γ. It was proved in [9] that, for r, an integer with $\frac{g+2}{3} \leqslant r \leqslant \lfloor \frac{g}{2} \rfloor$, it holds

$$o_{g,r} = \sum_{\Omega \in \mathcal{S}_{(\lfloor \frac{g}{2} \rfloor - r)}} \#C(\Omega, \lfloor \tfrac{g}{2} \rfloor - r + 1).$$

We will see now that, for q an integer with $\frac{g+1}{3} \leqslant q \leqslant \lfloor \frac{g-1}{2} \rfloor$, it holds

$$\varrho_{g,q} = \sum_{\Omega \in \mathcal{S}_{(\lfloor \frac{g-1}{2} \rfloor - q)}} \#C(\Omega, \lfloor \tfrac{g-1}{2} \rfloor - q + 1).$$

This proves that, for q, an integer with $\frac{g+2}{3} \leqslant q \leqslant \lfloor \frac{g-1}{2} \rfloor$, we have

$$\varrho_{g,q} = \begin{cases} o_{g,q} & \text{if } g \text{ is odd,} \\ o_{g,q+1} & \text{if } g \text{ is even.} \end{cases}$$

Theorem 3. *Let $g \in \mathbb{N}_0$, $g \geqslant 1$, and let q be an integer with $\frac{g+1}{3} \leqslant q \leqslant \lfloor \frac{g-1}{2} \rfloor$. Define $\omega = \lfloor \frac{g-1}{2} \rfloor - q$*
1. *If Ω is a numerical semigroup of genus ω and B is an Ω-closed set of size $\omega + 1$ and first element equal to 0 then*

$$\{2j : j \in \Omega\} \cup \{2j - 2\max(B) + 2g + 1 : j \in B\} \cup (2g + \mathbb{N}_0)$$

is a numerical semigroup of genus g and quasi-ordinarization number equal to q.

2. *All numerical semigroups of genus g and quasi-ordinarization number q can be uniquely written as*

$$\{2j : j \in \Omega\} \cup \{2j - 2\max(B) + 2g + 1 : j \in B\} \cup (2g + \mathbb{N}_0)$$

for a unique numerical semigroup Ω of genus ω and a unique Ω-closed set B of size $\omega + 1$ and first element equal to 0.

3. *The number $\rho_{g,q}$ of numerical semigroups of genus g and quasi-ordinarization number q depends only on ω. It is exactly*

$$\sum_{\Omega \in \mathcal{S}_\omega} \#C(\Omega, \omega + 1).$$

23

Proof.

1. Suppose that Ω is a numerical semigroup of genus w and B is an Ω-closed set of size $w+1$ and first element equal to 0. Let $X = \{2j : j \in \Omega\}$, $Y = \{2j - 2\max(B) + 2g + 1 : j \in B\}$, and $Z = (2g + \mathbb{N}_0)$.

   First of all, let us see that the complement $\mathbb{N}_0 \setminus (X \cup Y \cup Z)$ has g elements. Notice that all elements in X are even while all elements in Y are odd. So, X and Y do not intersect. Additionally, the unique element in $Y \cap Z$ is $2g + 1$. The number of elements in the complement will be

   $$\begin{aligned}\#\mathbb{N}_0 \setminus (X \cup Y \cup Z) &= 2g - \#\{x \in X : x < 2g\} - \#Y + 1 \\ &= 2g - \#\{s \in \Omega : s < g\} - \#B + 1 \\ &= 2g - w - \#\{s \in \Omega : s < g\}.\end{aligned}$$

   We know that all gaps of Ω are at most $2w - 1 = 2(\lfloor\frac{g-1}{2}\rfloor - q) - 1 \leq g - 1 - 2q - 1 < g$. So, $\#\{s \in \Omega : s < g\} = g - w$ and we conclude that $\#\mathbb{N}_0 \setminus (X \cup Y \cup Z) = g$.

   Before proving that $X \cup Y \cup Z$ is a numerical semigroup, let us prove that the number of non-zero elements in $X \cup Y \cup Z$, which are smaller than or equal to $g - 1$ is q. Once we prove that $X \cup Y \cup Z$ is a numerical semigroup, this will mean, by Lemma 3, that it has quasi-ordinarization number q. On the one hand, all elements in Y are larger than $g - 1$. Indeed, if λ is the enumeration of Ω (i.e., $\Omega = \{\lambda_0, \lambda_1, \dots\}$ with $\lambda_i < \lambda_{i+1}$), then $\max(B) \leq \lambda_w \leq 2w = 2\lfloor\frac{g-1}{2}\rfloor - 2q \leq g - 1 - 2\frac{g+1}{3} < \frac{g}{3}$. Now, for any $j \in B$, $2j - 2\max(B) + 2g + 1 > 2g - 2\max(B) > g$. On the other hand, all gaps of Ω are at most $2w - 1 = 2\lfloor\frac{g-1}{2}\rfloor - 2q - 1 < g - \frac{2(g+1)}{3} - 1 < \frac{g}{3} - 1$ and so all the even integers not belonging to X are less than g. So, the number of non-zero non-gaps of $X \cup Y \cup Z$ smaller than or equal to $g - 1$ is $\lfloor\frac{g-1}{2}\rfloor - w = q$.

   To see that $X \cup Y \cup Z$ is a numerical semigroup, we only need to see that it is closed under addition. It is obvious that $X + Z \subseteq Z$, $Y + Z \subseteq Z$, $Z + Z \subseteq Z$. It is also obvious that $X + X \subseteq X$ since Ω is a numerical semigroup and that $Y + Y \subseteq Z$ since, as we proved before, all elements in Y are larger than g.

   It remains to see that $X + Y \subseteq X \cup Y \cup Z$. Suppose that $x \in X$ and $y \in Y$. Then, $x = 2i$ for some $i \in \Omega$ and $y = 2j - 2\max(B) + 2g + 1$ for some $j \in B$. Then, $x + y = 2(i + j) - 2\max(B) + 2g + 1$. Since B is Ω-closed, we have that either $i + j \in B$ and so $x + y \in Y$ or $i + j > \max(B)$. In this case, $x + y \in Z$. So, $X + Y \subseteq Y \cup Z$.

2. First of all notice that, since the Frobenius number of a semigroup Λ of genus g is smaller than $2g$, it holds

   $$\Lambda \cap (2g + \mathbb{N}_0) = (2g + \mathbb{N}_0).$$

   For any numerical semigroup Λ, the set $\Omega = \{\frac{j}{2} : j \in \Lambda \cap (2\mathbb{N}_0)\}$ is a numerical semigroup. If Λ has a quazi-ordinarization number $q \geq \frac{g+1}{3}$ then, by Lemma 7,

   $$\Lambda \cap [0, g - 1] = (2\Omega) \cap [0, g - 1].$$

   The semigroup Ω has exactly $q + 1$ non-gaps between 0 and $\lfloor\frac{g-1}{2}\rfloor$ and $\omega = \lfloor\frac{g-1}{2}\rfloor - q$ gaps between 0 and $\lfloor\frac{g-1}{2}\rfloor$. We can use Lemma 8 with $n = \lfloor\frac{g+1}{2}\rfloor$ since

   $$2w + 2 = 2\left\lfloor\frac{g-1}{2}\right\rfloor - 2q + 2 \leq g - 1 - \frac{2(g+1)}{3} + 2 = \frac{g+1}{3},$$

   which implies $2w + 2 \leq \frac{g+1}{3} \leq \lfloor\frac{g+1}{2}\rfloor = n$. Then, the genus of Ω is ω and the Frobenius number of Ω is at most $\lfloor\frac{g+1}{2}\rfloor$. This means that all even integers larger than $g - 1$ belong to Λ.

Define $D = (\Lambda \cap [0, 2g]) \setminus 2\Omega$. That is, D is the set of odd non-gaps of Λ smaller than $2g$. We claim that $\bar{B} = \{\frac{i-1}{2} : i \in D \cup \{2g+1\}\}$ is a Ω-closed set of size $\omega + 1$. The size follows from the fact that the number of non-gaps of Λ between g and $2g$ is $g - q$ and that the number of even integers in the same interval is $\lceil \frac{g+1}{2} \rceil$. Suppose that $\lambda \in \Omega$ and $b \in \bar{B}$. Then, $b = \frac{j-1}{2}$ for some j in $D \cup \{2g+1\}$ and $b + \lambda = \frac{(j+2\lambda)-1}{2}$. If $\frac{(j+2\lambda)-1}{2} \geq \max(\bar{B}) = \frac{(2g+1)-1}{2}$, we are done. Otherwise, we have $j + 2\lambda \leq 2g$. Since Λ is a numerical semigroup and both $j, 2\lambda \in \Lambda$, it holds $j + 2\lambda \in \Lambda \cap [0, 2g]$. Furthermore, $j + 2\lambda$ is odd since j is also. So, $b + \lambda$ is either larger than $\max(\bar{B})$ or it is in $\bar{B}$. Then, $B = \bar{B} - \min(\bar{B})$ is a Λ-closed set of size $\omega + 1$ and first element zero.

3. The previous two points define a bijection between the set of numerical semigroups in $\mathcal{S}_g$ of quasi-ordinarization number q and the set $\{C(\Omega, \omega + 1) : \Omega \in \mathcal{S}_\omega\}$. Hence, $\varrho_{g,q} = \sum_{\Omega \in \mathcal{S}_\omega} \#C(\Omega, \omega + 1)$. □

Corollary 1. *Suppose that $\frac{g+2}{3} \leq q \leq \lfloor \frac{g-1}{2} \rfloor$. Then,*

$$\varrho_{g,q} = \begin{cases} o_{g,q} & \text{if } g \text{ is odd,} \\ o_{g,q+1} & \text{if } g \text{ is even.} \end{cases}$$

Define, as in [9], the sequence f_ω by $f_\omega = \sum_{\Omega \in \mathcal{S}_\omega} \#C(\Omega, \omega + 1)$. The first elements in the sequence, from f_0 to f_{15} are

ω	0	1	2	3	4	5	6	7	8	9	10	11	12	13	14	15
f_ω	1	2	7	23	68	200	615	1764	5060	14,626	41,785	117,573	332,475	933,891	2,609,832	7,278,512

We remark that this sequence appears in [5], where Bernardini and Torres proved that the number of numerical semigroups of genus 3ω whose number of even gaps is ω is exactly f_ω. It corresponds to the entry A210581 of The On-Line Encyclopedia of Integer Sequences [23].

We can deduce the values $\varrho_{g,q}$ using the values in the previous table together with Theorem 3 for any g, whenever $q \geq \max(\frac{g+2}{3}, \lfloor \frac{g-1}{2} \rfloor - 15)$.

The next corollary is a consequence of the fact that the sequence f_ω is increasing for ω between 0 and 15.

Corollary 2. *For any $g \in \mathbb{N}$ and any $q \geq \max(\frac{g}{3} + 1, \lfloor \frac{g}{2} \rfloor - 15)$, it holds $\varrho_{g,q} \geq \varrho_{g+1,q}$.*

If we proved that $f_\omega \leq f_{\omega+1}$ for any ω, this would imply $\varrho_{g,q} \leq \varrho_{g+1,q}$ for any $q > \frac{g}{3}$.

5. The Forest $\mathcal{F}_g$

Fix a genus g. We can define a graph in which the nodes are all semigroups of that genus and whose edges connect each semigroup to its quasi-ordinarization transform, if it is not itself. The graph is a forest $\mathcal{F}_g$ rooted at all ordinary and quasi-ordinary semigroups of genus g. In particular, the quasi-ordinarization transform defines, for each fixed genus and conductor, a tree rooted at the unique quasi-ordinary semigroup of that genus and conductor, given in Lemma 2. See $\mathcal{F}_4$ in Figure 3, $\mathcal{F}_6$ in Figure 4, and $\mathcal{F}_7$ in Figure 5.

In the forest $\mathcal{F}_g$, we know that the parent of a numerical semigroup that is not a root is its quasi-ordinarization transform. Let us analyze now, what the children of a numerical semigroup are. The next result is well known and can be found, for instance, in [2]. We use Λ^* to denote $\Lambda \setminus \{0\}$.

Lemma 9. *Suppose that Λ is a numerical semigroup and that $a \in \mathbb{N}_0 \setminus \Lambda$. The set $a \cup \Lambda$ is a numerical semigroup if and only if*

- $a + \Lambda^* \subseteq \Lambda^*$,
- $2a \in \Lambda$,

- $3a \in \Lambda$.

The elements $a \in \mathbb{N}_0 \setminus \Lambda$ such that $a + \Lambda \subseteq \Lambda$, are denoted *pseudo-Frobenius numbers* of Λ. The elements $a \in \mathbb{N}_0 \setminus \Lambda$ such that $\{2a, 3a\} \subseteq \Lambda$, are denoted *fundamental gaps* of Λ. The elements satisfying the three conditions will be called *candidates*.

Suppose that a numerical semigroup Λ with Frobenius number F has children in $\mathcal{F}_g$. Let $e_1, \ldots, e_r$ be the generators of Λ between the subconductor and $F - 1$. For $i = 1, \ldots, r$, let $c_1^i, \ldots, c_{k_i}^i$ be the candidates of $\Lambda \setminus \{e_i\}$. The children of Λ in $\mathcal{F}_g$ are the semigroups of the form $\Lambda \setminus \{e_i\} \cup \{c_j^i\}$, for $i = 1, \ldots, r$ and $j = 1, \ldots, k_i$.

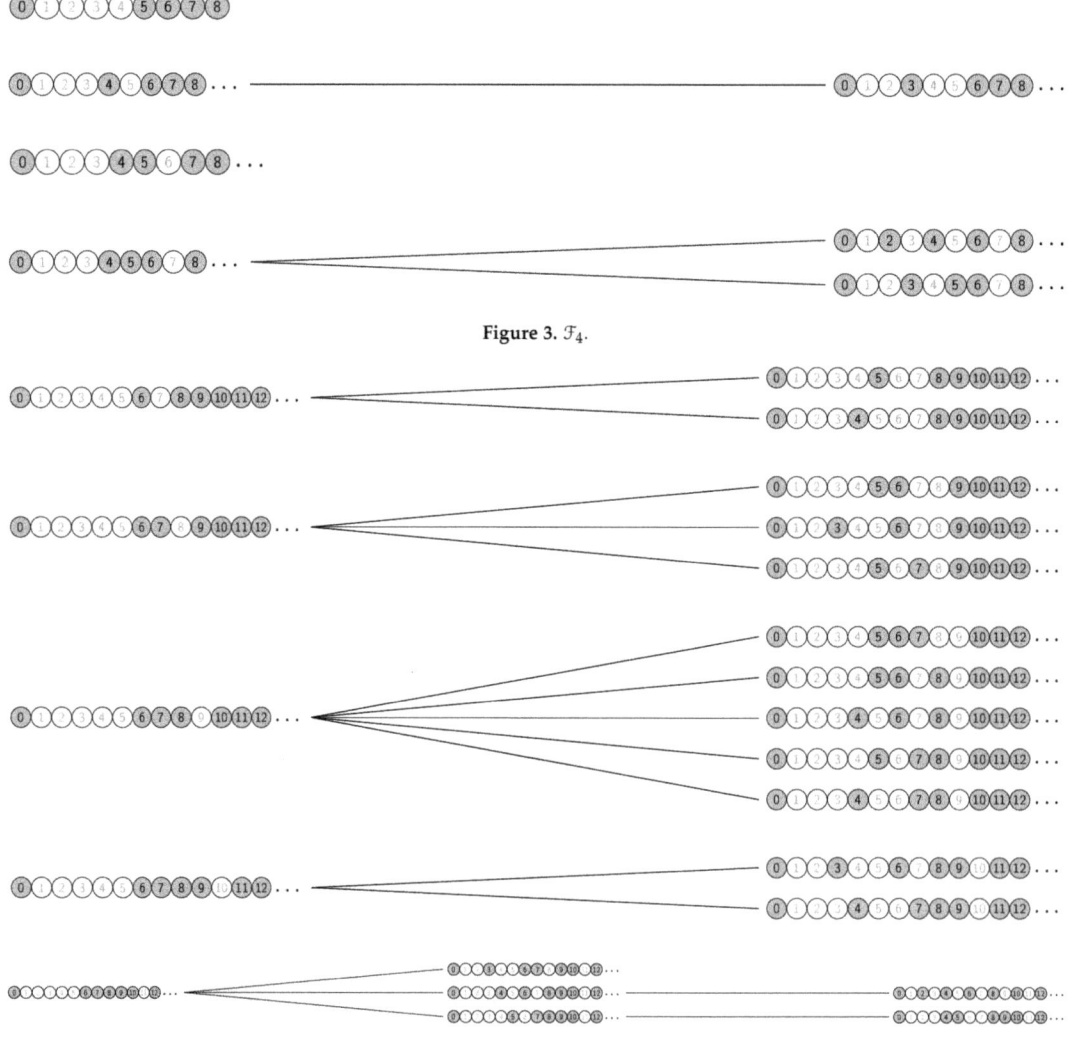

Figure 3. $\mathcal{F}_4$.

Figure 4. $\mathcal{F}_6$.

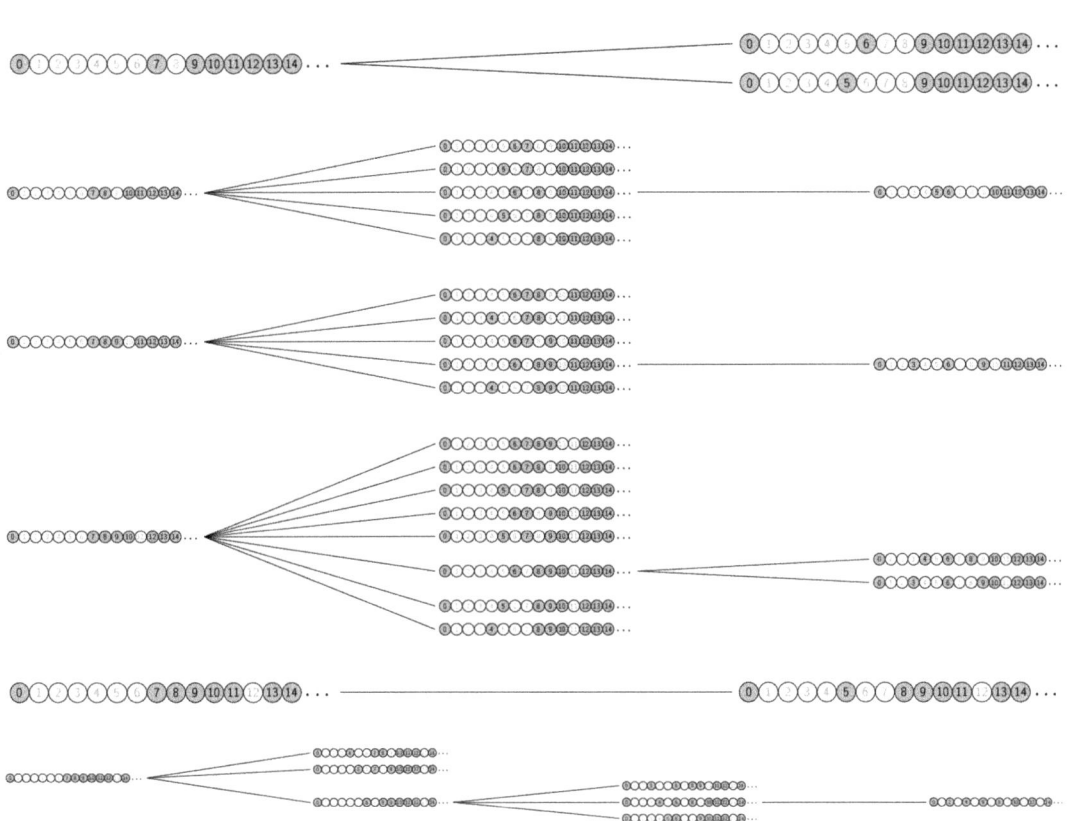

Figure 5. $\mathcal{F}_7$.

6. Relating $\mathcal{F}_g$, $\mathcal{T}_g$, and $\mathcal{T}$

Now, we analyze the relation between the kinship of different nodes in $\mathcal{F}_g$, $\mathcal{T}_g$, and $\mathcal{T}$. If two semigroups are children of the same semigroup Λ, then they are called *siblings*. If Λ_1 and Λ_2 are siblings, and Λ_3 is a child of Λ_2, then we say that Λ_3 is a *niece/nephew* of Λ_1.

Let $q(\Lambda)$ denote the quasi-ordinarization of Λ. The next lemmas are quite immediate from the definitions.

Lemma 10. *If Λ_1 is a child of Λ_2 in $\mathcal{T}$, then $q(\Lambda_1)$ is a niece/nephew of $q(\Lambda_2)$ in $\mathcal{T}$.*

As an example, $\Lambda_1 = \{0, 4, 5, 8, 9, 10, 12, \dots\}$ is a child of $\Lambda_2 = \{0, 4, 5, 8, \dots\}$ in $\mathcal{T}$, while $q(\Lambda_1) = \{0, 5, 7, 8, 9, 10, 12, \dots\}$ is a niece of $q(\Lambda_2) = \{0, 5, 6, 8, \dots\}$ in $\mathcal{T}$.

Lemma 11. *If Λ_1 and Λ_2 are siblings in $\mathcal{T}$, then they are siblings in $\mathcal{T}_g$ but not in $\mathcal{F}_g$.*

As an example, $\Lambda_1 = \{0, 5, 7, 9, 10, 11, 12, 14, \dots\}$ and $\Lambda_2 = \{0, 5, 7, 9, 10, 12, \dots\}$ are siblings in $\mathcal{T}$ and in $\mathcal{T}_7$ (see Figure 2), but they are not siblings in $\mathcal{F}_7$ (see Figure 5).

Lemma 12. *If Λ_1 and Λ_2 are siblings in $\mathcal{T}_g$, then $q(\Lambda_1)$ and $q(\Lambda_2)$ are siblings in $\mathcal{T}$.*

As an example, $\Lambda_1 = \{0, 3, 6, 9, 10, 12, \dots\}$ and $\Lambda_2 = \{0, 5, 6, 10, \dots\}$ are siblings in $\mathcal{T}_7$ (see Figure 2), and $q(\Lambda_1) = \{0, 6, 8, 9, 10, 12, \dots\}$ and $q(\Lambda_2) = \{0, 6, 8, 10, \dots\}$ are siblings in $\mathcal{T}$.

As a consequence of the previous two lemmas, we obtain this final lemma.

Lemma 13. *If Λ_1 and Λ_2 are siblings in $\mathcal{T}$, then $q(\Lambda_1)$ and $q(\Lambda_2)$ are siblings in $\mathcal{T}$.*

As an example, $\Lambda_1 = \{0, 5, 7, 9, 10, 11, 12, 14, \dots\}$ and $\Lambda_2 = \{0, 5, 7, 9, 10, 12, \dots\}$ are siblings in $\mathcal{T}$ and $q(\Lambda_1) = \{0, 7, 8, 9, 10, 11, 12, 14, \dots\}$ and $q(\Lambda_2) = \{0, 7, 8, 9, 10, 12, \dots\}$ are siblings in $\mathcal{T}$.

7. Conclusions

Quasi-ordinary semigroups are those semigroups that have all gaps except one in a row, while ordinary semigroups have all gaps in a row.

We defined a quasi-ordinarization transform that, applied repeatedly to a non-ordinary numerical semigroup stabilizes in a quasi-ordinary semigroup of the same genus.

From this transform, fixing a genus g, we can define a forest $\mathcal{F}_g$ whose nodes are all semigroups of genus g, whose roots are all ordinary and quasi-ordinary semigroups of that genus, and whose edges connect each non-ordinary and non-quasi-ordinary numerical semigroup to its quasi-ordinarization transform.

We conjectured that the number of numerical semigroups in $\mathcal{F}_g$ at a given depth is at most the number of numerical semigroups in $\mathcal{F}_{g+1}$ at the same depth. We provided a proof of the conjecture for the largest possible depths. Proving this conjecture for all depths would prove the conjecture that $n_{g+1} \geqslant n_g$. Hence, we expect our work to be a step toward the proof of the conjectured increasingness of the sequence n_g.

Author Contributions: All authors contributed equally. All authors have read and agreed to the published version of the manuscript.

Funding: This work was partly supported by the Catalan Government under grant 2017 SGR 00705 and by the Spanish Ministry of Economy and Competitivity under grant TIN2016-80250-R.

Institutional Review Board Statement: Not Applicable.

Informed Consent Statement: Not Applicable.

Data Availability Statement: Not Applicable.

Conflicts of Interest: The authors declare no conflict of interest.

References

1. Rosenfeld, V.R.; Nordahl, T.E. Semigroup theory of symmetry. *J. Math. Chem.* **2016**, *54*, 1758–1776. [CrossRef]
2. Rosales, J.; García-Sánchez, P. *Numerical Semigroups*; Volume 20 of Developments in Mathematics; Springer: New York, NY, USA, 2009.
3. Bras-Amorós, M. Tempered monoids of real numbers, the golden fractal monoid, and the well-tempered harmonic semigroup. *Semigroup Forum* **2019**, *99*, 496–516. [CrossRef]
4. Bras-Amorós, M. Fibonacci-like behavior of the number of numerical semigroups of a given genus. *Semigroup Forum* **2008**, *76*, 379–384. [CrossRef]
5. Bernardini, M.; Torres, F. Counting numerical semigroups by genus and even gaps. *Discret. Math.* **2017**, *340*, 2853–2863. [CrossRef]
6. Blanco, V.; Rosales, J. The set of numerical semigroups of a given genus. *Semigroup Forum* **2012**, *85*, 255–267. [CrossRef]
7. Blanco, V.; García-Sánchez, P.; Puerto, J. Counting numerical semigroups with short generating functions. *Int. J. Algebra Comput.* **2011**, *21*, 1217–1235. [CrossRef]
8. Bras-Amorós, M. Bounds on the number of numerical semigroups of a given genus. *J. Pure Appl. Algebra* **2009**, *213*, 997–1001. [CrossRef]
9. Bras-Amorós, M. The ordinarization transform of a numerical semigroup and semigroups with a large number of intervals. *J. Pure Appl. Algebra* **2012**, *216*, 2507–2518. [CrossRef]
10. Bras-Amorós, M.; Bulygin, S. Towards a better understanding of the semigroup tree. *Semigroup Forum* **2009**, *79*, 561–574. [CrossRef]

1. Bras-Amorós, M.; de Mier, A. Representation of numerical semigroups by Dyck paths. *Semigroup Forum* **2007**, *75*, 677–682. [CrossRef]
2. Bras-Amorós, M.; Fernández-González, J. The right-generators descendant of a numerical semigroup. *Math. Comp.* **2020**, *89*, 2017–2030. [CrossRef]
3. Elizalde, S. Improved bounds on the number of numerical semigroups of a given genus. *J. Pure Appl. Algebra* **2010**, *214*, 1862–1873. [CrossRef]
4. Fromentin, J.; Hivert, F. Exploring the tree of numerical semigroups. *Math. Comp.* **2016**, *85*, 2553–2568. [CrossRef]
5. Kaplan, N. Counting numerical semigroups by genus and some cases of a question of Wilf. *J. Pure Appl. Algebra* **2012**, *216*, 1016–1032. [CrossRef]
6. Kaplan, N. Counting numerical semigroups. *Am. Math. Mon.* **2017**, *124*, 862–875. [CrossRef]
7. Komeda, J. *On Non-Weierstrass Gap Sequqences*; Technical Report; Kanagawa Institute of Technology: Kanagawa, Japan, 1989.
8. Komeda, J. Non-Weierstrass numerical semigroups. *Semigroup Forum* **1998**, *57*, 157–185. [CrossRef]
9. O'Dorney, E. Degree asymptotics of the numerical semigroup tree. *Semigroup Forum* **2013**, *87*, 601–616. [CrossRef]
10. Zhao, Y. Constructing numerical semigroups of a given genus. *Semigroup Forum* **2010**, *80*, 242–254. [CrossRef]
11. Zhai, A. Fibonacci-like growth of numerical semigroups of a given genus. *Semigroup Forum* **2013**, *86*, 634–662. [CrossRef]
12. Bras-Amorós, M. On Numerical Semigroups and Their Applications to Algebraic Geometry Codes, in Thematic Seminar "Algebraic Geometry, Coding and Computing", Universidad de Valladolid en Segovia. 2007. Available online: http://www.singacom.uva.es/oldsite/seminarios/WorkshopSG/workshop2/Bras_SG_2007.pdf (accessed on 1 June 2021).
13. OEIS Foundation Inc. The On-Line Encyclopedia of Integer Sequences. 2021. Available online: http://oeis.org (accessed on 1 June 2021).
14. Rosales, J.C.; García-Sánchez, P.A.; García-García, J.I.; Jiménez Madrid, J.A. The oversemigroups of a numerical semigroup. *Semigroup Forum* **2003**, *67*, 145–158. [CrossRef]
15. Freĭman, G.A. On the addition of finite sets. *Dokl. Akad. Nauk SSSR* **1964**, *158*, 1038–1041.
16. Freĭman, G.A. *Nachala Strukturnoi Teorii Slozheniya Mnozhestv*; Kazanskii Gosudarstvennyi Pedagogiceskii Institut: Kazan, Russia, 1966.
17. Nathanson, M.B. *Additive Number Theory*; Volume 165 of Graduate Texts in Mathematics; Springer: New York, NY, USA, 1996.

Article

Cryptanalysis of a Group Key Establishment Protocol

Jorge Martínez Carracedo [1,†] and Adriana Suárez Corona [2,*,†]

1 School of Computing and Mathematics, Ulster University, Belfast BT37 0QB, UK; j.martinez-carracedo@ulster.ac.uk
2 Department of Mathematical Sciences, Universidad de León, 24071 León, Spain
* Correspondence: asuac@unileon.es
† These authors contributed equally to this work.

Abstract: In this paper, we analyze the security of a group key establishment scheme proposed by López-Ramos et al. This proposal aims at allowing a group of users to agree on a common key. We present several attacks against the security of the proposed protocol. In particular, an active attack is presented, and it is also proved that the protocol does not provide forward secrecy.

Keywords: cryptanalysis; group key establishment

Citation: Carracedo, J.M.; Corona, A.S. Cryptanalysis of a Group Key Establishment Protocol. *Symmetry* **2021**, *13*, 332. https://doi.org/10.3390/sym13020332

Academic Editors: Juan Alberto Rodríguez Velázquez and Alejandro Estrada-Moreno

Received: 20 January 2021
Accepted: 9 February 2021
Published: 17 February 2021

Publisher's Note: MDPI stays neutral with regard to jurisdictional claims in published maps and institutional affiliations.

Copyright: © 2021 by the authors. Licensee MDPI, Basel, Switzerland. This article is an open access article distributed under the terms and conditions of the Creative Commons Attribution (CC BY) license (https://creativecommons.org/licenses/by/4.0/).

1. Introduction

Secure multiparty communication is an important concern for many current applications that work over public insecure channels, such as the Internet. Wireless sensor networks, collaborative applications, multiparty voice and video conferences, etc. need to guarantee confidentiality, integrity and authentication in their communications.

Group key establishment (GKE) protocols are fundamental in that sense. They allow a set of participants to agree on a common secret key to be used afterwards with symmetric key cryptographic primitives.

In some settings all the nodes play an equivalent role, and thus the group protocol is somewhat symmetric. Nevertheless, there are other applications where some nodes are distinguished and one can assume they may have more computational power and resources, and thus, they are required to perform more computations.

Over recent decades, group key establishment protocols were widely discussed in the literature [1–7], and formal security models were proposed, indicating which attacks the adversary can perform and what a secure key establishment protocol is. What is typically required is that, after completion of the protocol, the intended users agree on a common key, whereas the adversary does not learn anything about it.

A standard technique to augment the security of a scheme is the use of compilers, which allows a modular design, going from passively secure solutions to authenticated ones [8], from 2-party to group solutions [9], or adding forward secrecy [10].

However, several protocols were found to be insecure after they were published, because the proposals do not provide security proofs or the proofs are not correct [11–13]. Other protocols were found to be insecure when considering active attacks [14].

Motivated by the works in López-Ramos et al. [14], in this paper, we analyze a group key establishment proposal by López-Ramos et al. [15] and present several attacks on the proposed protocols. In particular, we present here some active attacks against the protocols, proving they are insecure when considering active adversaries.

Contributions: We present several concrete attacks showing the security flaws of the protocols proposed in López-Ramos et al. [15]. In Section 2, we review the proposal of López Ramos et al. Then, in Section 3 we review a standard security model for group key exchange. We then present our attacks in Section 4.

2. The Protocol of López Ramos et al.

In this section, we describe Protocol 1 in López-Ramos et al. [15], which can be seen as an extension of the classical 2-party Diffie-Hellman key exchange. Four different protocols are presented, which are modifications of this first one. In particular, Protocol 2 computes the same session key, but publishing only one public key and sending a different message in Round 2. Protocol 3 describe the extra steps to be done if some participants leave the group and Protocol 4 deals with the case where some users join the group.

Initialization

Let $\{U_1, \ldots, U_n\}$ be the finite set of protocol participants, including U_{c_1}, who will act as controller. The users agree on a multiplicative cyclic group G of prime order p and on g, a generator of G.

Each user $U_i, 1 \leq i \leq n$ will have two random values, $r_i, x_i \in \mathbb{Z}_p^*$ as private keys and g^{r_i} and g^{x_i} will be their public keys.

Round 1

1. Each user U_i publishes his pair of public keys (g^{r_i}, g^{x_i}) (We assume that these keys are sent to the users, hence the adversary can potentially manipulate those values).
2. The group controller calculate $K_1 = g^{r_{c_1} \sum_{j=1, j \neq c_1}^{n} r_j}$, which will be the session key.
3. The group controller will choose a new pair of elements (r'_{c_1}, x'_{c_1}) that will be privately kept and will become his new private information at a later stage.

Round 2

Every user U_i, using the public information, computes $g^{\sum_{j \neq i, c_1} r_j}$ and sends this value to U_{c_1} (Notice that there is no need to send this information, since this value can be computed from the published public keys).

The group controller U_{c_1}, moreover, computes

$$Y_{1,i} = g^{-x_{c_1} x_i} \left(g^{r_{c_1} \sum_{j \neq c_1, i} r_j} \right) \quad \text{for} \quad i \in \{1, \ldots, n\} \setminus \{c_1\} \quad \text{and}$$

$$Y_{1,c_1} = K_1 g^{-r'_{c_1} r_{c_1}} g^{-x'_{c_1} x_{c_1}},$$

$$R_1 = g^{r_{c_1}} \quad \text{and} \quad S_1 = g^{x_{c_1}}.$$

He broadcasts $(Y_{1,1}, \ldots, Y_{1,c_1}, \ldots, Y_{1,n}, R_1, S_1)$

Key Computation

Once user U_i has received the second round message, he computes the common session key $K_1 := K_{1,i} = Y_{1,i} S_1^{x_i} R_1^{r_i}$.

The protocol is summarized in Figure 1.

Remark 1. *The subindex 1 in the session key K_1 indicates here that it is the first execution of the protocol. In Protocols 3 and 4 in López-Ramos et al. [15], this subindex changes when the participants involved in the protocol change, i.e., some participants leave or join the protocol, and thus, some extra computations are needed.*

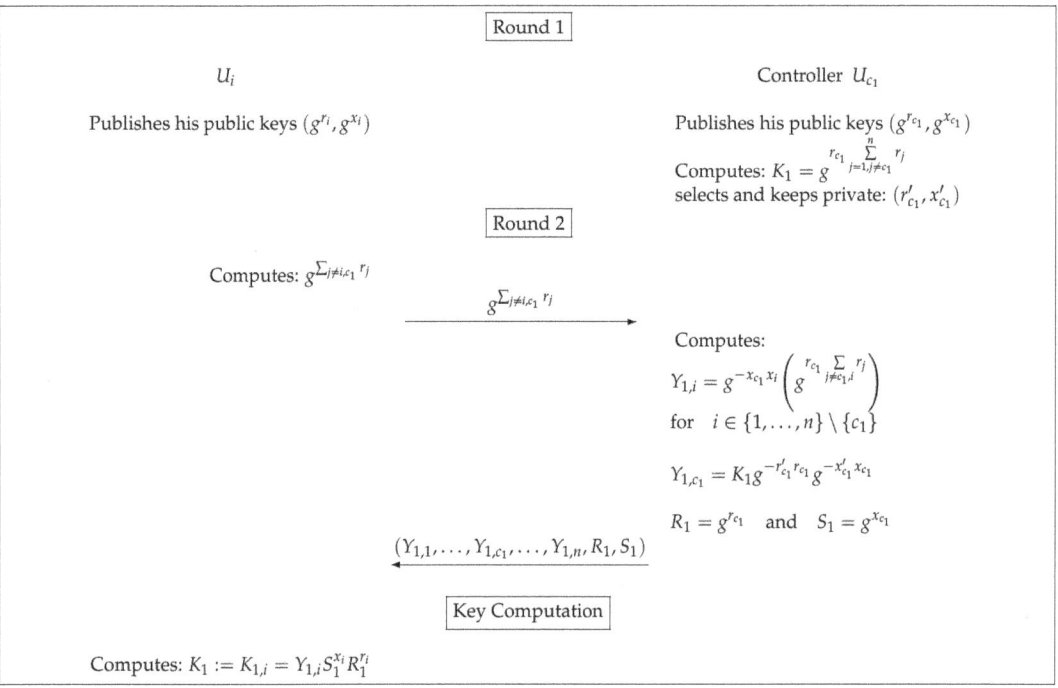

Figure 1. Protocol 1 of López Ramos et al.

3. Security Model

To formalize secure group key establishment, we use the somewhat standard Bohli et al.'s [5] security model, which builds on Jonathan Katz and Moti Yung [8].

Security Goals: Semantic Security and Authentication

Participants:

The (potential) protocol participants are modelled as probabilistic polynomial time (ppt) Turing machines in the finite set $\mathcal{U} = \{U_1, \ldots, U_n\}$. Each participant U_i in the set $\mathcal{U}$ is able to run a polynomial amount of protocol instances in parallel.

We will refer to instance s_i of principal U_i as $\Pi_i^{s_i}$ ($i \in \mathbb{N}$) and it has the following variables assigned:

pid$_i^{s_i}$: stores the identities of the parties user U_i aims at establishing a session key with (including U_i itself);

sid$_i^{s_i}$: is a variable storing a non-secret session identifier to the session key stored in sk$_i^{s_i}$;

acc$_i^{s_i}$: is a variable which indicates whether the session key in sk$_i^{s_i}$ was accepted;

term$_i^{s_i}$: is a variable which indicates whether the protocol execution has terminated;

used$_i^{s_i}$: is a variable which indicates whether this instance is taking part in a protocol run;

sk$_i^{s_i}$: this variable is initialized with a distinguished NULL value and will store the session key.

Communication network and adversarial capabilities:

We assume there exist arbitrary point to point connections among users and the network is non-private, fully asynchronous and in complete control of the adversary $\mathcal{A}$, who can eavesdrop, delay, delete, modify or insert messages. The adversary's capabilities are captured by the following *oracles*:

Send(U_i, s_i, M) : when querying this oracle, message M is sent to instance $\Pi_i^{s_i}$ of user $U_i \in \mathcal{U}$. The output will be the protocol message that the instance outputs after receiving message M. This oracle can also be used for the adversary $\mathcal{A}$ to initialize a protocol execution, by using the special message $M = \{U_{i_1}, \ldots, U_{i_r}\}$ to an unused instance $\Pi_j^{s_i}$. This oracle initializes a protocol run among $U_{i_1}, \ldots, U_{i_r} \in \mathcal{U}$. After such a query, $\Pi_i^{s_i}$ sets $\text{pid}_i^{s_i} := \{U_{i_1}, \ldots, U_{i_r}\}$, $\text{used}_i^{s_i} := \text{TRUE}$, and processes the first step of the protocol.

Execute($U_1, s_1, \ldots, U_r, s_r$) : if the instances $s_1, \ldots, s_r$ have not yet been used, this oracle will return a transcript of a complete execution of the protocol among the specified instances.

Reveal(U_i, s_i) : this oracle returns the session key stored in $\text{sk}_i^{s_i}$ if $\text{acc}_i^{s_i} = \text{TRUE}$ and a NULL value otherwise.

Corrupt(U_i) : this query returns U_i's long term secret key.

We can distinguish two types of adversaries. An adversary with access to all the oracles described above is considered to be *active*. If the adversary is not granted access to any of the Send oracles, then it is considered a *passive* adversary.

To define semantic security, we also allow the adversary to have access to a Test oracle, which can be queried only once. The query Test(U_i, s_i) can be made on input an instance $\Pi_i^{s_i}$ of user $U_i \in \mathcal{U}$ only if $\text{acc}_i^{s_i} = \text{TRUE}$. In that case, a bit $b \leftarrow \{0, 1\}$ is chosen uniformly at random; if $b = 0$, the oracle returns the session key stored in $\text{sk}_i^{s_i}$. Otherwise, the oracle outputs a uniformly at random chosen element from the space of session keys.

Security notions:

For the schemes to be useful, we need the group key establishments to be *correct*, i.e., without adversarial interference, the protocol would allow all users to compute the same key.

Definition 1 (correctness). *A group key establishment is correct if for all instances $\Pi_i^{s_i}, \Pi_j^{s_j}$ which accepted with $\text{sid}_i^{s_i} = \text{sid}_j^{s_j}$ and $\text{pid}_i^{s_i} = \text{pid}_j^{s_j}$, the condition $\text{sk}_i^{s_i} = \text{sk}_j^{s_j} \neq \text{NULL}$ is satisfied.*

To be more precise in the security definition, it is important to specify under which conditions the adversary can query the Test oracle. To do so, we first define the following notion of *partnering*:

Definition 2 (partnering). *Two terminated instances $\Pi_i^{s_i}$ and $\Pi_j^{s_j}$ are partnered if $\text{sid}_i^{s_i} = \text{sid}_j^{s_j}$, $\text{pid}_i^{s_i} = \text{pid}_j^{s_j}$ and $\text{acc}_{U_i}^{s_i} = \text{acc}_{U_j}^{s_j} = \text{TRUE}$.*

To avoid queries that would trivially allow the adversary to know the key, we restrict the instances that can be queried to the Test oracle, only allowing *fresh* instances:

Definition 3 (freshness). *We say an instance $\Pi_i^{s_i}$ is fresh if none of the following events has occurred:*

- *the adversary queried Reveal(U_j, s_j) for an instance $\Pi_j^{s_j}$ that is partnered with $\Pi_i^{s_i}$;*
- *the adversary queried Corrupt(U_j) for a user $U_j \in \text{pid}_i^{s_i}$ before a query of the form Send($U_l, s_l, *$);*

Remark 2. *The previous definition for freshness allows including the desired goal of* forward secrecy *in our definition of security given below: an adversary $\mathcal{A}$ is allowed to query Corrupt for all users and obtain their long term keys without violating freshness, if he does not send any message afterwards.*

Let $\text{Succ}_{\mathcal{A}}$ be the event that the adversary $\mathcal{A}$ queries the Test oracle with a fresh instance and makes a correct guess about the random bit b used by the Test oracle, we define the *advantage* of an adversary $\mathcal{A}$ attacking protocol P as

$$\text{Adv}_{\mathcal{A}}^{\text{ke}} = \text{Adv}_{\mathcal{A}}^{\text{ke}}(k) := \left| \Pr[\text{Succ}_{\mathcal{A}}] - \frac{1}{2} \right|.$$

Definition 4 (semantic security). *A group key establishment protocol is (semantically) secure, if $\text{Adv}_{\mathcal{A}}^{\text{ke}} = \text{Adv}_{\mathcal{A}}^{\text{ke}}(k)$ is negligible for every ppt adversary $\mathcal{A}$.*

4. Cryptanalysis of the Proposal of López-Ramos et al.

In this section, we describe several concrete attacks refuting the security results of López-Ramos et al. [15], where four different, but related, GKE protocols are described. The four protocols will be considered in this section. However, we will only explicitly attack Protocol 1, being the attacks to the others straightforwardly adapted.

4.1. Active Attack

Informally, since the protocol is not authenticated, we will describe here how an adversary can attack the protocol by mounting a Man-In-The-Middle attack. Users will end up sharing a key with the adversary, instead of with all the intended communication partners. We formalize the attack below.

Let us fix $\{U_1, ..., U_n\}$ the set of communication parties and let $\mathcal{A}$ be an active attacker able to supersede some parties in the set. We will distinguish two different cases: $\mathcal{A}$ shares a key with the group controller U_{c_1} and other with the rest of the users and $\mathcal{A}$ shares a key with any other party U_i, $i \neq c_1$, and a different key with the rest, including the controller.

If $\mathcal{A}$ tries to share a different key with the group controller U_{c_1} the adversary can build an attack by following the next steps:

1. The attacker $\mathcal{A}$ queries $\text{Send}(U_1, s_1, ..., U_n, s_n)$, to initiate a protocol instance. After this query, the first step of the protocol is executed. In particular, the adversary obtains every users' pairs of public keys (g^{r_i}, g^{x_i}), with $r_i, x_i \in \mathbb{Z}_p^*$.

2. The adversary $\mathcal{A}$ will delete the message $(g^{r_{c_1}}, g^{x_{c_1}})$ sent by the controller U_{c_1} to the rest of the users and delete the public keys (g^{r_1}, g^{x_1}) sent by user U_1 to U_{c_1}.

3. The adversary $\mathcal{A}$ generates its private keys $a_{c_1}, b_{c_1} \in \mathbb{Z}_p^*$ and public keys $(g^{a_{c_1}}, g^{b_{c_1}})$ and queries $\text{Send}(U_i, s_i, (g^{a_{c_1}}, g^{b_{c_1}}))$, for all $i \in \{1, ..., n\} \setminus \{c_1\}$. the adversary $\mathcal{A}$ generates its private keys $a_1, b_1 \in \mathbb{Z}_p^*$ and public keys (g^{a_1}, g^{b_1}) and queries $\text{Send}(U_{c_1}, s_{c_1}, (g^{a_1}, g^{b_1}))$.

   Notice that every user U_i, $i \neq 1, c_1$, after receiving that message, will compute and send the value $g^{\sum_{j=1, j \neq c_1}^{n} r_j}$ and therefore this value will be output by the Send oracle.

   The controller U_{c_1}, after receiving that message, will compute and send the value $g^{a_1 + \sum_{j=2, j \neq c_1}^{n} r_j}$ and therefore this value will be output by the Send oracle.

4. The adversary $\mathcal{A}$ will compute the session key $Q_1 = g^{a_{c_1}(\sum_{j=1, j \neq c_1}^{n} r_j)}$ and the values $T_1 = g^{a_{c_1}}$ and $V_1 = g^{b_{c_1}}$, along with the keying values

$$Z_{1,i} = g^{-b_{c_1} x_i} g^{a_{c_1}(\sum_{j=1, j \neq c_1, i}^{n} r_j)}, \quad i \neq c_1$$

$$Z_{1,c_1} = Q_1 g^{-a'_{c_1} a_{c_1}} g^{-b'_{c_1} b_{c_1}}.$$

5. The adversary $\mathcal{A}$ will query $\text{Send}(U_i, s_i, (Z_{1,1}, ..., Z_{1,n}, T_1, V_1))$ oracle, for all $i \in \{1, ..., n\} \setminus \{c_1\}$.

6. The adversary $\mathcal{A}$ will compute the session key $K_1 = g^{r_{c_1}(\sum_{j=1,j\neq c_1}^{n} r_j)}$ and the values $R_1 = g^{r_{c_1}}$ and $S_1 = g^{x_{c_1}}$, along with the keying values

$$Y_{1,i} = g^{-x_{c_1}x_i}g^{r_{c_1}(a_1+\sum_{j=2,j\neq c_1,i}^{n} r_j)}, \quad i \neq c_1$$

$$Y_{1,c_1} = K_1 g^{-r'_{c_1}r_{c_1}}g^{-x'_{c_1}x_{c_1}}.$$

7. The adversary $\mathcal{A}$ will query $Send(\mathcal{U}_1, s_1, (Y_{1,1}, \ldots, Y_{1,n}, S_1, T_1))$ oracle.

Please note that after receiving this last message, users $\{\mathcal{U}_1, \ldots, \mathcal{U}_n\} \setminus \{\mathcal{U}_{c_1}\}$, following the protocol, will compute $Q_{1,i} = Z_{1,i}T_1^{x_i}V_1^{r_i}$. Please note that $Q_{1,i} = Q_1$ for every $i \neq c_1$.
On the other hand, the group controller $\mathcal{U}_{c_1}$ will compute $K_1 = Y_{1,1}S_1^{b_1}R_1^{a_1}$.
Therefore, after this attack, the adversary has established a shared key Q_1 with the set of parties $\{\mathcal{U}_1, \ldots, \mathcal{U}_n\} \setminus \{\mathcal{U}_{c_1}\}$ and the key K_1 with the group controller $\mathcal{U}_{c_1}$, where

$$Q_1 = g^{a_{c_1} \sum_{j=1,j\neq c_1}^{n} r_j} \quad \text{and} \quad K_1 = g^{r_{c_1}(a_1+\sum_{j=2,j\neq c_1}^{n} r_j)}.$$

Consequently, all the users will believe they are establishing a common key when they are not. Moreover, the adversary can decrypt the messages sent encrypted with both keys and forward the communication between the users that do not share a key.

This attack is outlined in Figure 2.

If $\mathcal{A}$ tries to compute a different key with any user different from the group controller, we can assume without loss of generality that $\mathcal{A}$ is sharing it with $\mathcal{U}_1$. The adversary $\mathcal{A}$ can build an attack following the subsequent steps:

1. The attacker $\mathcal{A}$ queries $Send(\mathcal{U}_1, s_1, \ldots, \mathcal{U}_n, s_n)$, to initiate a protocol instance. After this query, the first step of the protocol is executed. In particular, the users send their public keys and thus, the adversary obtains (g^{r_i}, g^{x_i}), with $r_i, x_i \in \mathbb{Z}_p^*$ for all the participants $\{\mathcal{U}_1, \ldots, \mathcal{U}_n\}$.
2. The adversary $\mathcal{A}$ will delete the message $(g^{r_{c_1}}, g^{x_{c_1}})$ sent by the controller $\mathcal{U}_{c_1}$ to user $\mathcal{U}_1$ and the message (g^{r_1}, g^{x_1}) sent by user $\mathcal{U}_1$ to the rest of the participants.
3. The adversary $\mathcal{A}$, will choose random values $a_1, b_1, a_{c_1}, b_{c_1} \in \mathbb{Z}_p^*$, and queries $Send(\mathcal{U}_i, s_i, (g^{a_1}, g^{b_1}))$, for all $i \in \{2, \ldots, n\}$, including c_1.

Notice that every user $\mathcal{U}_i, i \neq 1, c_1$ and the adversary $\mathcal{A}$, after receiving that message, will compute $g^{(a_1+\sum_{j=2,j\neq c_1}^{n} r_j)}$ and therefore this value will be output by the Send oracle.

Moreover, the group controller $\mathcal{U}_{c_1}$ will calculate the session key $Q_1 = g^{r_{c_1}(a_1+\sum_{j=2,j\neq c_1}^{n} r_j)}$ and he will send $R_1 = g^{r_{c_1}}$ and $S_1 = g^{x_{c_1}}$, along with the keying values

$$Z_{1,1} = g^{-x_{c_1}b_1}g^{(r_{c_1}\sum_{j=2,j\neq c_1}^{n} r_j)},$$

$$Z_{1,i} = g^{-x_{c_1}x_i}g^{r_{c_1}(a_1+\sum_{j=2,j\neq c_1}^{n} r_j)} \quad i \neq 1, c_1,$$

$$Z_{1,c_1} = Q_1 g^{-r'_{c_1}r_{c_1}}g^{-x'_{c_1}x_{c_1}}.$$

These values will also be part of the output of the Send oracle.
Please note that after receiving this message every user $\mathcal{U}_i, i \neq 1$, can compute the key $Q_1 = Z_{1,i}S_1^{x_i}R_1^{r_i}$ that will be shared with the adversary $\mathcal{A}$.

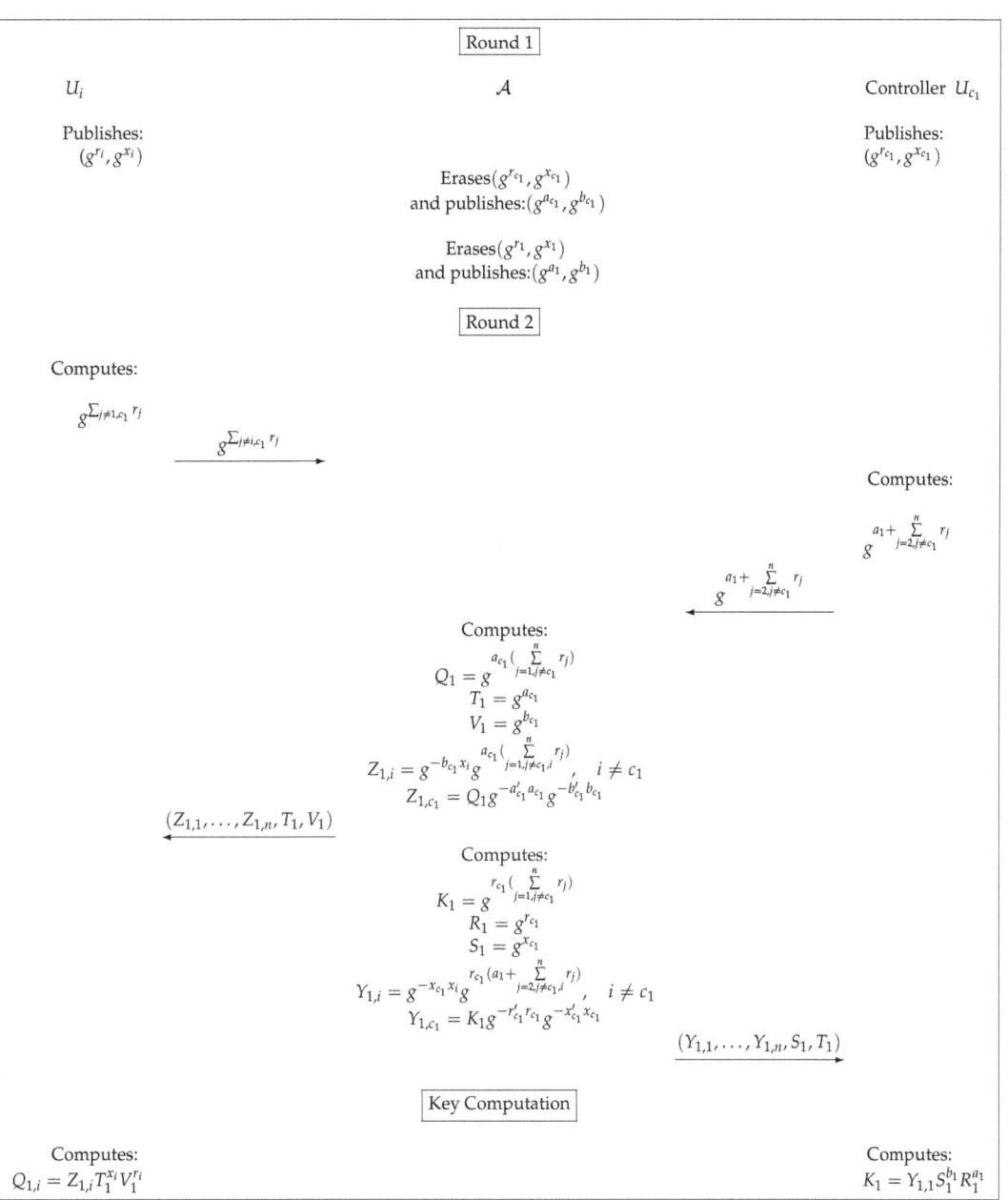

Figure 2. Active attack on Protocol 1 of López Ramos et al.

4. The attacker $\mathcal{A}$ will delete the message sent by U_{c_1} to the superseded user U_1, and queries $\text{Send}(U_1, s_1, (W_{1,1}, \ldots, W_{1,n}, T_1, V_1))$, where

$$W_{1,i} = g^{-b_{c_1} x_i} g^{a_{c_1}(\sum_{j=1, j \neq i, c_1}^{n} r_j)},$$

$$W_{1,c_1} = K_1 g^{-a'_{c_1} a_{c_1}} g^{-b'_{c_1} b_{c_1}},$$

$$T_1 = g^{a_{c_1}} \quad \text{and} \quad V_1 = g^{b_{c_1}}.$$

37

Please note that user U_1, after receiving these last messages, can compute the key $K_1 = W_{1,1} T_1^{x_1} V_1^{r_1}$ which is shared with the adversary $\mathcal{A}$.

5. With the information received, the users, following the protocol, will compute the subsequent keys:

    (a) The superseded user U_1 will compute $K_1 = W_{1,1} T_1^{x_1} V_1^{r_1}$.

    (b) Every user U_i, $i \neq 1$ computes $Q_{1,i} = Y_{1,i} S_1^{x_i} R_1^{r_i}$.

    (c) Adversary $\mathcal{A}$ computes $Q_{1,1} = Z_{1,1} S_1^{b_1} R_1^{a_1}$ and $K_1 = W_{1,1} T_1^{x_1} V_1^{r_1}$.

Therefore, the adversary $\mathcal{A}$ has established a shared key Q_1 with the set of parties $\{U_2, ..., U_n\}$. On the other hand, both U_1 and the adversary $\mathcal{A}$ share the common key K_1.

Remark 3. *While in López-Ramos et al. [15] four different protocols were described, in the previous lines only Protocol 1 was attacked.*

In Protocol 2, authors try to share the computational requirements in a more even way among the parties by slightly modifying which values every participant sends to the group controller and the computations that this user has to perform. However, the only private information for every user is the tuple (r_i, x_i) as in Protocol 1. Thus, an attack can be built analogously by following the steps described above.

In Protocol 3, authors assume that the group controller has changed. The new group controller, by using two private elements (r'_{c_t}, x'_{c_t}) makes a transformation of the key. The next steps of Protocol 3 follows the description of Protocol 1. Therefore, an attack can be built following the previous description.

In Protocol 4, new users take part in the round with new private elements (r_t, x_t). Therefore a new key has to be computed by the group controller using those new elements. Once more, subsequent steps of Protocol 4 follows the description of Protocol 1 and an attack can be constructed analogously.

4.2. Forward Secrecy

We will informally describe how a passive adversary who corrupts a participant $U_i \in \{U_1, \ldots, U_n\}$ involved in a protocol run will be able to compute the shared session key. Therefore, the protocol does not provide forward secrecy.

Let $\mathcal{A}$ be a probabilistic polynomial time adversary (modelled as a Turing machine). He may perform an attack by following the next steps:

1. The attacker $\mathcal{A}$ queries Corrupt(U_i), obtaining the private keys r_i and x_i.
2. Afterwards, he queries, Execute($U_1, s_1, \ldots, U_r, s_r$), obtaining a protocol transcript. In particular, he gets the values $Y_{1,i}$, R_1 and S_1.
3. The adversary now can compute the key as user U_i would do according to the protocol description: $K_1 = Y_{1,i} S_1^{x_i} R_1^{r_i}$.
4. The adversary now queries Test(U_j, s_j) on any user instance involved in the above execution. Since he knows the key established, he wins the game with probability one.

Please note that session s_j of user U_j remains fresh, since, the adversary has not made any Send or Reveal query, so the attack is legitimate.

Remark 4. *In Protocols 3 and 4 in López-Ramos et al. [15], it is described how to proceed when participants may join or leave the group. However, when a participant leaves, the only user changing his private and public keys is the new controller. This means that the rest of the users will have the same private and public key used for previous instances. Therefore, when corrupting any user that is not the new controller, one will obtain their private keys and mount the attack described above. Protocol 2, can also be attacked in the same way, just changing the computations to obtain the session key according to the protocol description.*

Remark 5. *As observed in Theorem 2.4 in López-Ramos et al. [15], the keying messages sent to establish the key can be seen as ElGamal-like encryptions of the key K_1 under a different key for* each user. In that sense, the protocol can be interpreted as a key transport protocol, which cannot be forward secret.

Remark 6. Countermeasures: *If a security proof of the protocols in López-Ramos et al. [15] is provided for passive adversaries, and one consider the private keys as random nonces to be used only in one instance of the protocol, one could then apply the compiler in Katz and Yung [8] to avoid active attacks, generating long term keys for each user to compute digital signatures on all the exchanged messages to guarantee authentication. In that case, if the keys are nonces,* Corrupt *oracle queries would return the signing private keys, thus, corrupted users would not be able to compute the session keys and forward secrecy would also be granted.*

5. Conclusions

As demonstrated above, the protocol proposed by López Ramos et al. [15] does not offer security guarantees. The paper does not provide a rigorous security proof in any standard security model using provable security techniques. The proofs provided are too schematic. If a compiler for authentication is used and the private keys are ephemeral, some attacks could not be applicable. Nevertheless, a security proof should be provided.

Author Contributions: Individual contributions to this article: conceptualization, J.M.C. and A.S.C.; methodology, J.M.C. and A.S.C.; validation, J.M.C. and A.S.C.; formal analysis, J.M.C. and A.S.C.; software, J.M.C. and A.S.C.; investigation, J.M.C. and A.S.C.; resources, J.M.C. and A.S.C.; writing—original draft preparation, J.M.C. and A.S.C.; writing–review and editing, J.M.C. and A.S.C.; project administration, J.M.C. and A.S.C.; funding acquisition, J.M.C. and A.S.C. All authors have read and agreed to the published version of the manuscript.

Funding: This research was funded in part through research project MTM2017-83506-C2-2-P by the Spanish MICINN.

Institutional Review Board Statement: Not applicable.

Informed Consent Statement: Not applicable.

Data Availability Statement: Not applicable.

Conflicts of Interest: The authors declare no conflict of interest.

References

1. Bellare, M.; Rogaway, P. *Entity Authentication and Key Distribution*; CRYPTO, Lecture Notes in Computer Science; Springer: Berlin/Heidelberg, Germany, 1993; Volume 773, pp. 232–249.
2. Bellare, M.; Pointcheval, D.; Rogaway, P. *Authenticated Key Exchange Secure against Dictionary Attacks*; EUROCRYPT, Lecture Notes in Computer Science; Springer: Berlin/Heidelberg, Germany, 2000; Volume 1807, pp. 139–155.
3. Boyd, C.; Mathuria, A. *Protocols for Authentication and Key Establishment*; Information Security and Cryptography; Springer: Berlin/Heidelberg, Germany, 2003.
4. Burmester, M.; Desmedt, Y. A secure and scalable Group Key Exchange system. *Inf. Process. Lett.* **2005**, *94*, 137–143. [CrossRef]
5. Bohli, J.; Vasco, M.I.G.; Steinwandt, R. Secure group key establishment revisited. *Int. J. Inf. Sec.* **2007**, *6*, 243–254. [CrossRef]
6. Boyd, C.; Davies, G.T.; Gjøsteen, K.; Jiang, Y. *Offline Assisted Group Key Exchange*; ISC, Lecture Notes in Computer Science; Springer: Berlin/Heidelberg, Germany, 2018; Volume 11060, pp. 268–285.
7. Vasco, M.I.G.; del Pozo, A.L.P.; Corona, A.S. Group key exchange protocols withstanding ephemeral-key reveals. *IET Inf. Secur.* **2018**, *12*, 79–86. [CrossRef]
8. Katz, J.; Yung, M. *Scalable Protocols for Authenticated Group Key Exchange*; CRYPTO, Lecture Notes in Computer Science; Springer: Berlin/Heidelberg, Germany, 2003; Volume 2729, pp. 110–125.
9. Abdalla, M.; Bohli, J.; Vasco, M.I.G.; Steinwandt, R. *(Password) Authenticated Key Establishment: From 2-Party to Group*; TCC, Lecture Notes in Computer Science; Springer: Berlin/Heidelberg, Germany, 2007; Volume 4392, pp. 499–514.
10. Neupane, K.; Steinwandt, R.; Corona, A.S. *Group Key Establishment: Adding Perfect Forward Secrecy at the Cost of One Round*; CANS; Springer: Berlin/Heidelberg, Germany, 2012; Volume 7712, pp. 158–168.
11. Vasco, M.I.G.; Robinson, A.; Steinwandt, R. Cryptanalysis of a Proposal Based on the Discrete Logarithm Problem Inside S_n. *Cryptography* **2018**, *2*, 16. [CrossRef]
12. Steinwandt, R.; Corona, A.S. Cryptanalysis of a 2-party key establishment based on a semigroup action problem. *Adv. Math. Commun.* **2011**, *5*, 87–92. [CrossRef]

13. Vasco, M.I.G.; del Pozo, A.L.P.; Corona, A.S. Pitfalls in a server-aided authenticated group key establishment. *Inf. Sci.* **2016**, *363*, 1–7. [CrossRef]
14. Baouch, M.; López-Ramos, J.A.; Torrecillas, B.; Schnyder, R. An active attack on a distributed Group Key Exchange system. *Adv. Math. Commun.* **2017**, *11*, 715–717. [CrossRef]
15. López-Ramos, J.A.; Rosenthal, J.; Schipani, D.; Schnyder, R. An Application of Group Theory in Confidential Network Communications. *Math. Methods Appl. Sci.* **2016**. [CrossRef]

Article

From Total Roman Domination in Lexicographic Product Graphs to Strongly Total Roman Domination in Graphs

Ana Almerich-Chulia [1,†], Abel Cabrera Martínez [2,*,†], Frank Angel Hernández Mira [3,†] and Pedro Martin-Concepcion [1,†]

1. Department of Continuum Mechanics and Theory of Structures, Universitat Politecnica de Valencia, Camino de Vera s/n, 46022 Valencia, Spain; analchu@mes.upv.es (A.A.-C.); pmartin@mes.upv.es (P.M.-C.)
2. Departament d'Enginyeria Informàtica i Matemàtiques, Universitat Rovira i Virgili, Av. Països Catalans 26, 43007 Tarragona, Spain
3. Centro de Ciencias de Desarrollo Regional, Universidad Autónoma de Guerrero, Los Pinos s/n, Colonia El Roble, Acapulco 39400, Mexico; fmira8906@gmail.com
* Correspondence: abel.cabrera@urv.cat
† These authors contributed equally to this work.

Abstract: Let G be a graph with no isolated vertex and let $N(v)$ be the open neighbourhood of $v \in V(G)$. Let $f : V(G) \to \{0,1,2\}$ be a function and $V_i = \{v \in V(G) : f(v) = i\}$ for every $i \in \{0,1,2\}$. We say that f is a strongly total Roman dominating function on G if the subgraph induced by $V_1 \cup V_2$ has no isolated vertex and $N(v) \cap V_2 \neq \emptyset$ for every $v \in V(G) \setminus V_2$. The strongly total Roman domination number of G, denoted by $\gamma_{tR}^s(G)$, is defined as the minimum weight $\omega(f) = \sum_{x \in V(G)} f(x)$ among all strongly total Roman dominating functions f on G. This paper is devoted to the study of the strongly total Roman domination number of a graph and it is a contribution to the Special Issue "Theoretical Computer Science and Discrete Mathematics" of Symmetry. In particular, we show that the theory of strongly total Roman domination is an appropriate framework for investigating the total Roman domination number of lexicographic product graphs. We also obtain tight bounds on this parameter and provide closed formulas for some product graphs. Finally and as a consequence of the study, we prove that the problem of computing $\gamma_{tR}^s(G)$ is NP-hard.

Keywords: strongly total Roman domination; total Roman domination; total domination; lexicographic product graph

1. Introduction

Let G be a simple graph with no isolated vertex. Given a vertex $v \in V(G)$, $N(v)$ and $N[v]$ denote the *open neighbourhood* and the *closed neighbourhood* of v in G, respectively. The *order*, *minimum degree* and *maximum degree* of G will be denoted by $n(G)$, $\delta(G)$ and $\Delta(G)$, respectively. As usual, given a set $D \subseteq V(G)$ and a vertex $v \in D$, the *external private neighbourhood* and the *internal private neighbourhood* of v with respect to D is defined to be $epn(v, D) = \{u \in V(G) \setminus D : N(u) \cap D = \{v\}\}$ and $ipn(v, D) = \{u \in D : N(u) \cap D = \{v\}\}$, respectively.

Domination in graphs is a classical research topic that has experienced rapid growth since its introduction. A set $D \subseteq V(G)$ is said to be a *dominating set* of G if $N(v) \cap D \neq \emptyset$ for every $v \in V(G) \setminus D$. Let $\mathcal{D}(G)$ be the set of dominating sets of G. The *domination number* of G is defined to be the following.

$$\gamma(G) = \min\{|D| : D \in \mathcal{D}(G)\}.$$

We define a $\gamma(G)$-set as a set $D \in \mathcal{D}(G)$ with $|D| = \gamma(G)$. The same agreement will be assumed for optimal parameters associated with other characteristic functions or sets of

a graph. For more information on domination and its variants in graphs, we suggest the books [1–4].

An important domination variant in graph, which may be the most studied, is the total domination number. A *total dominating set* of G is a set $D \in \mathcal{D}(G)$ such that $N(v) \cap D \neq \emptyset$ for every $v \in D$. Let $\mathcal{D}_t(G)$ be the set of total dominating sets of G. The *total domination number* of G is defined to be the following.

$$\gamma_t(G) = \min\{|D| : D \in \mathcal{D}_t(G)\}.$$

More information on total domination in graphs can be found in the works [5–7]. Let G be a graph with no isolated vertex and $f : V(G) \to \{0,1,2\}$ a function. For every $i \in \{0,1,2\}$, we define $V_i = \{v \in V(G) : f(v) = i\}$. We will use the unified notation $f(V_0, V_1, V_2)$ to identify the function f with the subsets V_0, V_1, V_2 associated with it. Given a set $X \subseteq V(G)$, we define $f(X) = \sum_{x \in X} f(x)$ and, particularly, we define the *weight* of f as $\omega(f) = f(V(G)) = |V_1| + 2|V_2|$. One of the domination topics widely studied by research is the Roman domination, which is a domination variant arising from some historical roots coming from the ancient Roman Empire [8]. A function $f(V_0, V_1, V_2)$ is a *Roman dominating function* on G if $N(v) \cap V_2 \neq \emptyset$ for every vertex $v \in V_0$. The *Roman domination number* of G denoted by $\gamma_R(G)$ is the minimum weight among all Roman dominating functions on G. For more information on Roman domination in graphs, we suggest the referenced works [9–12].

One of the classical variants of Roman domination is the so-called total Roman domination. This article deals precisely with this style of domination. A *total Roman dominating function* (TRDF) on a graph G with no isolated vertex is a Roman dominating function $f(V_0, V_1, V_2)$ such that $V_1 \cup V_2 \in \mathcal{D}_t(G)$. The minimum weight among all TRDFs on G is the *total Roman domination number* of G and is denoted as $\gamma_{tR}(G)$. This concept was introduced in 2013 by Liu and Chang [13] and formally presented and deeply studied three years later by Abdollahzadeh Ahangar et al. [14]. Subsequently, several researchers have continued with the study of this parameter. For instance, in [15–17], some combinatorial results were presented. In [18–21], constructive characterizations in trees related with this domination parameter were provided. In [22–25], studies of the total Roman domination number on graph products were carried out. In particular, we want to highlight the following closed formula provided in [25] for the case of lexicographic product graphs.

For any graph G with no isolated vertex and any nontrivial graph H, the total Roman domination number of the lexicographic product graph $G \circ H$ is given by the following [25]:

$$\gamma_{tR}(G \circ H) = \begin{cases} 2\gamma_t(G) & \text{if } \gamma(H) \geq 2, \\ \xi(G) & \text{if } \gamma(H) = 1, \end{cases} \quad (1)$$

where $\xi(G) = \min\{|A| + 2|B| : B \in \mathcal{D}(G) \text{ and } A \cup B \in \mathcal{D}_t(G)\}$. As it can be observed, the authors [25] showed that the behavior of $\gamma_{tR}(G \circ H)$ depends on two domination parameters of graphs, namely the well-known total domination number and the incipient parameter $\xi(G)$. In that regard, the authors exposed some results on this last parameter and they raised the challenge of conducting a detailed study of the topic.

In this paper, we continue with the study of this novel parameter although it will be carried out by considering a different approach. In Section 2 we define a new variant of total Roman domination, namely strongly total Roman domination number and denoted by $\gamma_{tR}^s(G)$. We then show that this variant is an appropriate framework to investigate the parameter $\xi(G)$ of a graph. Section 3 is devoted to providing closed formulas for some product graphs. As a consequence of the study, we conclude this section by showing that the problem of computing $\gamma_{tR}^s(G)$ is NP-hard. Finally, in Section 4 we obtain tight bounds on the strongly total Roman domination number of a graph and we discuss the tightness of these bounds.

We assume that the reader is familiar with the basic concepts and terminology of graph domination. If this is not the case, then we suggest the textbooks [1,4]. For the remainder of the article, definitions will be introduced whenever a concept is required.

2. Strongly Total Roman Dominating Functions

The concept of a total Roman dominating function on a graph is associated with the "total domination" property, i.e., this kind of functions requires that each vertex has a neighboring vertex with a positive label assigned to it. However, some vertices have a "special property", which in some cases others do not have. In particular, vertices with label zero must always have a neighbor with label two, but it is not always the case that a vertex with label one satisfies this property. In relation to the above situation, we introduce a "stronger" version of the standard total Roman domination below.

A *strongly total Roman dominating function* (STRDF) on a graph G with no isolated vertex is a total Roman dominating function $f(V_0, V_1, V_2)$ with the additional property that V_2 is a dominating set of G. The minimum weight among all STRDFs on G is the *strongly total Roman domination number* of G and is denoted $\gamma_{tR}^s(G)$.

To illustrate this concept, we consider the graph G shown in Figure 1. For this example, $\gamma_{tR}(G) < \gamma_{tR}^s(G)$.

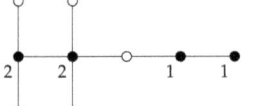 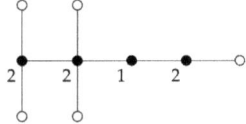

Figure 1. The function on the left is a $\gamma_{tR}(G)$-function, while the function on the right is a $\gamma_{tR}^s(G)$-function.

Now, we proceed to show that this new domination variant is an appropriate framework to investigate the parameter $\xi(G)$.

Theorem 1. *For any graph G with no isolated vertex,*

$$\gamma_{tR}^s(G) = \xi(G).$$

Proof. Let $f(V_0, V_1, V_2)$ be a $\gamma_{tR}^s(G)$-function. By definition we have that $V_2 \in \mathcal{D}(G)$ and $V_1 \cup V_2 \in \mathcal{D}_t(G)$. Therefore, the following obtains.

$$\xi(G) = \min\{|A| + 2|B| : B \in \mathcal{D}(G) \text{ and } A \cup B \in \mathcal{D}_t(G)\} \leq |V_1| + 2|V_2| = \gamma_{tR}^s(G).$$

On the other side, let $A', B' \subseteq V(G)$ such that $B' \in \mathcal{D}(G)$, $A' \cup B' \in \mathcal{D}_t(G)$ and $\xi(G) = |A'| + 2|B'|$. Notice that the function $f'(V_0', V_1', V_2')$, defined by $V_2' = B'$, $V_1' = A'$ and $V_0' = V(G) \setminus (A' \cup B')$, is a STRDF on G. Hence, $\gamma_{tR}^s(G) \leq w(f') = |A'| + 2|B'| = \xi(G)$, which completes the proof. □

To end this section and as a consequence of previous theorem, we show the basic results given in [25] for the strongly total Roman domination number.

Theorem 2. Ref. [25] *For any graph G with no isolated vertex,*

$$\max\{\gamma_{tR}(G), \gamma_t(G) + \gamma(G)\} \leq \gamma_{tR}^s(G) \leq \min\{3\gamma(G), 2\gamma_t(G)\}.$$

Furthermore,

(i) $\gamma_{tR}^s(G) = \gamma_{tR}(G)$ *if and only if there exists a $\gamma_{tR}(G)$-function $f(V_0, V_1, V_2)$ such that V_2 is dominating set of G.*

(ii) $\gamma_{tR}^s(G) = \gamma_t(G) + \gamma(G)$ *if and only if there exists a $\gamma_t(G)$-set that contains some $\gamma(G)$-set.*

3. Exact Formulas for Some Graph Products and Computational Complexity

In order to show the tightness of several bounds and relationships, in this section we obtain the strongly total Roman domination number concerning a well-know families of graphs. We emphasize that we will use the notation K_n, $K_{1,n-1}$, $K_{r,n-r}$ and W_n for complete graphs, star graphs, complete bipartite graphs and the wheel graphs of order n, respectively.

The join graph $G + H$ of the graphs G and H is the graph with vertex set $V(G + H) = V(G) \cup V(H)$ and edge set $E(G + H) = E(G) \cup E(H) \cup \{uv : u \in V(G), v \in V(H)\}$.

Theorem 3. *For any graphs G and H,*

$$\gamma_{tR}^s(G + H) = \begin{cases} 3 & \text{if } \gamma(G) = 1 \text{ or } \gamma(H) = 1, \\ 4 & \text{otherwise.} \end{cases}$$

Proof. We first notice that $\gamma_t(G + H) = 2$. Now, we observe that $\gamma(G + H) = 1$ if and only if $\gamma(G) = 1$ or $\gamma(H) = 1$. Therefore, by Theorem 2 we deduce that $\gamma_{tR}^s(G + H) = 3$ if and only if $\gamma(G) = 1$ or $\gamma(H) = 1$, which completes the proof. □

The following corollary is an immediate consequence of the theorem above.

Corollary 1. *The following equalities hold for any integer $n \geq 3$.*
(i) $\gamma_{tR}^s(K_{1,n-1}) = \gamma_{tR}^s(W_n) = \gamma_{tR}^s(K_n) = 3$.
(ii) *If $r \in \mathbb{Z}$ such that $n - r \geq r \geq 2$, then $\gamma_{tR}^s(K_{r,n-r}) = 4$.*

Let G be a graph with no isolated vertex and H is any graph. The corona product graph $G \odot H$ is defined as the graph obtained from G and H, by taking one copy of G and $|V(G)|$ copies of H and joining by an edge every vertex from the ith-copy of H with the ith-vertex of G. Next, we study the strongly total Roman domination number of any corona product graph.

Theorem 4. *For any graph G with no isolated vertex and any graph H,*

$$\gamma_{tR}^s(G \odot H) = 2n(G).$$

Proof. First, we notice that $\gamma_t(G \odot H) = \gamma(G \odot H) = n(G)$. Hence, Theorem 2 leads to the equality $\gamma_{tR}^s(G \odot H) = 2n(G)$. Therefore, the proof is complete. □

Let G be a graph with no isolated vertex and H a nontrivial graph. The *lexicographic product* of G and H is the graph $G \circ H$ for which the vertex set is $V(G \circ H) = V(G) \times V(H)$ and two vertices $(u, v), (x, y) \in V(G \circ H)$ are neighbors if and only if $ux \in E(G)$ or $u = x$ and $vy \in E(H)$.

Theorem 5. *Ref. [26] For any graph G with no isolated vertex and any nontrivial graph H,*

$$\gamma_t(G \circ H) = \gamma_t(G).$$

We next show that the strongly total Roman domination number and the total Roman domination number coincide for lexicographic product graphs.

Theorem 6. *For any graph G with no isolated vertex and any nontrivial graph H,*

$$\gamma_{tR}^s(G \circ H) = \begin{cases} 2\gamma_t(G) & \text{if } \gamma(H) \geq 2, \\ \gamma_{tR}^s(G) & \text{if } \gamma(H) = 1. \end{cases}$$

Proof. If $\gamma(H) \geq 2$, then the result immediately follows by applying Equation (1) and Theorems 2 and 5, i.e., we have the following.

$$2\gamma_t(G) = \gamma_{tR}(G \circ H) \leq \gamma_{tR}^s(G \circ H) \leq 2\gamma_t(G \circ H) = 2\gamma_t(G).$$

From this moment on, we assume that $\gamma(H) = 1$. By Equation (1) and Theorems 1 and 2 we deduce that $\gamma_{tR}^s(G) = \xi(G) = \gamma_{tR}(G \circ H) \leq \gamma_{tR}^s(G \circ H)$. We only need to prove that $\gamma_{tR}^s(G \circ H) \leq \gamma_{tR}^s(G)$. Let $f(V_0, V_1, V_2)$ be a $\gamma_{tR}^s(G)$-function and $\{v\}$ a $\gamma(H)$-set. Notice that the function $g(W_0, W_1, W_2)$, defined by $W_2 = V_2 \times \{v\}$, $W_1 = V_1 \times \{v\}$ and $W_0 = V(G \circ H) \setminus (W_1 \cup W_2)$, is a STRDF on $G \circ H$. Hence, $\gamma_{tR}^s(G \circ H) \leq |W_1| + 2|W_2| = |V_1| + 2|V_2| = \gamma_{tR}^s(G)$, which completes the proof. □

As shown in [27], the general optimization problem of computing the total domination number of a graph with no isolated vertex is NP-hard. Therefore, by Theorem 6 (considering the case $\gamma(H) \geq 2$) we immediately obtain the analogous result for the strongly total Roman domination number.

Theorem 7. *The problem of computing the strongly total Roman domination number of a graph with no isolated vertex is NP-hard.*

4. Primary Combinatorial Results

The first result of this section provides bounds for the strongly total Roman domination number in terms of the order of a graph. For this purpose, we need to recall the following well-known result.

Theorem 8. Ref. [5] *If G is a connected non-complete graph of order at least three, then G has a $\gamma_t(G)$-set D such that every vertex $v \in D$ satisfies $epn(v, D) \neq \emptyset$ or is adjacent to a vertex $v' \in ipn(v, D)$ satisfying $epn(v', D) \neq \emptyset$.*

Theorem 9. *For any connected graph G of order at least three,*

$$3 \leq \gamma_{tR}^s(G) \leq n(G).$$

Furthermore,

(i) $\gamma_{tR}^s(G) = 3$ *if and only if* $\gamma(G) = 1$.
(ii) $\gamma_{tR}^s(G) = 4$ *if and only if* $\gamma_t(G) = \gamma(G) = 2$.

Proof. The lower bound is straightforward. Now, we proceed to prove the upper bound. If G is isomorphic to a complete graph, then $\gamma_{tR}^s(G) = 3 \leq n(G)$, as desired. From this moment, we assume that G is different of a complete graph. Let D be a $\gamma_t(G)$-set which satisfies Theorem 8 and $\overline{D} = V(G) \setminus D$. Now, we consider the following sets.

$$D_e = \{v \in D : epn(v, D) \neq \emptyset\} \text{ and } \overline{D_e} = \{v \in \overline{D} : N(v) \cap D_e \neq \emptyset\}.$$

Let us define $f'(V_0', V_1', V_2')$ as a function of minimum weight among all functions $f(V_0, V_1, V_2)$ on G satisfying the following conditions.

(a) $V_1' \cup V_2' = D$.
(b) $D_e \subseteq V_2'$.
(c) $N(v) \cap V_2' \neq \emptyset$ for every vertex $v \in \overline{D} \setminus \overline{D_e}$.

By (a), it is straightforward that $V_1' \cup V_2' \in \mathcal{D}_t(G)$. By (b) and (c) we deduce that every vertex in $V_0' = \overline{D}$ has a neighbor in V_2'. Now, let $v \in V_1'$. By definition, $v \in D \setminus D_e$ and thus Theorem 8 results in $N(v) \cap D_e \neq \emptyset$. Hence, $N(v) \cap V_2' \neq \emptyset$ by (b). This implies that $V_2' \in \mathcal{D}(G)$. Therefore, f' is a STRDF on G and thus $\gamma_{tR}^s(G) \leq w(f')$.

We only need to prove that $w(f') \leq n(G)$. Let $v \in \overline{D} \setminus \overline{D_e}$. By definition, we have that $N(v) \cap D \subseteq D \setminus D_e$ and $|N(v) \cap D| \geq 2$. Hence, by (a) and (c) we deduce that

$N(v) \cap V_2' \setminus D_e \neq \emptyset$. Thus, the minimality of f' results in $|V_2' \setminus D_e| \leq |\overline{D} \setminus \overline{D_e}|$ and it is straightforward that $|V_2' \cap D_e| \leq |D_e| \leq |\overline{D_e}|$. Therefore, the following

$$\begin{aligned} \omega(f') &= |V_1'| + 2|V_2'| \\ &= |D| + |V_2' \setminus D_e| + |V_2' \cap D_e| \\ &\leq |D| + |\overline{D} \setminus \overline{D_e}| + |\overline{D_e}| \\ &= |D| + |\overline{D}| \\ &= n(G), \end{aligned}$$

is as required. Hence, the proof of the upper bound is complete.

We then proceed to prove (i). By Theorem 2 we deduce that $\gamma_{tR}^s(G) = 3$ if and only if $\gamma(G) = 1$. Hence, (i) follows.

Finally, we proceed to prove (ii). If $\gamma_t(G) = \gamma(G) = 2$, then Theorem 2 leads to $\gamma_{tR}^s(G) = 4$. Conversely, if $\gamma_{tR}^s(G) = 4$, then by (i) we deduce that $\gamma(G) \geq 2$. Thus, Theorem 2 results in $\gamma_t(G) = \gamma(G) = 2$. Therefore, (ii) follows. □

The upper bound above is tight. For instance, it is achieved for the graph G given in Figure 1. Moreover and as an immediate consequence of Theorems 2 and 9, it is also achieved for the graphs G with $\gamma_{tR}^s(G) = n(G)$. This family is defined in [14].

We continue by providing additional upper bounds. As shown in Theorem 2, the strong total Roman domination number is bounded from above by $3\gamma(G)$. Since $\gamma_R(G) \leq 2\gamma(G)$, the next result improves this upper bound for any graph G with no isolated vertex. We need to introduce the following result.

Theorem 10. Ref. [9] *Let $f(V_0, V_1, V_2)$ be a $\gamma_R(G)$-function on a graph G with no isolated vertex such that $|V_1|$ is minimum. Then the following conditions hold.*

(a) $N(v) \subseteq V_0$ *for every vertex* $v \in V_1$.
(b) $N[x] \cap N[y] = \emptyset$ *for any two different vertices* $x, y \in V_1$.

Theorem 11. *For any graph G with no isolated vertex,*

$$\gamma_{tR}^s(G) \leq \gamma_R(G) + \gamma(G).$$

Proof. Let $f(V_0, V_1, V_2)$ be a $\gamma_R(G)$-function such that $|V_1|$ is minimum. Hence, conditions (a) and (b) of Theorem 10 are satisfied. Now, we consider a function $g'(W_0', W_1', W_2')$ of minimum weight among all functions $g(W_0, W_1, W_2)$ on G such that the following conditions are satisfied:

(i) $V_2 \subseteq W_2$.
(ii) $N(v) \cap W_2 \neq \emptyset$ for every vertex $v \in V_1$.
(iii) $N(v) \cap (W_1 \cup W_2) \neq \emptyset$ for every vertex $v \in V_2$.

We proceed to prove that g' is a STRDF on G. By definitions of f and g', it is straightforward that $W_1' \cup W_2' \in \mathcal{D}_t(G)$. Now, let $v \in V(G) \setminus W_2'$. By (i) we deduce that $v \in V_0 \cup V_1$. Moreover, if $v \in V_0$, then $N(v) \cap W_2' \neq \emptyset$ because $N(v) \cap V_2 \neq \emptyset$. Otherwise, if $v \in V_1$, then (ii) results in $N(v) \cap W_2' \neq \emptyset$. Hence, $W_2' \in \mathcal{D}(G)$, which implies that g' is a STRDF on G, as desired. Thus, $\gamma_{tR}^s(G) \leq \omega(g')$.

Now, let D be a $\gamma(G)$-set. Hence, $N[v] \cap D \neq \emptyset$ for every $v \in V(G)$. In addition, notice that $N(v) \cap (W_2' \setminus V_2) \neq \emptyset$ for every $v \in V_1$. Thus, by (ii) and (iii), conditions (a) and

(b) of Theorem 10 and the minimality of g', we deduce that $2|W_2' \setminus V_2| + |W_1'| \leq |V_1| + |D|$. From the inequalities above we obtain

$$\begin{aligned}\gamma_{tR}^s(G) &\leq \omega(g') \\ &= |W_1'| + 2|W_2'| \\ &= |W_1'| + 2|W_2' \setminus V_2| + 2|W_2' \cap V_2| \\ &\leq |V_1| + |D| + 2|V_2| \\ &= \gamma_R(G) + \gamma(G).\end{aligned}$$

Therefore, the proof is complete. □

The bound above is tight. For instance, it is achieved for any graph $G = G_1 + G_2$ such that $\gamma(G_1) = 1$. In this case, Theorem 3 results in $\gamma_{tR}^s(G) = 3 = \gamma_R(G) + \gamma(G)$.

The following characterization is an immediate consequence of Theorem 11 and the well-known inequality $\gamma_R(G) \leq 2\gamma(G)$.

Theorem 12. *Let G be a graph with no isolated vertex. Then $\gamma_{tR}^s(G) = 3\gamma(G)$ if and only if $\gamma_{tR}^s(G) = \gamma_R(G) + \gamma(G)$ and $\gamma_R(G) = 2\gamma(G)$.*

From Theorem 4 and the fact that $\gamma_R(G_1 \odot G_2) = 2\gamma(G_1 \odot G_2) = 2n(G_1)$ we deduce that $\gamma_{tR}^s(G_1 \odot G_2) = \gamma_R(G_1 \odot G_2) = 2\gamma(G_1 \odot G_2)$ for any graph G_1 with no isolated vertex and any nontrivial graph G_2. This previous equality chain shows that the condition $\gamma_R(G) = 2\gamma(G)$ is a necessary condition but is not sufficient to satisfy $\gamma_{tR}^s(G) = 3\gamma(G)$.

We continue the study by providing a new upper bound, which improves the classical inequality $\gamma_{tR}^s(G) \leq 2\gamma_t(G)$. We need to introduce some concepts and tools. For any $\gamma_t(G)$-set D, let $D^* \subseteq D$ be a set of minimum cardinality such that $D^* \in \mathcal{D}(G)$. Observe that D^* is not necessarily a $\gamma(G)$-set. For instance, for the graph G given in Figure 2 we have that $\gamma_t(G) = 4$ and $\gamma(G) = 3$. However, the set D of black-colored vertices is the only $\gamma_t(G)$-set; moreover, the only dominating set that is a subset of D is D itself.

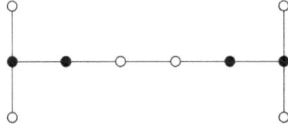

Figure 2. The set of black-colored vertices is the only $\gamma_t(G)$-set.

We define $K_G(D) = D \setminus D^*$ as the *kernel* of D. The maximum cardinality among all kernels $K_G(D)$ from all $\gamma_t(G)$-sets D is the *kernel* of G and it is denoted by $k(G)$. For instance, $k(G_1 \odot G_2) = 0$ and also if $\gamma(G_1 + G_2) = 1$, then $k(G_1 + G_2) = 1$.

Theorem 13. *For any graph G with no isolated vertex,*

$$\gamma_{tR}^s(G) \leq 2\gamma_t(G) - k(G).$$

Proof. Let D be a $\gamma_t(G)$-set such that $k(G) = |K_G(D)|$. Let $D^* \subseteq D \cap \mathcal{D}(G)$ be the set such that $K_G(D) = D \setminus D^*$. Notice that the function $f(V_0, V_1, V_2)$, defined by $V_2 = D^*$, $V_1 = K_G(D)$ and $V_0 = V(G) \setminus D$, is a STRDF on G. Hence,

$$\begin{aligned}\gamma_{tR}^s(G) &\leq \omega(f) \\ &= |V_1| + 2|V_2| \\ &= |K_G(D)| + 2|D^*| \\ &= 2|D| - |K_G(D)| \\ &= 2\gamma_t(G) - k(G).\end{aligned}$$

Therefore, the proof is complete. □

The following result provides a necessary condition for the graphs G satisfying $\gamma_{tR}^s(G) = 2\gamma_t(G)$.

Theorem 14. *Let G be a graph of order at least three with no isolated vertex. If $\gamma_{tR}^s(G) = 2\gamma_t(G)$, then $epn(v, D) \neq \emptyset$ for every $\gamma_t(G)$-set D and $v \in D$.*

Proof. If there exist a $\gamma_t(G)$-set D and a vertex $v \in D$ such that $epn(v, D) = \emptyset$, then $|K_G(D)| \geq 1$ because $D \setminus \{v\} \in \mathcal{D}(G)$. Hence, $k(G) \geq 1$ and Theorem 13 results in $\gamma_{tR}^s(G) < 2\gamma_t(G)$, which completes the proof. □

The following results provide lower bounds for the strongly total Roman domination number in terms of order, maximum degree and total domination number of a graph.

Theorem 15. *For any graph G with every component of order at least three,*

$$\gamma_{tR}^s(G) \geq \gamma_t(G) + \frac{n(G) - \gamma_t(G)}{\Delta(G) - 1}.$$

Proof. Let $f(V_0, V_1, V_2)$ be a $\gamma_{tR}^s(G)$-function. As $V_1 \cup V_2 \in \mathcal{D}_t(G)$, we deduce that

$$|V_2| = \omega(f) - (|V_1| + |V_2|) \leq \gamma_{tR}^s(G) - \gamma_t(G).$$

Now, it is easy to deduce that $|V_0| \leq (\Delta(G) - 1)|V_2|$ because $V_2 \in \mathcal{D}(G)$. Hence,

$$\begin{aligned}\gamma_{tR}^s(G) &= |V_1| + 2|V_2| \\ &= n(G) - |V_0| + |V_2| \\ &\geq n(G) - (\Delta(G) - 1)|V_2| + |V_2| \\ &= n(G) - (\Delta(G) - 2)|V_2| \\ &\geq n(G) - (\Delta(G) - 2)(\gamma_{tR}^s(G) - \gamma_t(G)).\end{aligned}$$

Therefore, we deduce that $\gamma_{tR}^s(G) \geq \gamma_t(G) + \frac{n(G) - \gamma_t(G)}{\Delta(G) - 1}$, which completes the proof. □

In order to show a class of graphs satisfying the equality in the previous bound, we consider the corona product graphs $K_2 \odot H$. For these graphs we obtain that

$$\gamma_{tR}^s(K_2 \odot H) = 4 = \gamma_t(K_2 \odot H) + \frac{n(K_2 \odot H) - \gamma_t(K_2 \odot H)}{\Delta(K_2 \odot H) - 1},$$

because $\gamma_{tR}^s(K_2 \odot H) = 4$ by Theorem 4, $\gamma_t(K_2 \odot H) = 2$, $n(K_2 \odot H) = 2n(H) + 2$ and $\Delta(K_2 \odot H) = n(H) + 1$.

In [15], the authors showed that $\gamma_{tR}(G) \geq \frac{2n(G)}{\Delta(G)}$ for any graph G with no isolated vertex. The following result is a direct consequence of this previous inequality and Theorems 2, 5 and 15.

Theorem 16. *For any graph G with no isolated vertex,*

$$\gamma_{tR}^s(G) \geq \left\lceil \frac{2n(G)}{\Delta(G)} \right\rceil.$$

Furthermore, if $\gamma_t(G) = \frac{n(G)}{\Delta(G)}$, then the previous bound is achieved.

The next theorem shows another relationship between our parameter and the order, maximum degree and total domination number of a graph. This result improves the bound given in the previous theorem whenever $\gamma_t(G) \geq \frac{2n(G)}{\Delta(G)}$.

Theorem 17. *For any graph G with no isolated vertex,*

$$\gamma_{tR}^s(G) \geq \left\lceil \frac{2n(G) + \gamma_t(G)}{\Delta(G) + 1} \right\rceil.$$

Proof. Let $f(V_0, V_1, V_2)$ be a $\gamma_{tR}^s(G)$-function. As $V_1 \cup V_2 \in \mathcal{D}_t(G)$, we deduce that

$$\gamma_t(G) \leq |V_1| + |V_2| = \omega(f) - |V_2| = \gamma_{tR}^s(G) - |V_2|.$$

Now, notice that the following is the case:

$$\begin{aligned}
\Delta(G)\gamma_{tR}^s(G) &= \Delta(G)\omega(f) \\
&= \Delta(G) \sum_{x \in V(G)} f(x) \\
&\geq \sum_{x \in V(G)} |N(x)| f(x) \\
&= \sum_{x \in V(G)} \sum_{u \in N(x)} f(u) \\
&\geq 2|V_0| + 2|V_1| + |V_2| \\
&= 2n(G) - |V_2|.
\end{aligned}$$

From previous inequality chains we deduce the following:

$$2n(G) + \gamma_t(G) \leq \Delta(G)\gamma_{tR}^s(G) + |V_2| + \gamma_{tR}^s(G) - |V_2| = (\Delta(G) + 1)\gamma_{tR}^s(G).$$

Therefore, $\gamma_{tR}^s(G) \geq \left\lceil \frac{2n(G) + \gamma_t(G)}{\Delta(G) + 1} \right\rceil$, as desired. □

The bound above is tight. For instance, it is achieved for any graph G such that $\Delta(G) = n(G) - 1$.

5. Conclusions and Open Problems

In this article we introduced the concept of strongly total Roman domination number and showed that this parameter is an appropriate framework to study the total Roman domination number of lexicographic product graphs. Moreover, we obtained new tight bounds and provided exact formulas for some product graphs. As a consequence of this study, we showed that the problem of computing $\gamma_{tR}^s(G)$ is NP-hard.

We next propose some open problems which we consider to be interesting:

(i) Since the optimization problem of finding $\gamma_{tR}^s(G)$ is NP-hard, it would be interesting to compute the value of this parameter for other families of graphs.
(ii) We propose the problem of characterizing the graphs satisfying the following equalities:
  (a) $\gamma_{tR}^s(G) = n(G)$;
  (b) $\gamma_{tR}^s(G) = \gamma_R(G) + \gamma(G)$;
  (c) $\gamma_{tR}^s(G) = 2\gamma_t(G) - k(G)$.

Author Contributions: The work was organized and led by A.C.M. All authors contributed equally to this work. All authors have read and agreed to the published version of the manuscript.

Funding: This research received no external funding.

Conflicts of Interest: The authors declare no conflict of interest.

References

1. Haynes, T.W.; Hedetniemi, S.T.; Slater, P.J. *Domination in Graphs: Volume 2: Advanced Topics*; Chapman & Hall/CRC Pure and Applied Mathematics, Taylor & Francis: Abingdon, UK, 1998.
2. Haynes, T.W.; Hedetniemi, S.T.; Henning, M.A. Topics in domination in graphs. In *Developments in Mathematics*; Springer: Cham, Switzerland, 2020; Volume 64.
3. Haynes, T.W.; Hedetniemi, S.T.; Henning, M.A. Structures of domination in graphs. In *Developments in Mathematics*; Springer: Cham, Seitzerland, 2021; Volume 66.
4. Haynes, T.W.; Hedetniemi, S.T.; Slater, P.J. *Fundamentals of Domination in Graphs*; Chapman and Hall/CRC Pure and Applied Mathematics Series; Marcel Dekker, Inc.: New York, NY, USA, 1998.
5. Henning, M.A. Graphs with large total domination number. *J. Graph Theory* **2000**, *35*, 21–45. [CrossRef]
6. Henning, M. A survey of selected recent results on total domination in graphs. *Discret. Math.* **2009**, *309*, 32–63. [CrossRef]
7. Henning, M.; Yeo, A. *Total Domination in Graphs*. Springer Monographs in Mathematics; Springer: New York, NY, USA, 2013.
8. Stewart, I. Defend the Roman Empire! *Sci. Am.* **1999**, *281*, 136–138. [CrossRef]
9. Cockayne, E.J.; Dreyer, P.A., Jr.; Hedetniemi, S.M.; Hedetniemi, S.T. Roman domination in graphs. *Discret. Math.* **2004**, *278*, 11–22. [CrossRef]
10. Chambers, E.W.; Kinnersley, B.; Prince, N.; West, D.B. Extremal problems for Roman domination. *SIAM J. Discret. Math.* **2009**, *23*, 1575–1586. [CrossRef]
11. Chellali, M.; Jafari Rad, N.; Sheikholeslami, S.M.; Volkmann, L. Roman Domination in Graphs. In *Topics in Domination in Graphs. Developments in Mathematics*; Springer: Cham, Switzerland, 2020; Volume 64.
12. Henning, M.A. A characterization of Roman trees. *Discuss. Math. Graph Theory* **2002**, *22*, 325–334. [CrossRef]
13. Liu, C.-H.; Chang, G.J. Roman domination on strongly chordal graphs. *J. Comb. Optim.* **2013**, *26*, 608–619. [CrossRef]
14. Abdollahzadeh Ahangar, H.; Henning, M.A.; Samodivkin, V.; Yero, I.G. Total Roman domination in graphs. *Appl. Anal. Discret. Math.* **2016**, *10*, 501–517. [CrossRef]
15. Abdollahzadeh Ahangar, H.; Amjadi, J.; Sheikholeslami, S.M.; Soroudi, M. On the total Roman domination number of graph. *Ars Combin.* **2020**, *150*, 225–240.
16. Cabrera Martínez, A.; Cabrera García, S.; Carrión García, A. Further results on the total Roman domination of graphs. *Mathematics* **2020**, *8*, 349. [CrossRef]
17. Pushpam, P.R.L.; Sampath, P. On total Roman domination in graphs. In *Theoretical Computer Science and Discrete Mathematics*; Lecture Notes in Comput. Sci.; Springer: Cham, Switzerland, 2017; Volume 10398, pp. 326–331.
18. Cabrera Martínez, A.; Martínez Arias, A.; Menendez Castillo, M. A characterization relating domination, semitotal domination and total Roman domination in trees. *Commun. Comb. Optim.* **2021**, *6*, 197–209.
19. Amjadi, J.; Nazari-Moghaddam, S.; Sheikholeslami, S.M.; Volkmann, L. Total Roman domination number of trees. *Australas. J. Combin.* **2017**, *69*, 271–285.
20. Cabrera García, S.; Cabrera Martínez, A.; Hernández Mira, F.A.; Yero, I.G. Total Roman {2}-domination in graphs. *Quaest. Math* **2021**, *44*, 411–434. [CrossRef]
21. Amjadi, J.; Sheikholeslami, S.M.; Soroudi, M. On the total Roman domination in trees. *Discuss. Math. Graph Theory* **2019**, *39*, 519–532. [CrossRef]
22. Cabrera Martínez, A.; Cabrera García, S.; Carrión García, A.; Hernández Mira, F.A. Total Roman domination number of rooted product graphs. *Mathematics* **2020**, *8*, 1850. [CrossRef]
23. Cabrera Martínez, A.; Cabrera García, S.; Peterin, I.; Yero, I.G. Dominating the direct product of two graphs through total Roman strategies. *Mathematics* **2020**, *8*, 1438. [CrossRef]
24. Campanelli, N.; Kuziak, D. Total Roman domination in the lexicographic product of graphs. *Discret. Appl. Math.* **2019**, *263*, 88–95. [CrossRef]
25. Cabrera Martínez, A.; Rodríguez-Velázquez, J. A. Closed formulas for the total Roman domination number of lexicographic product graphs. *ARS Math. Contemp.* **2021**, in press. [CrossRef]
26. Cabrera Martínez, A.; Rodríguez-Velázquez, J.A. Total protection of lexicographic product graphs. *Dis. Math. Graph Theory* **2020**, in press. [CrossRef]
27. Laskar, R.; Pfaff, J.; Hedetniemi, S.; Hedetniemi, S. On the algorithmic complexity of total domination. *SIAM J. Alg. Discr. Meth.* **1984**, *5*, 420–425. [CrossRef]

Article

Modeling and Optimization for Multi-Objective Nonidentical Parallel Machining Line Scheduling with a Jumping Process Operation Constraint

Guangyan Xu [1], Zailin Guan [1], Lei Yue [2,*], Jabir Mumtaz [3] and Jun Liang [1]

1. School of Mechanical Science and Engineering, Huazhong University of Science and Technology, Wuhan 430074, China; xuguangyan@hust.edu.cn (G.X.); zlguan@hust.edu.cn (Z.G.); liangjun@alumni.hust.edu.cn (J.L.)
2. School of Mechanical and Electrical Engineering, Guangzhou University, Guangzhou 510000, China
3. College of Mechanical and Electronic Engineering, Wenzhou University, Wenzhou 325000, China; jabir.mumtaz@me.uol.edu.pk
* Correspondence: leileiyok@gzhu.edu.cn

Citation: Xu, G.; Guan, Z.; Yue, L.; Mumtaz, J.; Liang, J. Modeling and Optimization for Multi-Objective Nonidentical Parallel Machining Line Scheduling with a Jumping Process Operation Constraint. *Symmetry* **2021**, *13*, 1521. https://doi.org/10.3390/sym13081521

Academic Editors: Juan Alberto Rodríguez Velázquez and Alejandro Estrada-Moreno

Received: 13 July 2021
Accepted: 13 August 2021
Published: 18 August 2021

Publisher's Note: MDPI stays neutral with regard to jurisdictional claims in published maps and institutional affiliations.

Copyright: © 2021 by the authors. Licensee MDPI, Basel, Switzerland. This article is an open access article distributed under the terms and conditions of the Creative Commons Attribution (CC BY) license (https://creativecommons.org/licenses/by/4.0/).

Abstract: This paper investigates the nonidentical parallel production line scheduling problem derived from an axle housing machining workshop of an axle manufacturer. The characteristics of axle housing machining lines are analyzed, and a nonidentical parallel line scheduling model with a jumping process operation (NPPLS-JP), which considers mixed model production, machine eligibility constraints, and fuzzy due dates, is established so as to minimize the makespan and earliness/tardiness penalty cost. While the physical structures of the parallel lines in the NPPLS-JP model are symmetric, the production capacities and process capabilities are asymmetric for different models. Different from the general parallel line scheduling problem, NPPLS-JP allows for a job to transfer to another production line to complete the subsequent operations (i.e., jumping process operations), and the transfer is unidirectional. The significance of the NPPLS-JP model is that it meets the demands of multivariety mixed model production and makes full use of the capacities of parallel production lines. Aiming to solve the NPPLS-JP problem, we propose a hybrid algorithm named the multi-objective grey wolf optimizer based on decomposition (MOGWO/D). This new algorithm combines the GWO with the multi-objective evolutionary algorithm based on decomposition (MOEA/D) to balance the exploration and exploitation abilities of the original MOEA/D. Furthermore, coding and decoding rules are developed according to the features of the NPPLS-JP problem. To evaluate the effectiveness of the proposed MOGWO/D algorithm, a set of instances with different job scales, job types, and production scenarios is designed, and the results are compared with those of three other famous multi-objective optimization algorithms. The experimental results show that the proposed MOGWO/D algorithm exhibits superiority in most instances.

Keywords: nonidentical parallel production lines; axle housing machining; mixed model production; eligibility constraint; fuzzy due date; grey wolf optimizer

1. Introduction

Flow shop scheduling problems are the most common and widely studied problems in the manufacturing industry, and the examined problems are usually simplified versions of the real flow shop scheduling problem so as to reduce the difficulty of modeling and solving these problems. These oversimplified schemes often cannot perfectly solve such scheduling problems in the actual environment. In real-world manufacturing situations, some special constraints or uncertainties must usually be considered and handled, such as sequence-dependent setup times [1–3], the kinds of parallel machines under study [4,5], machine eligibility constraints [6–9], resource constraints [10,11], and fuzzy stochastic demand [12,13]. Considerations of these additional constraints and uncertainties make the developed scheduling models closer to real production scenarios, but also increase their scheduling complexity. Because of product iteration requirements and the diversified needs of customers, production systems must often address multivariety production on multiple production lines. This is very common in manufacturing enterprises such as those in the automobile industry and household appliance industry, as well as for construction machinery manufacturers. When coping with these situations, the equipment configurations of multiple lines may be different for meeting multivariety production. In this paper, we study this nonidentical parallel production line scheduling problem. To improve the machine utilization and shorten the waiting times of the jobs to be processed, a jumping process operation is often used. This means that if a certain process operation of a job is finished, it can move to another production line to complete the subsequent process operations. This jumping process operation is unidirectional. To solve this kind of scheduling problem, nonidentical parallel line scheduling with a jumping process operation (NPPLS-JP) is proposed; this is also in essence a parallel production line scheduling problem. Notably, the proposed NPPLS-JP problem is an NP-hard problem because of its complexity.

This NPPLS-JP problem is derived from the axle housing machining workshop of an axle manufacturer. Axle housing is an important part of axle production; it usually adopts make-to-stock (MTS) production and make-to-order (MTO) assembly. The machining of an axle housing is shown in Figure 1. The axle manufacturer adopts nonidentical parallel production lines for axle housing machining. There are two parallel production lines (A and B) in the axle housing machining workshop, and each production line is installed linearly, as shown in Figure 1. The physical structures of the two production lines are symmetrical. Each production line contains five stages corresponding to five operations, and the parallel machines at any stage of each line are identical. The corresponding stages of production lines A and B have similar functions, but the configurations of the machines in the different lines are different in order to meet the needs of multivariety mixed production. In this production situation, the production load of each stage on any line is not easy to balance via a simple scheduling scheme.

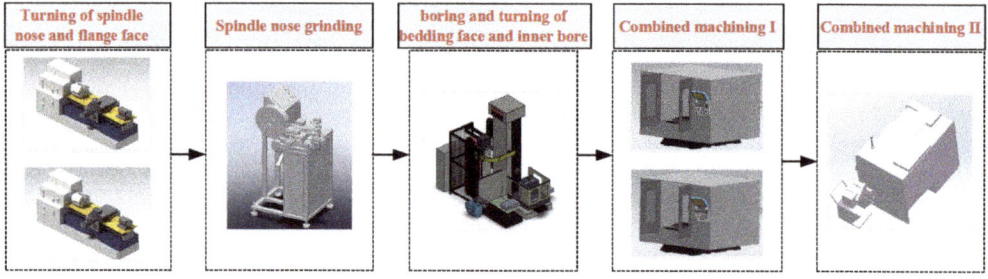

Figure 1. The composition of an axle housing machining line.

According to the different vehicle models, there are eight types of axle housing products that can be processed on line A and line B with multivariety mixed model production; all types of products are processed through five operations, as shown in Figure 1. Two types of axle housing products have machine eligibility constraints, and the operation "combined machining I" can only be performed at the corresponding machines in production line B; the others can finish processing on any line independently. For different types of axle housing, the different processing times required for the same operation on the same machine and the different mixing ratios of the various axle housing types increase the complexity of the scheduling problems. It is difficult to balance all the stages of each production line, so a jumping process operation is adopted to address this problem by allowing a job to be processed on two production lines. For example, when the operation "combined machining I" is finished on line A, the job is transferred to line B to complete the subsequent machining operations; this is called the jumping process operation, and the stage "combined machining I" is called the jumping process point. A jumping process operation is unidirectional, which means that the axle housings processed on line A are allowed to be transferred to production line B, but not the other way around. Appropriate jumping process operations can reduce waiting times, improve the utilization rate of equipment, balance the production capacity of each stage, and ensure the due date of each order.

From the above description of a production system, it can be seen that the axle housing machining line scheduling problem has the following characteristics: (1) The configurations of the multiple production lines are similar but not the same. (2) Mixed multivariety production is adopted to organize production. (3) Several jobs with special types have machine eligibility constraints. (4) The jumping process operation is unidirectional in the production process. This problem can be regarded as a variant and extension of the flow shop scheduling problem or general parallel production line scheduling problems, and it involves four key decision-making processes, namely: job sequencing decisions, parallel line decisions, parallel machine decisions, and job jumping process operation decisions. It is obvious that the NPPLS-JP problem proposed in this paper is a rather complex scheduling optimization problem.

Because of the fierce competition in the market and the diversified needs of customers, the NPPLS-JP problem is widespread in manufacturing environments and has an important impact on the manufacturing efficiency of production systems. However, there is no relevant research on this topic in the existing literature, so the study in this paper is of exploratory and practical significance. In this paper, we establish a scheduling model for the NPPLS-JP problem, which involves multivariety mixed model production, multiline scheduling, and machine eligibility constraints. The objectives are to minimize the makespan and earliness/tardiness penalty in a production cycle. In the NPPLS-JP model, to more closely approximate the actual production environment, the due date of a production order denoted by the fuzzy earliness/tardiness penalty model [14], and a hybrid algorithm combining the grey wolf optimization algorithm and the multi-objective evolutionary algorithm based on decomposition (MOEA/D) are proposed to solve the NPPLS-JP problem.

The remainder of this paper is organized as follows. Section 2 reviews the literature relevant to multivariety mixed model production, parallel line scheduling, and multi-objective optimization. Section 3 gives a general statement of the NPPLS-JP problem and establishes a production-based, order-oriented, multi-objective scheduling model. Section 4 proposes the multi-objective grey wolf optimizer based on decomposition (MOGWO/D) for NPPLS-JP and describes the procedures in detail. Section 5 tests the performance of the proposed MOGWO/D algorithm by comparing it with three other famous multi-objective algorithms based on a set of designed test instances, and the experimental results are analyzed. Section 6 summarizes the research content and discusses the direction of future research.

2. Literature Review

Mixed model production refers to the production of a variety of products on a single production line to increase the flexibility of the line and meet the multivariety and small-batch production demands. For multiproduct demands, mixed model production is widely adopted by manufacturing enterprises, especially in assembly workshops [15,16]. Because of the widespread application of mixed model production, many scholars are devoted to research in this field and have made many achievements [17–21].

Mcmullen et al. [22] studied the mixed model scheduling problem with the consideration of a setup time. They presented a bean search heuristic method to obtain an efficient front. Leu and Hwang [23] proposed a resource-constrained mixed-production flow shop scheduling system for mixed precast production task problems and developed a search method based on a genetic algorithm (GA) to minimize the output makespan under resource constraints and mixed production. Wang et al. [24] studied final assembly line scheduling, which considers order scheduling and mixed model sequencing simultaneously, and combined the original artificial bee colony (ABC) algorithm with some steps of the GA and Pareto optimality to solve this problem. Bahman et al. [25] constructed a mixed-integer linear programming model with a tighter linear relaxation for a realistic automotive industry assembly line, including a set of specific requirements involving moving workers and limited workspace. Alghazi and Kurz [26] proposed a mixed model line-balancing integer program for mixed model assembly lines with the aim of minimizing the number of chemical workers; a constraint programming model was established to address larger assembly line balancing problems.

Parallel line scheduling problems are very common in both mass production and multiproduct production. Parallel line production can enhance the stability and flexibility of the production system and improve production efficiency. When one production line breaks down, not all production activities are stopped. All parallel lines may have the same number of processing stages, the pieces of equipment in the same stage are similar and can complete the same production processes, and every line can substitute for all other lines to produce all or some types of the desired products [27]. However, in some situations, the configurations of the machines in different lines are nonidentical; this situation is more convenient for the production of multiple products. For example, two types of equipment in different production lines can independently complete the operation of milling surfaces, but the machining accuracies are different.

Haq et al. [28] studied the line scheduling problem with multiple parallel processing in job shops. Each job can only be processed on a particular line and is not allowed to move between parallel lines. Meyr and Mann et al. [29] introduced a new solution approach to determine the lot-sizing and scheduling problem for parallel production lines with the consideration of scarce capacity; sequence-dependent setup times; and the deterministic, dynamic demands of multiple products. Mumtaz et al. [30] investigated the multiple assembly line scheduling problem for a printed circuit board (PCB) assembly and developed a hybrid spider monkey optimization approach with an improved replacement strategy to solve it. Rajeswari et al. [27] presented parallel flow line scheduling with a dual objective to minimize the tardiness and earliness of jobs. All parallel flow lines had similar sets, and the authors developed a hybrid algorithm that used a GA and particle swarm optimization (PSO) to incorporate greedy randomized adaptive search to address the problem. Ebrahimipour et al. [31] proposed linear programming and a bagged binary knapsack to address the multiple production line scheduling problem. Mumtaz et al. [32] developed a mixed-integer programming model for the multilevel planning and scheduling problem of parallel PCB assembly lines, and a hybrid spider monkey optimization (HSMO) algorithm was proposed.

Multi-objective optimization refers to a situation where more than one conflicting objective is to be optimized simultaneously. It is often impossible to obtain an optimal solution as in a single-objective optimization problem, but a set of tradeoff solutions, namely, nondominated solutions, can be used to choose the most suitable solution according to the actual requirements. Therefore, it is more applicable to the actual situation, which requires the consideration of multiple indicators affecting decision making. As it is conducive to obtaining an ideal decision making effect, multi-objective optimization has a wider application field and more practical value. MOEAs are global optimization algorithms based on populations that simulate the evolution process of natural organisms. Since the whole solution set can be obtained in one run, MOEAs have become some of the mainstream algorithms in multi-objective optimization. According to their selection mechanisms, MOEAs can be classified into three classes [33]: Pareto-based algorithms, indicator-based algorithms, and decomposition-based algorithms. The Pareto-based method was first proposed by Goldberg et al. [34] and has been widely studied since then. Many classical multi-objective optimization algorithms are based on the Pareto relation, and many have been proposed based on the Pareto dominance relationship, such as the famous

SPEA [35], SPEA2 [36], and NSGA-II [37]. The indicator-based method uses performance evaluation indicators to guide the search process and the choice of solutions [38,39]. The MOEA/D was first proposed by Zhang and Hui in 2007 [40], and it is the most representative multi-objective optimization algorithm based on decomposition. Different from the classical multi-objective optimization algorithms, it decomposes an input multi-objective optimization problem (MOP) into a series of single-objective optimization subproblems by using a set of uniformly distributed weight vectors and optimizing these subproblems simultaneously. Since the MOEA/D was proposed, it has attracted increasing attention from scholars, improvements and applications for the MOEA/D are constantly emerging, and it has become one of the best multi-objective optimization algorithms.

Li and Landa-Silva et al. [41] proposed evolutionary multi-objective simulated annealing (EMOSA), which incorporates a simulated annealing algorithm and introduces an adaptive search strategy. The experimental results showed that the algorithm obtained a good effect in terms of solving the multi-objective knapsack problem and multi-objective salesman problem. Tan et al. [42] developed a multi-objective meme algorithm based on decomposition, which integrates a simplified quadratic approximation (SQA) into the MOEA/D as a local search operator to balance its local and global search strategies. Wang et al. [43] designed a multi-objective particle swarm optimization algorithm based on decomposition (MPSO/D). This algorithm adopts relevant measures to ensure that only one solution is present in each subregion in oder to maintain solution diversity, and the fitness is calculated by the crowding distance. Ke and Zhang et al. [44] proposed the MOEA/D-ACO algorithm, which incorporates ant colony optimization (ACO) into the MOEA/D; they then tested the performance of the proposed algorithm in 12 instances and obtained good results. Alhindi et al. [45] developed a hybrid algorithm called MOEA/D-GLS, which integrated guided local search (GLS) with the MOEA/D to promote the exploitation ability of the original MOEA/D. The experimental results showed that the proposed MOEA/D-GLS was superior to the original MOEA/D. Zhang et al. [46] proposed MOEA/D-EGO to address expensive MOPs. In this method, the input problem is decomposed into several subproblems, and a prediction model is established for each subproblem based on the evaluated points in order to reduce modeling costs and improve the prediction quality. Wang et al. [47] proposed adaptive replacement strategies by adjusting the problem size dynamically for the MOEA/D. This approach can balance the diversity and convergence of the MOEA/D.

However, according to the no free lunch theorem [48], no algorithm can solve all of the optimization problems in all of the fields. Because of the continuous emergence of new optimization problems, the existing algorithms cannot solve these new optimization problems well, so new algorithms or improved algorithms are needed. The MOEA/D algorithm exhibits good diversity in solving MOPs; its characteristics include simplicity, few parameters, and better result distributions. In this paper, the MOGWO/D, which incorporates the GWO into the MOEA/D, is proposed to solve the NPPLS-JP problem.

3. Problem Description and Mathematical Modeling

3.1. Problem Definition and Assumption

Suppose that O orders are processed in L production lines, the job types for each order are the same and those for different orders may be different, and the operations of each type of job are predetermined and similar. All production lines have the same number of stages, and the machine configurations may be different. If parallel machines exist in some stages of the production line, they are identical parallel machines. Some types of jobs may have machine eligibility constraints, namely some job operations must be carried out on specific machines at certain stages of some production lines. For any production line l, if s is set as the jumping process operation point, it means that after processing is completed in stage s, the job can be transferred to line l' to continue the processing of the subsequent operations ($l \neq l'$ and $l, l' \in \{1, 2, \cdots, L\}$); the jumping process operation is unidirectional.

The scheduling objectives are to minimize the makespan and the earliness/tardiness penalty cost.

In addition, there are usually several complicated constraints and perturbations in the real-world production environment. To prevent the loss of generality and reduce the computational complexity of the scheduling model, some modeling assumptions are given, as follows.

(1) The type, quantity, and due date of each order are known.
(2) The jobs in each order are of the same type.
(3) All the machines in the production system are available at the beginning.
(4) The processing times and setup times of the jobs on each machine do not overlap.
(5) Each job can be processed on only one machine at any time, each machine can process only one job at any time, and operations cannot be interrupted.
(6) For each job, the jumping operation can only occur once.
(7) The jumping process operation point is singular, fixed, and unidirectional.
(8) The setup time and machine breakdown time are ignored.

Meanwhile, to solve the NPPLS-JP problem more conveniently, the concept of a virtual production line is introduced. This means that if there is a jumping process operation point in the manufacturing process of a job of a certain type, all the stages in two different production lines that can complete the processing of jobs of this type are regarded as a new production line, namely, a virtual production line.

In a real-world production environment, a breach of the order due dates is not always unacceptable. In general, a few occurrences of tardiness are allowed, to achieve the smallest due date penalty cost across the total orders. Here, the fuzzy due date is used to deal with this situation. Trapezoidal fuzzy due date and triangular fuzzy due date are two common fuzzy due dates that have been investigated in the literature regarding scheduling problems [14,49–51]. Most researchers choose the type of fuzzy due date depending on the research background and problem characteristics. In our research, neither early nor late completion were the best solutions for the automobile industries. Completing the production order in advance will increase the inventory cost, while the order delay will lead to customer penalty loss. Finishing and delivering the orders in a given period is the most feasible result. Therefore, the trapezoidal fuzzy due date and earliness/tardiness penalty cost model was adopted in this scheduling problem according to the real-world production requirement.

As shown in Figure 2a–b, (a) is the trapezoidal fuzzy due date and earliness/tardiness penalty cost model, (b) is the corresponding satisfaction model of the fuzzy due date. Model (a) shows each order has a corresponding trapezoidal fuzzy due date that is denoted by a trapezoidal fuzzy number $d_o = (d_o^1, d_o^2, d_o^3, d_o^4)$, and the earliness/tardiness penalty cost coefficients are α_o and β_o, respectively. A completion time before d_o^1 it means that the orders are produced prematurely, and additional inventory costs are generated; if the completion time comes after d_o^4, the production order seriously violates the due date requirement. These two cases are both unacceptable, so the satisfaction is 0, and the maximal earliness/tardiness penalty is imposed. If the completion time is in the time interval $[d_o^1, d_o^2]$ or $[d_o^3, d_o^4]$, the due date penalty costs decrease and increase linearly, respectively, and the corresponding due date satisfaction is just the opposite. Only in the time interval $[d_o^2, d_o^3]$ are the completion times reasonable, and the due date penalty cost in this case is 0. Therefore, the production orders should be arranged optimally in terms of time according to different due dates and earliness/tardiness penalty costs.

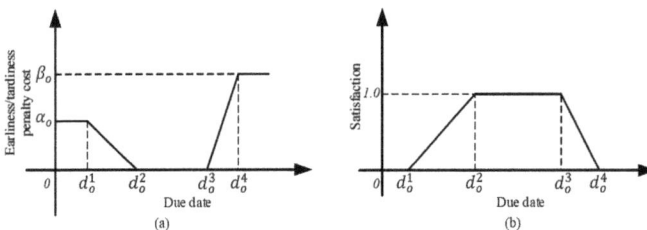

Figure 2. The earliness/tardiness penalty cost and satisfaction model. (**a**) The earliness/tardiness penalty cost; (**b**) The satisfaction model.

3.2. Mathematical Modeling

A mathematical model for the proposed NPPLS-JP problem is established using the above notations, and the two objective functions, which minimize the makespan and earliness/tardiness penalty cost, are formulated as Equations (1) and (2), respectively.

$$\text{Min} \quad f_1 = \max\{C_o\} \tag{1}$$

$$\text{Min} \quad f_2 = \sum_{o=1}^{O} P_o \tag{2}$$

where the calculation formulas of the completion time and due earliness/tardiness penalty for order o are formulated as Equations (3) and (4), respectively:

$$C_o = \max\{u_{i,o} C_i\} \; i \in J \tag{3}$$

$$P_o = \begin{cases} \alpha_o & C_o < d_o^1 \\ \alpha_o \cdot \frac{d_o^2 - C_o}{d_o^2 - d_o^1} & d_o^1 \le C_o < d_o^2 \\ 0 & d_o^2 \le C_o \le d_o^3 \\ \beta_o \cdot \frac{C_o - d_o^3}{d_o^4 - d_o^3} & d_o^3 \le C_o < d_o^4 \\ \beta_o & C_o \ge d_o^4 \end{cases} \quad o \in O \tag{4}$$

The constraints are as follows:

$$\sum_{l=1}^{L} \sum_{s=1}^{(s_l-1)} (X_{i,s,l} Y_{i,s,s'}) = 1 \; i \in J; \; s' = (s+1) \tag{5}$$

$$\sum_{l=1}^{L} \sum_{s=(s_l+1)}^{(S-1)} (X_{i,s,l} Y_{i,s,s'}) = 1 \; i \in J; \; s' = (s+1) \tag{6}$$

$$Y_{i,s_l,(s_l+1)} = 0, 1 \; i \in J \tag{7}$$

$$\sum_{l=1}^{L} (X_{i,s,l} X_{i,s',l}) = X_{i,s,s'} \; i \in J; \; s \in \{1, 2, \cdots, (S-1)\}; \; s' = (s+1) \tag{8}$$

$$\sum_{k=1}^{M_{sl}} X_{k,i,s,l} = X_{i,s,l} \; i \in J; \; s \in S; \; l \in L \tag{9}$$

$$X_{i,s,l} \le \sum_{o=1}^{O} \sum_{t=1}^{N_t} \sum_{l=1}^{L} x_{i,o} x_{o,t} x_{t,s,l} \; i \in J; \; s \in S; \; l \in L \tag{10}$$

$$\sum_{l=1}^{L} \sum_{k=1}^{M_{sl}} X_{k,i,s,l} = 1 \; i \in J; \; s \in S \tag{11}$$

$$Z_{k,i,i',s,l} + Z_{k,i',i,s,l} \le X_{k,s,i,l} \; i, i' \in J; \; i \ne i'; \; s \in S; \; k \in M_{s,l}; \; l \in L \tag{12}$$

$$Z_{k,i,i',s,l} + Z_{k,i',i,s,l} \leq X_{k,s,i',l} \quad i,i' \in J; i \neq i'; s \in S; k \in M_{s,l}; l \in L \tag{13}$$

$$C_{k,s,i',l'} X_{k,s,i',l'} + M(1 - Z_{k,i,i',s,l}) \geq C_{k,s,i,l} X_{k,s,i,l} + \sum_{o=1}^{O} \sum_{t=1}^{N_t} (pt_{t,k,s} x_{o,t} x_{i,o}) X_{k,s,i',l'}$$
$$i,i' \in J; i \neq i'; s \in S; k \in \{M_{s,l} \cup M_{s,l'}\}; l, l' \in L \tag{14}$$

$$C_{k',s',i,l'} X_{k',s',i,l'} + M(1 - X_{k',s',i,l'}) \geq C_{k,s,i,l} X_{k,s,i,l} + \sum_{o=1}^{O} \sum_{t=1}^{N_t} (pt_{t,k',s'} x_{o,t} x_{i,o}) X_{k',s',i,l'}$$
$$s \in \{1,2,\cdots,S-1\}; s' = (s+1); k \in M_{s,l}; k' \in M_{s',l'}; l, l' \in L \tag{15}$$

Among the above constraints, constraints (5)–(7) together define the jumping process operation. Constraint (5) defines the operations before the jumping process operation point (including the operation on the jumping process operation stage), which can only be completed in one production line. Constraint (6) restricts all of the operations after the jumping process operation point for each job that can only be processed on the same production line. Constraint (7) states the jumping process operation for any job may or may not occur after completing another jumping process operation, and the three constraints together guarantee that the jumping process operation can only occur once at most. Constraint sets (8) and (9) define the relationships between several decision variables. Constraint (10) states that the processing of each job operation must satisfy the machine eligibility constraints. Constraint (11) ensures that each operation of a job can only be processed on one machine. Constraint sets (12) and (13) together restrict the processing sequence of two jobs on a machine to only one possible result. Constraint (14) guarantees that one machine can only process one job at a time, which means that the completion time of the current job is longer than the sum of the completion time of the immediate predecessor job and the processing time of the current job. Constraint (15) ensures that a job is processed by only one machine at a time, that is, the completion time of the job operation is greater than the sum of the completion time of the immediate predecessor operation and the processing time of the current operation.

4. Proposed MOGWO/D Algorithm

The MOEA/D provides a general framework that allows any single objective to be applied to the subproblems of a MOP [41]. Compared with the other multi-objective optimization algorithms, such as the Pareto-based optimization algorithms, MOEA/D has less computational complexity, and its results have better diversity. In this section, we present a hybrid algorithm that combines GWO with MOEA/D, and uses the mechanism of searching for prey in the GWO algorithm to enforce a balance between exploration and exploitation. According to the characteristics of the NPPLS-JP problem, problem-specific encoding and decoding rules are given, and some main procedures in the proposed MOGWO/D are also stated in detail.

4.1. Original GWO

The GWO was inspired by the leadership hierarchy and hunting behaviors of grey wolves [52]. In GWO, initial populations are used to simulate the grey wolf group, which is divided into four hierarchies; the solutions with the best, second, and third fitness values are α, β, and δ, respectively, are utilized to find the optimal solution by simulating the hunting process of grey wolves. In the process of hunting, the location of the prey is unknown. Therefore, to simulate the hunting behavior of grey wolves and the prey behavior from the perspective of mathematical modeling, suppose that α, β, and δ are closest to the potential position of the prey. Under the guidance of α, β, and δ, the position vector is updated to approximate the optimal solutions in the search space.

The main procedure of wolf hunting includes encircling prey and hunting, and the mathematical models for grey wolves approaching and encircling their prey are as follows:

$$\vec{D} = \left| \vec{C} \cdot \vec{X}_p(t) - \vec{X}(t) \right| \tag{16}$$

$$\vec{X}(t+1) = \vec{X}_p(t) - \vec{A} \cdot \vec{D} \tag{17}$$

$$\vec{A} = 2\vec{a} \cdot \vec{r}_1 - \vec{a} \tag{18}$$

$$\vec{C} = 2\vec{r}_2 \tag{19}$$

In Equations (16) and (17), $\vec{X}_p$ indicates the position vector of the prey and $\vec{X}$ is the position vector of the grey wolf. $\vec{A}$ and $\vec{C}$ are coefficient vectors, and the calculation methods are shown in Equations (18) and (19). By changing the value of the vector $\vec{A}$, the search process can be guided. When $|\vec{A}| > 1$, α, β, and δ diverge from each other, which is good for global search; when $|\vec{A}| < 1$, α, β, and δ converge to the prey, which contributes to the local search. The parameter $\vec{C}$ is generated randomly to help grey wolves jump out of the local optima.

In Equations (18) and (19), $\vec{r}_1$ and $\vec{r}_2$ are randomly generated in $[0, 1]$, and the values of the parameter $\vec{a}$ linearly decrease from 2 to 0 over the course of the iterations. During the process of hunting, the position vectors are updated using the following equations:

$$\vec{X}_1 = \vec{X}_\alpha - \vec{A}_1 \cdot \left| \vec{C}_1 \cdot \vec{X}_\alpha - \vec{X} \right| \tag{20}$$

$$\vec{X}_2 = \vec{X}_\beta - \vec{A}_2 \cdot \left| \vec{C}_2 \cdot \vec{X}_\beta - \vec{X} \right| \tag{21}$$

$$\vec{X}_3 = \vec{X}_\delta - \vec{A}_3 \cdot \left| \vec{C}_3 \cdot \vec{X}_\delta - \vec{X} \right| \tag{22}$$

$$\vec{X}(t+1) = \frac{\vec{X}_1 + \vec{X}_2 + \vec{X}_3}{3} \tag{23}$$

In Equations (20)–(23), $\vec{X}_\alpha$, $\vec{X}_\beta$, and $\vec{X}_\delta$ are the position vectors of α, β, and δ, respectively, and $\vec{X}$ denotes the current position vector that needs to be updated.

4.2. MOGWO/D Algorithm Framework

The proposed MOGWO/D is a hybrid algorithm that integrates GWO into MOEA/D. Similar to the original MOEA/D, the MOGWO/D algorithm decomposes the input multi-objective problem into a series of single-objective scalar optimization subproblems by utilizing a set of uniformly distributed weight vectors and a scalar function. Here, we use the Tchebycheff method to construct each subproblem; then, subproblem i can be described as follows [40]:

$$\text{Minimize } g^{te}(x|\lambda_i, z^*) = \max_{1 \le j \le m} \left\{ \lambda_i^j | f_i(x) - z_m^* | \right\} \tag{24}$$

where z_i^* is the i-th component of reference point $(z_1^*, z_2^*, \cdots, z_m^*)^T$, $z_i^* = \min\{f_i(x) | x \in \Omega\}$, $i = 1, 2, \cdots, m$, $\lambda_i = (\lambda_i^1, \lambda_i^2, \cdots, \lambda_i^m)^T$. The purpose is to minimize each single-objective function $g^{te}(x_i|\lambda_i, z^*)$, and each subproblem uses the approach of the GWO to update its position vectors. It is worth noting that finding accurate reference points is difficult and time-consuming work, so the best objective values $z = (z_1, z_2, \cdots, z_m)^T$ method is used in the initial population as the initial reference point, and to update the reference point over the course of iterations by generations.

As the normalization method for objectives is conducive to increase the uniformness of the obtained solutions when the input objectives are disparately scaled [40], we used a simple normalization method to replace f_i and obtained a normalization Tchebycheff approach, as follows.

$$\text{Minimize } g^{te}(x|\lambda_i, z^*) = \max_{1 \leq j \leq m} \left\{ \lambda_i^j \left| \frac{f_i(x) - z_i^*}{z_i^{nad} - z_i^*} \right| \right\} \quad (25)$$

where z^* is the reference point and $z^{nad} = \{z_1^{nad}, z_2^{nad}, \cdots, z_m^{nad}\}$ is the nadir point in the objective space. In our calculation, z^* is replaced by z in Step 2.3 of Algorithm 1, and the maximum value of $f_i(x)$ in the current population is the substitute for z_i^{nad}. This calculation strategy can meet the needs of the algorithm.

Algorithm 1 MOGWO/D

Input:
A multiobjective problem;
A stopping criterion;
A set of uniformly spread weight vectors $\{\lambda^1, \lambda^2, \cdots, \lambda^N\}$;
N: population size (equal to the number of the weight vectors or subproblems);
T: neighborhood size;
T': the number of position vectors in the neighborhood to be updated of a subproblem (where $T' < T$).

Output:
External population, EP for short.

Step 1) Initialization:

Step 1.1) Set EP = ∅;

Step 1.2) Generate a set of uniformly distributed weight vectors $\{\lambda^1, \lambda^2, \cdots, \lambda^N\}$, calculate the Euclidean distances of any pair of weight vectors, for
$\forall i = 1, 2, \cdots, N$, defines a set $B(i) = \{i_1, i_2, \cdots, i_T\}$, $\lambda^{i_1}, \lambda^{i_2}, \cdots, \lambda^{i_T}$ are T closest weight vectors of the weight vecto λ^i.

Step 1.3) Randomly generate an initial population $\{x^1, x^2, \cdots, x^N\}$ or use a problem-specific approach. The objective of each position vector is calculated and labeled as FV^i, $FV^i = F(x^i)$, $i = 1, 2, \cdots, N$.

Step 1.4) Initialize
$z = (z_1, z_2, \cdots, z_m)^T$, $z_i = \min\{f_i(x^1), f_i(x^2), \cdots, f_i(x^N)\}$, $i = 1, 2, \cdots m$.

Step 1.5) Calculate $g^{te}(x^j|\lambda^i, z^*)$ for each $j \in B(i)$, and the three best position vectors are labeled as x_α, x_β and x_δ, respectively corresponding to weight vector λ^i.

Step 2) Update:
for $i = 1, 2, \cdots, N$, do

Step 2.1) Randomly select T' indexes $k_1, k_2, \cdots, k_{T'}$ from $B(i)$, then yield a set of new position vectors $x_{k_1}, x_{k_2}, \cdots, x_{k_{T'}}$ according to the Equations (20)–(23) by the guidance of $x_\alpha^i x_\beta^i$, and x_δ^i, set $PS(i) = \{x_{k_1}, x_{k_2}, \cdots, x_{k_{T'}}\}$.

Step 2.2) Update of x_α^i, x_β^i and x_δ^i. Comparing the value of $g^{te}(x'|\lambda^i, z)$ with $g^{te}(x_\alpha^i|\lambda^i, z)$, $g^{te}(x_\beta^i|\lambda^i, z)$ and $g^{te}(x_\delta^i|\lambda^i, z)$, $x' \in PS(i)$, then update x_α^i, x_β^i and x_δ^i with the three best position vectors of all.

Step 2.3) Update of z. For each $j = 1, 2, \cdots, m$, if $f_j(x_\alpha^i) < z_j$, set $z_j = f_j(x_\alpha^i)$.

Step 2.4) Update of neighborhood. For each
$j \in B(i)$, if $(x_\alpha^i, |\lambda^j, z|) \leq g^{te}(x^j|\lambda^j, z)$, set $x^j = x_\alpha^i$, and update of $FV_j = f(x_\alpha^i)$.

Step 2.5) Update of EP. Add $f(x_\alpha^i)$ to EP if no vectors in EP dominate $f(x_\alpha^i)$; if the number of vectors exceeds the EP capacity, the kth nearest neighbor method is used as a truncation strategy. If the vectors in EP are dominated by $f(x_\alpha^i)$, remove from EP.

Step 3) Stopping criterion:
If the stopping criteria is satisfied, stop running and output EP. Otherwise, return to Step 2.

Similar to the original MOEA/D, the MOGWO/D algorithm (Algorithm 1) also optimizes a number of scalar optimization subproblems simultaneously in one iteration, thus improving the optimization efficiency of the proposed algorithm.

4.3. Generate a Set of Uniform Weight Vectors

In Step 1.2 of Algorithm 1, a simplex-lattice design [53] is adopted to generate a set of uniformly distributed weight vectors $\lambda^i = (\lambda_1^i, \lambda_2^i, \cdots, \lambda_m^i)$, $i \in N$, and m is the dimensionality of the objective space. For each λ^i, $\sum_{j=1}^{m} \lambda_j^i = 1$, and $\lambda_j^i \in \left\{0, \frac{1}{H}, \frac{2}{H}, \cdots, \frac{H}{H}\right\}$, H is a predetermined positive integer determined according to the sizes of the problems, so, a total of C_{H+m-1}^{m-1} weight vectors are obtained. For each λ^i, the Euclidean distance to any weight vector is calculated, defining a set $B_{(i)} = \{i_1, i_2, \cdots, i_T\}$, in which $\lambda^{i_1}, \lambda^{i_2}, \cdots, \lambda^{i_T}$ are the indexes of the T closest weight vectors to λ^i; then, $B_{(i)}$ is called neighborhood of λ^i (including λ^i itself, as λ^i is the closest weight vector to itself, of which the Euclidean distance is 0). At the same time, the response of x^i to λ^i also generates a neighborhood, and each individual in the neighborhood corresponds to each weight vector determined by $B(i)$.

4.4. Encoding and Decoding

In the proposed MOGWO/D algorithm, the encoding method is similar to the original GWO, and the initial population is randomly generated from a uniform distribution. All position vectors in the initial population are continuous, but the scheduling solutions to the proposed combinatorial optimization problem are not, so a decoding approach is needed to convert the continuous position vectors to the scheduling solutions.

In the proposed NPPLS-JP problem, each position vector needs to include two pieces of information, a job permutation and a production line sequence, and the production line sequence corresponds to the job permutation. Suppose that there are O orders in a planning cycle; if the number of jobs in the order o is N_o, then there are a total of $N = \sum_{o=1}^{O} N_o$ jobs in this planning cycle, and each position vector in the population is represented as $x^i = [x_1^i, x_2^i, \cdots, x_N^i, |x_{(N+1)}^i, \cdots, x_{2N}^i]$. For convenience of expression, the first N position values are marked as Part 1, which corresponds to the job permutation, and the last N position values are marked as Part 2, which corresponds to the production line sequence.

Part 1 and Part 2 are decoded independently. The decoding methods for Part 1 and Part 2 are different because the machining operations of the jobs of some types have machine eligibility constraints. The selection of the production line involves the decoded information of Part 1, that is, the decoding process of Part 2 depends on the obtained jobs' permutation. To discretize the continuous position vectors, the ranked-order value (ROV) rule [54] is used in the decoding processes of Part 1 and Part 2.

For each position value in Part 1 of the position vector, the ROV rule is used to generate ROVs according to the position values in ascending order. If identical position values exist, the ROVs increase from left to right, and then the ROV permutation is obtained. Then, the N_1 smallest values are picked and all are assigned a value of V_1. V_1 is the order number of order 1, and N_1 is the size of order 1; similarly, ROVs $(N_1 + 1)$ to N_2 are picked and assigned V_2. V_2 is the order number of order 2, and N_2 is the size of order 2. In the same way, the job permutation is obtained.

For Part 2, first, as in Part 1, the ROV sequence is obtained according to the position values in Part 2 in ascending order. The next step is different from Part 1. For each ROV in Part 2, the same ROV in Part 1 is found, the corresponding order number is obtained, and then its job type is determined. Then, the job type in the first column of the given line selection information table is found, and the corresponding number of optional lines in the second column is obtained. Next, the remainder of the current ROV divided by the number of available lines is obtained. Finally, the corresponding production line number in the third column is found according to the calculated remainder, and the production line number is assigned to the position in Part 2 corresponding to current ROV. In the same way, the production line sequence is obtained. This decoding method for Part 2 can prevent increases in the calculation cost due to invalid solutions caused by the selection of unusable production lines.

A simple example, as shown in Table 1, is used to demonstrate the decoding rules. The sizes of the three orders are two, three, and one, and the total number of jobs is 6six. The detailed decoding process for the example is shown in Figure 3.

Table 1. The data used in the example.

Order No.	Number of Jobs	Job Type
1	2	3
2	3	1
3	1	2

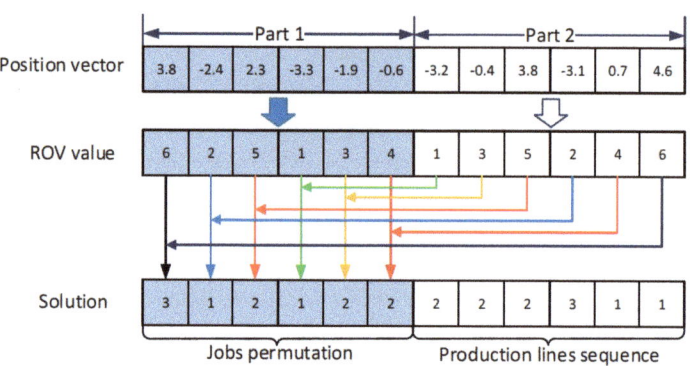

Figure 3. An example of decoding.

In this example, there are a total of five jobs. Each position value of the position vector is taken from the uniform distribution U[−6, 6] so that the position vector x = [3.8, −2.4, 2.3, −3.3, −1.9, −0.6, −3.2, −0.4, 3.8, −3.1, 0.7, 4.6] can be obtained, where Part 1 is [3.8, −2.4, 2.3, −3.3, −1.9, −0.6], and Part 2 is [−3.2, −0.4, 3.8, −3.1, 0.7, 4.6]. For the decoding process, the production line selection information used in Part 2 is given in Table 2.

Table 2. The line selection information of the example.

Job Types	Available Quantity	Optional Lines		
		0	1	2
1	2	1	2	
2	3	1	2	3
3	3	1	2	3

4.5. Updating the Position Vectors

The Tchebycheff approach is adopted to decompose the input MOP into N single-objective optimization subproblems and to optimize them simultaneously. The neighborhood of each subproblem is defined based on the distances between their weight vectors. Adjacent subproblems have similar approximate solutions, so each subproblem is optimized, and only the information of neighboring subproblems is used. For every generation, T' position vectors corresponding to the subproblems are updated, where T' is a positive integer smaller than the neighborhood size, and these T' position vectors are randomly selected from the neighborhood corresponding to the subproblem.

Each selected position vector of each subproblem is updated, and this information is used in its neighborhood. First, position vectors are updated according to Equations (20)–(23) through the guidance of the current three best position vectors x_α, x_β, and x_δ, and the position vector set $PS = \{x_1, x_2, \cdots, x_{T'}\}$ of the subproblem is obtained. Second, calculating the g^{te} value of every position vector in the $GS(i) = PS \cup \{x_\alpha, x_\beta, x_\delta\}$ corresponding to λ^i. Then, x_α, x_β, and x_δ are updated with the best, second best, and third best position vectors in the $GS(i)$ corresponding to the weight vector λ^i, where x_α is the optimal solution to the current subproblem. After that, the reference point is updated by calculating the objectives with x_α; if $f_j(x_\alpha) < z_j$, then $z_j = f_j(x_\alpha)$. Third, the neighborhood of the current subproblem is updated with respect to each position vector to the weight vectors of the neighborhood; for each $j \in B(i)$, if $g^{te}(x_\alpha|\lambda^j, z^*) \le g^{te}(x_j|\lambda^j, z^*)$, $x_j = x_\alpha$ and $FV_j = f(x_\alpha)$ simultaneously. The pseudocode for implementing the update process in one iteration is shown in Algorithm 2.

Algorithm 2 The update process of one iteration

While(stopping condition is not satisfied){
//Main loop
for($i = 1; i \le N; i++$)
{//optimize N subproblems simultaneously
idxes = getRandoms(T', $B(i)$);
selectedPop = getIndividuals(neighborhood, idxes);
$PS(i)$ = updateIndividuls$\Big($selectedPop, $x_\alpha, x_\beta, x_\delta\Big)$;
sortedPop = sort($PS(i)$);
x_α = sortedPop(0);
x_β = sortedPop(1);
x_δ = sortedPop(2);
updateZ(x_α); //update the reference point;
updateNeighborhood(x_α);
updateEP(x_α);
}
}

4.6. Updating the External Population (EP)

After obtaining the best position vector x_α of every subproblem in each generation, the EP needs to be updated. If the condition that $F(x_\alpha)$ is not dominated by the individuals in the EP, $F(x_\alpha)$ is added to the EP, and the individuals that are dominated by $F(x_\alpha)$ are removed.

In the search process of the algorithm, excess individuals are added to the EP. Too many nondominated individuals are not of great significance for solving practical problems, but they increase the difficulty of the data analysis. Therefore, a special truncation strategy is used to maximize the retention of nondominated solution characteristics while maintaining the appropriate EP size. In this strategy, the kth nearest neighbor method [36] is used here to evaluate the individuals in the EP, and the calculation approach for the kth nearest neighbor distance is as follows.

$$D(i) = \frac{1}{\sigma_i^k + 2} \tag{26}$$

$$k = \sqrt{|P| + |EP|} \tag{27}$$

where $D(i)$ is the inverse function of the Euclidean distance from individual i to its k-th nearest neighbor, which is used to reflect the density information of the objective space. σ_i^k is the kth nearest Euclidean distance of individual i, where the value of k is calculated using Equation (27), and $|P|$ and $|EP|$ are the population size and external population size, respectively. The smaller the $D(i)$ value is, the more scattered the solutions are.

5. Computational Experiments and Results Analysis

NPPLS-JP is a novel multiline scheduling problem derived from a real-world manufacturing workshop; it is an extension of regular flow shop scheduling problems that has no related research. Therefore, there are no benchmarks available for the proposed MOGWO/D algorithm. In this section, test experiments are designed to assess the performance of the proposed MOGWO/D algorithm by comparing the results obtained on the proposed NPPLS-JP problem with those of three other famous multi-objective optimization algorithms, i.e., NSGA-II [37], MOGWO [55], and MOPSO [56], in terms of three metrics. The results illustrate the effectiveness of the proposed MOGWO/D algorithm.

5.1. Evaluation Metrics

Different from single objective optimization problems, MOPs involve the simultaneous optimization of multiple conflicting objectives. An improvement of one objective results in the deterioration of another objective, so MOP algorithms usually obtain a set of tradeoff solutions in terms of the desired objectives, namely, nondominated solutions. There is no absolute optimum among these solutions, and fitness functions cannot evaluate their effectiveness, so a set of metrics is needed to evaluate the performance of multi-objective algorithms for solving MOPs. If the obtained Pareto front is closer to the Pareto optimal front, covering the extreme regions as much as possible, and the nondominated solutions are uniformly distributed in the obtained Pareto front, this means that the obtained results have better convergence and distribution effects. In this paper, the following three metrics are used:

(1) Generational distance (GD) [57]. The GD is the most common multi-objective indicator for convergence. It is used to calculate the mean Euclidean distance between the obtained Pareto front and the Pareto optimal front. The calculation formula for the GD is as follows.

$$GD = \frac{\sqrt{\sum_{i=1}^{|OF|} d_i^2}}{|OF|} \tag{28}$$

where d_i is the Euclidean distance from point i of the obtained Pareto front to the closest point in the Pareto optimal front, and $|OF|$ is the number of nondominated solutions in the obtained Pareto front; therefore, GD denotes the mean value of the closest distance from each point in the obtained Pareto front to the Pareto optimal front. A smaller GD value indicates that the obtained Pareto front is closer to the Pareto optimal front; namely, the obtained Pareto front has good convergence. When GD equals zero, the obtained Pareto front is located at the Pareto optimal front.

(2) Inversed generational distance (IGD) [58]. This metric is a variant of the GD and is a comprehensive performance indicator. This metric represents the mean Euclidean distance from the points in the Pareto optimal front to the obtained Pareto front. The formulation of the IGD is as follows.

$$IGD = \frac{\sqrt{\sum_{i=1}^{|PF|} D_i^2}}{|PF|} \tag{29}$$

where $|PF|$ denotes the number of points in the Pareto optimal front and D_i is the Euclidean distance from point i in the Pareto optimal front to the closest point in the obtained Pareto front. A smaller IGD value indicates better convergence and diversity for the obtained Pareto front. In our experiments, the nondominated solutions obtained from all independent runs of the four algorithms on each instance are regarded as the Pareto optimal front of that instance.

(3) Spread (Δ) [37]. Δ is the diversity metric of the multi-objective optimization that can measure the distribution and spread of solutions. Δ is calculated as follows.

$$\Delta = \frac{\sum_{j=1}^{m} d_j^e + \sum_{i=1}^{n} |d_i - \bar{d}|}{\sum_{j=1}^{m} d_j^e + n\bar{d}} \quad (30)$$

where m is the number of objectives and n is the number of solutions in the obtained Pareto front. d_j^e is the minimum Euclidean distance from the nondominated solutions in the obtained Pareto front to the extreme point j of the Pareto optimal front, and d_i is the Euclidean distance of the closest pairwise points in the obtained Pareto front, and $\bar{d}$ is the average value of d_i. A smaller value of Δ represents a better distribution and increased diversity. The calculation of Δ is simple and does not require knowledge of the whole Pareto optimal front, it uses only the extreme objectives of the Pareto optimal front.

5.2. Instance Generation

In the ideal situation, the proposed MOGWO/D algorithm is suitable for all kinds of problems that meet the model definition of NPPLS-JP with different numbers of parallel production lines and jobs, different mixing ratios of job types, and different configurations of production lines. Here, by combining the production system and customers' demand data to generate a set of instances, the performance of the proposed MOGWO/D algorithm is tested. Therefore, we evaluated the effectiveness of the proposed algorithm in different production scenarios by varying the order quantity and order capacity. In the experiment in this section, we only considered the case with two parallel production lines, and gave three production scenarios with different configurations, four different numbers of orders, and three different order capacities; thus, a tally of 3 × 4 × 3 = 36 instances were generated by the combination of the three factors.

The three production scenarios are shown in Figure 4a–c, each with two nonidentical parallel production lines, and each production line consisting of five stages corresponding to five operations of eight types of jobs. The three production scenarios consist of two general flow lines, a general flow production line and a hybrid flow production line, and two hybrid flow lines. The machines of each production scenario are taken from Table 3. Notably, the machine configuration for the first line in Scenario (b) is the same as that of the first line in Scenario (a), and the machines configuration for the second line in Scenario (c) is the same as that of the second line in Scenario (b). All of the parallel machines are identical parallel machines in each line of any production scenario. Table 3 also gives the processing time of each operation of the eight types of jobs on each machine. "_" indicates that jobs of the current type cannot be processed on the corresponding machines. Therefore, if the next operation of a job cannot be processed on the current line, the jumping process operation will occur. The production line selection information for decoding is shown in Table 4.

Table 3. The optional machines and the processing times for each type of job.

Job Types	Stage 1				Stage 2				Stage 3				Stage 4				Stage 5			
	M_1	M_2	M_3	M_4	M_5	M_6	M_7	M_8	M_9	M_{10}	M_{11}	M_{12}	M_{13}	M_{14}	M_{15}	M_{16}	M_{17}	M_{18}	M_{19}	M_{20}
1	41	41	30	30	39	39	34	34	42	42	31	31	—	—	26	26	49	49	23	23
2	40	40	31	31	42	42	34	34	40	40	32	32	—	—	23	23	50	50	23	23
3	40	40	32	32	38	38	32	32	41	41	32	32	47	47	26	26	52	52	24	24
4	39	39	32	32	41	41	32	32	39	39	29	29	48	48	26	26	52	52	24	24
5	42	42	34	34	40	40	34	34	39	39	30	30	—	—	24	24	51	51	24	24
6	40	40	35	35	39	39	31	31	42	42	31	31	—	—	23	23	46	46	22	22
7	39	39	32	32	40	40	32	32	40	40	33	33	54	54	24	24	54	54	24	24
8	43	43	31	31	42	42	31	31	43	43	34	34	57	57	26	26	56	56	26	26

Table 4. Production line selection information for the experiments.

Type	Available Quantity	Optional Lines		
		0	1	2
1	2	2	3	-
2	2	2	3	-
3	3	1	2	3
4	3	1	2	3
5	3	2	3	-
6	3	2	3	-
7	3	1	2	3
8	3	1	2	3

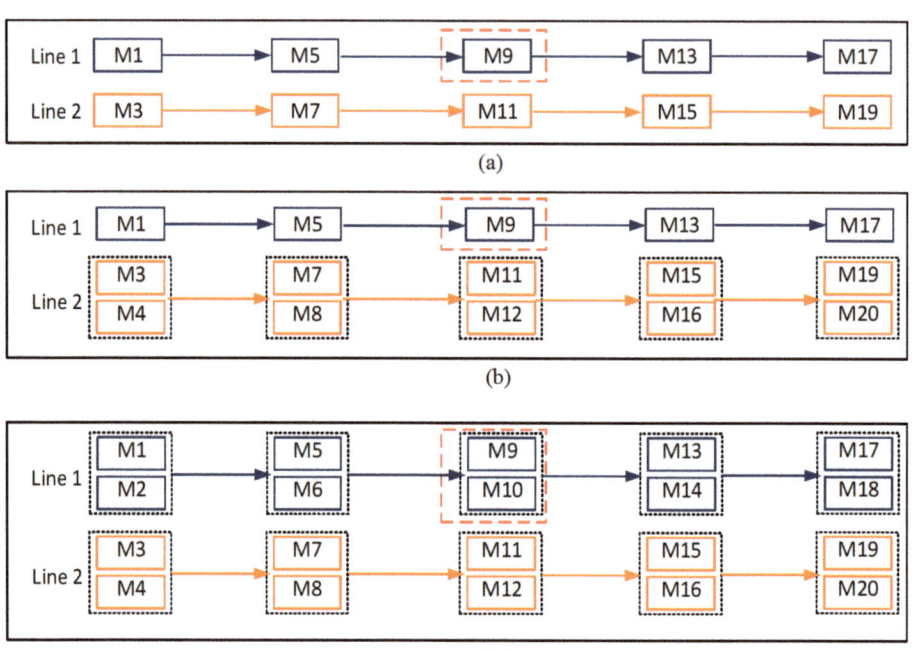

Figure 4. The three different production scenarios. (**a**) Scenario 1; (**b**) Scenario 2; (**c**) Scenario 3.

The number of orders $O = \{2, 4, 6, 8\}$ and the capacity of order o is N_o, where $N_o = \{10, 20, 30\}$. As the jobs in each order are of the same type, O represents not only the number of orders, but also the number of job types in the production cycle, and both O and N_o determine the total number of jobs. Each instance is denoted in the form "x_y_z", where x represents the production scenario taken from the three scenarios described earlier in Figure 4a–c, y is the total number of orders (corresponding to the first y types in Table 1), and z is the number of jobs in each order. For example, 1_4_10 represents the instance in scenario 1 that includes four orders, and there are 10 jobs in each order.

For each order in the instances, the trapezoidal fuzzy number for the trapezoidal fuzzy due date is generated as follows. First, a uniform distribution $U[0.8 \times N_J, 3 \times N_J]$ is given, where $N_J = N_o \times \sum_{s=1}^{S} pt_s$, N_o is the number of jobs in order o, S is the number of operations in the job, and pt_s is the maximum processing time for each job operation in order o at any available machine, as shown in Table 3. Then, four integers are randomly taken from the uniform distribution $U[0.8 \times N_J, 3 \times N_J]$. The last four time points are sorted in ascending order, as required for the trapezoidal fuzzy due date. The earliness/tardiness penalty coefficients of all orders are set to 5 and 15, respectively.

There are 36 instances to be tested using the proposed MOGWO/D algorithm and the three compared algorithms. For experimentation, the four algorithms are all coded in Java, and all instances are run on an HP Pavilion m4 notebook PC with a Windows 10 Professional 64-bit operating System, 8 GB of RAM, and an Intel Core i5 CPU at 2.60 GHz.

5.3. Parameter Settings

From the 36 instances generated above, we can see that the jobs in each instance have eight different scales: 20, 40, 60, 80, 120, 160, 180, and 240. For instances with different job sizes, different population sizes and numbers of iterations should be set so as to obtain the best optimization performance for the algorithms; thus, four different population sizes are given in Table 5. For fairness, MOGWO and MOPSO used the same encoding and decoding method as the proposed MOGWO/D algorithm. Because of the different mechanisms, the encoding and decoding methods of NSGA-II are slightly different. In the NSGA-II algorithm, Part 1 and Part 2 use a permutation encoding scheme [59] to obtain independent encodings, both of which are natural number permutations. The decoding scheme is similar to that of the proposed MOGWO/D algorithm; the only difference is that the individuals in the population are discrete, and these discrete values are natural number permutations. They can be regarded as ROVs, and then decoded according to the decoding rules in Section 4. Unlike the other three algorithms, the encodings in NSGA-II do not need to be discretized first.

Table 5. The parameter settings for the four algorithms.

MOGWO/D	NSGA-II	MOGWO	MOPSO
Population size (N): 60 and iterations: 300 (20 and 40 jobs)	Population size (N): 60 and iterations: 300 (20 and 40 jobs)	Population size (N): 60 and iterations: 300 (20 and 40 jobs)	Population size (N): 60 and iterations: 300 (20 and 40 jobs)
Population size (N): 100 and iterations: 500 (60 and 80 jobs)	Population size (N): 100 and iterations: 500 (60 and 80 jobs)	Population size (N): 100 and iterations: 500 (60 and 80 jobs)	Population size (N): 100 and iterations: 500 (60 and 80 jobs)
Population size (N): 240 and iterations: 800 (120 and 160 jobs)	Population size (N): 240 and iterations: 800 (120 and 160 jobs)	Population size (N): 240 and iterations: 800 (120 and 160 jobs)	Population size (N): 240 and iterations: 800 (120 and 160 jobs)
Population size (N): 300 and iterations: 1000 (180 and 240 jobs)	Population size (N): 300 and iterations: 1000 (180 and 240 jobs)	Population size(N): 300 and iterations: 1000 (180 and 240 jobs)	Population size (N): 300 and iterations: 1000 (180 and 240 jobs)
The external population size: N	Crossover rate: 0.9	The external population size: N	The external archive size: N
The neighborhood size: 20 (for all instances)	Mutation rate: 0.1		The inertia weight $w_o = 0.4$
$T' = 6$ (for all instances)			The acceleration coefficients $c_1 = c_2 = 2.0$

In addition, in NSGA-II, a binary tournament is used to as a selection operator for Part 1 and Part 2, the partial mapped crossover (PMX) and swap operation are used as the crossover operator of the two parts, and the insert operation is used as the mutation operator of both parts. The other parameter settings of the four algorithms are also shown in Table 5.

Each instance is run 30 times independently for the proposed MOGWO/D algorithm and the other three comparison algorithms.

5.4. Experimental Results Analysis

Table 6 presents the means and standard deviations of the three metrics for the MOGWO/D algorithm—NSGA-II, MOGWO, and MOPSO. By comparing the GD, IGD, and Spread values of the four multi-objective algorithms, it can be seen in Table 6 that the proposed MOGWO/D algorithm has better results in the vast majority of instances. Specifically, with regard to the convergence metric GD, the MOGWO/D algorithm achieves the best results for 33 instances; it is inferior to MOGWO on the "2_8_10" and "2_8_20" instances and inferior to MOPSO on "3_4_30", but the differences are very small. For the spread metric, MOGWO/D algorithm achieves the optimal scheme in 34 instances; it is only inferior to MOGWO on "2_8_10" and inferior to NSGA-II on "3_4_30", but still better than MOPSO. Regarding the comprehensive metric IGD, 34 instances obtained the best metrics values with the MOGWO/D algorithm, only instance "2_ 8_10" was slightly better when using MOGWO, and NSGA-II is superior to MOGWO/D algorithm on "3_4_30". Furthermore, the standard deviations achieved for the three metrics for these instances by the MOGWO/D algorithm were better than those of the other three comparison algorithms in the vast majority of instances.

Table 6. Means and deviations of the three metrics obtained by MOGWO/D, NSGA-II, MOGWO, and MOPSO.

Problems	MOGWO/D			NSGA-II			MOGWO			MOPSO		
	GD	IGD	Δ	GD	IGD	Δ	GD	IGD	Δ	GD	IGD	Δ
1_2_10	3.11E-2	3.93E-3	2.62E-1	7.52E-2	1.20E-2	7.96E-1	6.52E-2	1.02E-2	7.56E-1	8.50E-2	1.33E-2	7.70E-1
	6.73E-3	8.93E-4	4.68E-2	8.07E-3	1.62E-3	8.51E-2	7.80E-3	2.01E-3	/6.50E-	1.06E-2	1.67E-3	7.80E-2
1_4_10	1.38E-2	4.36E-3	3.57E-1	3.32E-2	1.47E-2	8.15E-1	2.86E-2	1.18E-2	8.62E-1	3.82E-2	1.69E-2	7.71E-1
	5.11E-3	1.03E-3	6.08E-2	1.05E-2	3.66E-3	1.00E-1	9.12E-3	3.19E-3	9.77E-2	1.32E-2	3.93E-3	7.77E-2
1_6_10	7.45E-3	2.09E-3	6.24E-1	5.82E-2	8.47E-3	7.61E-1	2.96E-2	4.17E-3	7.50E-1	8.15E-2	1.08E-2	6.43E-1
	9.70E-3	6.04E-4	6.19E-2	6.83E-3	1.19E-2	5.64E-1	3.16E-2	3.61E-3	5.55E-1	1.29E-1	1.68E-3	5.35E-1
1_8_10	1.70E-3	1.05E-3	3.90E-1	8.44E-3	4.70E-3	8.74E-1	4.62E-3	2.88E-3	8.31E-1	1.27E-2	6.14E-3	6.97E-1
	8.40E-4	4.09E-4	2.95E-2	3.99E-3	3.14E-3	2.23E-1	1.72E-3	1.26E-3	5.83E-2	1.06E-2	5.77E-3	4.15E-1
1_2_20	2.36E-2	1.74E-2	3.81E-1	3.46E-2	4.84E-2	1.01E+00	3.94E-2	4.60E-2	1.02E+00	2.54E-2	5.02E-2	9.70E-1
	3.96E-2	8.09E-3	1.95E-2	4.16E-2	5.90E-3	7.97E-2	4.96E-2	9.25E-3	1.44E-1	2.30E-2	3.68E-3	4.40E-2
1_4_20	1.12E-2	4.20E-3	3.64E-1	2.20E-2	1.02E-2	7.68E-1	2.00E-2	9.37E-3	7.80E-1	2.48E-2	1.11E-2	7.60E-1
	5.60E-3	2.48E-3	4.85E-2	8.89E-3	4.67E-3	4.50E-2	8.78E-3	4.78E-3	5.88E-2	1.04E-2	4.81E-3	8.10E-2
1_6_20	3.49E-2	1.94E-2	4.68E-1	3.59E+01	4.37E-2	9.65E-1	3.83E+01	4.08E-2	9.36E-1	3.74E+01	4.69E-2	9.96E-1
	1.56E-2	4.78E-3	7.30E-2	1.32E+02	1.05E-2	1.97E-1	1.43E+02	1.04E-2	2.23E-1	1.27E+02	1.32E-2	2.16E-1
1_8_20	3.02E+01	3.22E-3	5.40E-1	8.49E+01	7.62E-3	8.48E-1	7.42E+01	7.19E-3	9.86E-1	9.41E+01	9.05E-3	7.21E-1
	6.97E+01	2.23E-3	9.90E-2	1.69E+02	6.33E-3	5.76E-1	1.67E+02	5.32E-3	4.56E-1	1.69E+02	7.71E-3	6.13E-1
1_2_30	4.34E-2	5.87E-3	1.93E-1	1.51E-1	1.69E-2	7.50E-1	1.16E-1	1.45E-2	7.12E-1	1.77E-1	1.89E-2	7.55E-1
	2.94E-2	1.03E-3	7.06E-2	3.79E-2	3.32E-3	1.02E-1	3.69E-2	2.81E-3	9.66E-2	3.80E-2	3.85E-3	1.00E-1
1_4_30	3.26E-2	1.08E-2	3.92E-1	5.55E-2	2.68E-2	8.50E-1	5.43E-2	2.41E-2	8.14E-1	5.57E-2	2.82E-2	8.27E-1
	1.98E-2	7.54E-3	5.74E-2	2.93E-2	1.55E-2	6.21E-2	3.06E-2	1.54E-2	7.18E-2	2.82E-2	1.54E-2	6.66E-2
1_6_30	4.64E+01	1.60E-1	4.35E-1	6.90E+01	3.61E-1	8.93E-1	5.37E+01	3.44E-1	8.81E-1	1.23E+02	3.75E-1	9.42E-1
	2.02E+02	9.72E-2	6.76E-2	3.00E+02	2.10E-1	1.23E-1	2.33E+02	2.10E-1	1.44E-1	4.29E+02	2.12E-1	1.51E-1
1_8_30	1.76E+02	3.22E+00	6.02E-1	3.62E+02	1.07E+01	1.28E+00	2.77E+02	8.59E+00	1.29E+00	3.94E+02	1.50E+01	1.27E+00
	1.35E+02	4.78E+00	7.64E-2	1.71E+02	1.10E+01	6.85E-1	1.74E+02	1.05E+01	1.12E-1	1.84E+02	1.06E+01	5.52E-2
2_2_10	1.69E-3	1.05E-3	3.76E-1	6.66E-3	5.14E-3	1.02E+00	4.83E-3	3.34E-3	8.82E-1	9.05E-3	7.88E-3	1.14E+00
	3.88E-4	1.91E-4	3.07E-2	9.63E-4	8.87E-4	8.09E-2	8.66E-4	4.83E-4	7.77E-2	1.57E-3	2.14E-3	6.75E-2
2_4_10	7.71E-3	9.17E-4	4.50E-1	3.38E-2	5.39E-3	8.93E-1	2.75E-2	3.31E-3	9.12E-1	4.15E-2	7.50E-3	9.21E-1
	5.79E-3	4.21E-4	5.34E-2	2.28E-2	2.90E-3	9.24E-2	1.53E-2	6.50E-4	9.76E-2	3.02E-2	5.06E-3	8.74E-2
2_6_10	1.60E-1	3.14E-2	5.50E-1	2.21E-1	6.49E-2	8.99E-1	2.05E-1	5.64E-2	8.87E-1	2.37E-1	7.13E-2	9.23E-1
	1.30E-1	5.50E-3	1.04E-1	1.05E-1	9.55E-3	1.09E-1	1.18E-1	8.56E-3	9.91E-2	1.04E-1	9.18E-3	1.13E-1
2_8_10	1.04E-3	3.60E-3	6.16E-1	1.03E-3	1.04E-3	5.81E-2	1.37E-4	3.20E-4	5.46E-2	7.52E-4	2.21E-3	4.89E-2
	2.20E-4	1.69E-3	5.64E-2	4.49E-3	4.53E-3	2.53E-1	5.96E-4	1.40E-3	2.38E-1	3.28E-3	9.64E-3	2.13E-1
2_2_20	1.26E-3	7.85E-4	3.51E-1	2.74E-2	1.50E-2	1.11E+00	9.37E-3	3.80E-3	9.47E-1	3.72E-2	2.48E-2	1.28E+00
	4.72E-4	1.52E-4	2.68E-2	2.04E-2	7.67E-3	1.18E-1	4.89E-3	5.08E-4	9.58E-2	2.59E-2	8.00E-3	2.26E-1
2_4_20	6.34E-3	2.90E-3	4.98E-1	2.15E-2	9.74E-3	1.05E+00	1.55E-2	6.88E-3	1.03E+00	2.68E-2	9.90E-3	9.77E-1
	4.19E-3	1.71E-3	5.70E-2	1.48E-2	6.84E-3	1.52E-1	7.67E-3	2.89E-3	1.40E-1	1.37E-2	6.18E-3	2.52E-1
2_6_20	1.67E-2	8.88E-3	5.86E-1	2.19E-2	1.60E-2	9.59E-1	1.86E-2	1.45E-2	9.07E-1	2.81E-2	1.85E-2	9.74E-1
	6.75E-3	3.72E-3	4.58E-2	7.23E-3	5.75E-3	5.46E-2	5.82E-3	5.77E-3	6.54E-2	9.63E-3	6.00E-3	7.60E-2
2_8_20	1.89E-2	1.88E-2	6.05E-1	2.26E-2	3.12E-2	9.34E-1	2.16E-2	2.95E-2	9.08E-1	2.56E-2	3.28E-2	9.84E-1
	9.36E-3	9.15E-3	6.20E-2	9.38E-3	1.41E-2	8.92E-2	8.77E-3	1.38E-2	9.37E-2	1.17E-2	1.43E-2	1.13E-1
2_2_30	1.07E-3	9.50E-4	4.10E-1	8.16E-3	9.67E-3	1.12E+00	5.01E-3	5.72E-3	9.80E-1	1.22E-2	1.46E-2	1.24E+00
	5.40E-4	4.77E-4	4.66E-2	4.11E-3	8.56E-3	2.01E-1	3.22E-3	5.86E-3	1.80E-1	4.87E-3	8.92E-3	1.63E-1
2_4_30	1.75E-2	1.65E-2	4.80E-1	2.22E-2	2.91E-2	8.30E-1	2.08E-2	2.78E-2	8.44E-1	2.35E-2	3.04E-2	8.43E-1
	1.63E-2	9.82E-3	6.57E-2	1.55E-2	2.00E-2	2.98E-1	1.47E-2	1.97E-2	3.17E-1	1.63E-2	2.03E-2	3.02E-1
2_6_30	9.06E-3	5.89E-3	6.22E-1	1.24E-2	9.63E-3	9.98E-1	1.10E-2	9.27E-3	9.70E-1	1.39E-2	9.81E-3	1.01E+00
	6.63E-3	3.02E-3	7.22E-2	7.18E-3	4.56E-3	8.25E-1	7.36E-3	4.69E-3	8.23E-1	7.24E-3	4.29E-3	8.61E-2
2_8_30	1.40E-2	2.55E-2	6.39E-1	1.54E-2	4.10E-2	1.00E+00	1.55E-2	4.00E-2	1.03E+00	1.62E-2	4.18E-2	1.02E+00
	5.97E-3	9.17E-3	4.74E-2	5.50E-3	1.45E-2	7.58E-2	5.57E-3	1.45E-2	7.57E-2	5.85E-3	1.47E-2	6.46E-2

Table 6. Cont.

Problems	MOGWO/D			NSGA-II			MOGWO			MOPSO		
	GD	IGD	Δ	GD	IGD	Δ	GD	IGD	Δ	GD	IGD	Δ
3_2_10	3.81E-3	2.44E-3	2.93E-1	7.36E-3	7.38E-3	6.54E-1	6.53E-3	6.16E-3	6.26E-1	8.96E-3	8.81E-3	6.70E-1
	5.73E-4	1.96E-4	3.01E-2	6.19E-4	7.85E-4	6.62E-2	6.45E-4	4.02E-4	4.89E-2	9.92E-4	1.10E-3	4.24E-2
3_4_10	1.75E-3	4.21E-3	6.30E-1	1.80E-2	3.63E-2	1.03E + 00	1.48E-2	2.00E-2	1.13E + 00	1.68E-2	4.94E-2	1.01E + 00
	1.65E-3	9.72E-4	5.37E-2	1.96E-2	1.55E-2	1.44E-1	1.58E-2	1.23E-2	1.05E-1	1.66E-2	4.51E-3	6.04E-2
3_6_10	3.05E-3	2.02E-3	5.03E-1	3.45E-2	1.69E-2	9.37E-1	1.35E-2	8.79E-3	1.01E + 00	4.13E-2	1.33E-2	6.96E-1
	2.82E-3	1.20E-3	7.36E-2	2.52E-2	1.63E-2	3.19E-1	8.74E-3	5.07E-3	2.55E-1	4.68E-2	1.35E-2	4.67E-1
3_8_10	8.48E-4	8.06E-4	4.09E-1	2.30E-2	1.55E-2	1.09E + 00	4.49E-3	3.82E-3	1.01E + 00	3.23E-2	1.80E-2	8.86E-1
	4.96E-4	4.08E-4	3.25E-2	1.70E-2	1.20E-2	2.93E-1	2.69E-3	2.88E-3	1.40E-1	3.31E-2	1.52E-2	5.31E-1
3_2_20	3.22E-3	9.19E-4	3.59E-1	2.72E-2	1.32E-2	1.14E + 00	1.40E-2	4.41E-3	9.26E-1	3.86E-2	2.67E-2	1.29E + 00
	1.27E-3	1.41E-4	3.27E-2	8.04E-3	4.61E-3	9.75E-2	4.02E-3	1.22E-3	9.68E-2	1.02E-2	4.68E-3	1.31E-1
3_4_20	4.37E-3	9.39E-4	3.66E-1	1.32E-2	3.46E-3	8.24E-1	9.38E-3	2.58E-3	8.42E-1	1.89E-2	4.74E-3	8.89E-1
	2.73E-3	1.84E-4	7.22E-2	6.55E-3	6.67E-4	1.14E-1	5.05E-3	4.39E-4	1.11E-1	9.37E-3	8.16E-4	1.16E-1
3_6_20	6.68E-3	3.27E-3	5.18E-1	1.88E-2	9.60E-3	1.07E + 00	1.47E-2	8.03E-3	1.05E + 00	2.63E-2	1.29E-2	1.08E + 00
	4.66E-3	1.89E-3	4.42E-2	9.13E-3	4.32E-3	1.18E-1	7.62E-3	3.77E-3	7.74E-2	1.91E-2	5.64E-3	1.27E-1
3_8_20	2.79E-3	1.74E-3	4.18E-1	9.66E-3	6.78E-3	9.36E-1	8.24E-3	5.52E-3	9.39E-1	1.63E-2	1.28E-2	1.01E + 00
	2.21E-3	1.28E-3	4.86E-2	4.76E-3	4.90E-3	2.91E-1	5.74E-3	6.58E-3	1.19E-1	1.59E-2	1.33E-2	2.97E-1
3_2_30	3.92E-3	3.23E-3	2.34E-1	1.19E-2	1.08E-2	5.27E-1	1.00E-2	8.67E-3	5.25E-1	1.39E-2	1.34E-2	5.42E-1
	1.48E-3	3.54E-4	4.10E-2	5.13E-3	7.11E-4	1.04E-1	6.26E-3	6.82E-4	1.67E-1	3.99E-3	8.41E-4	1.09E-1
3_4_30	3.24E-2	1.14E-2	1.30E-1	3.05E-2	2.14E-3	1.31E-1	4.24E-3	4.28E-3	1.79E-1	2.39E-3	3.05E-3	1.49E-1
	8.88E-3	2.88E-3	6.68E-2	9.15E-3	6.46E-3	3.94E-1	9.13E-3	9.65E-3	4.35E-1	9.50E-3	9.34E-3	4.47E-1
3_6_30	8.38E-3	1.36E-2	5.02E-1	1.76E-2	1.76E-2	4.33E-1	3.87E-2	3.30E-2	7.28E-1	2.12E-2	2.43E-2	4.22E-1
	3.06E-3	5.69E-3	5.45E-2	3.95E-2	3.05E-2	5.44E-1	6.29E-2	3.14E-2	4.95E-1	4.61E-2	3.89E-2	5.25E-1
3_8_30	1.76E-2	1.20E-2	4.68E-1	3.21E-2	2.64E-2	1.02E + 00	2.79E-2	2.50E-2	1.04E + 00	4.39E-2	2.84E-2	1.02E + 00
	9.12E-3	4.95E-3	5.92E-2	1.45E-2	1.03E-2	1.18E-1	1.23E-2	9.90E-3	1.10E-1	3.48E-2	1.08E-2	1.38E-1
Statistics	33/30	34/34	35/33	0/1	1/0	0/1	2/3	1/1	1/0	1/2	0/1	0/2

It is clear that the MOGWO/D algorithm has a better optimization performance for solving NPPLS-JP problems, and meets the needs to solve such problems. This may be because of the better balance of the MOGWO/D algorithm between exploitation and exploration, as well as the strategy in which the three best solutions x_α, x_β, and x_δ can be obtained from different levels of the grey wolves' social hierarchy. The experimental results show that the proposed algorithm is superior to other multi-objective optimization algorithms in solving the NPPLS-JP problem.

To observe the experimental results more intuitively, Figure 5a–j gives the convergence curve of each instance in the instance set, for which the order capacity is 20, and each curve is the best running result according to the comprehensive performance metric IGD over 30 independent runs for the proposed MOGWO/D algorithm and the three other comparison algorithms. It can be seen clearly that the convergence curves of the proposed algorithm are better than those of the comparison algorithms through these curve graphs. Among them, only on instances "3_2_20" and "3_8_20" is the proposed algorithm similar to the comparison algorithms, and the other instances all yield much better convergence curves than those of the comparison algorithms, which proves the effectiveness of the proposed MOGWO/D algorithm in solving the NPPLS-JP problem.

The NPPLS-JP problem proposed in this paper comes from the actual demand of an axle housing machining workshop. This scheduling demand exists widely in a multi-variety mixed production environment, so the proposed model and method are of great significance for production practice. This research has a substantive impact on improving the production efficiency of the workshop, and can significantly enhance the production management level of enterprises, so as to increase the market competitiveness.

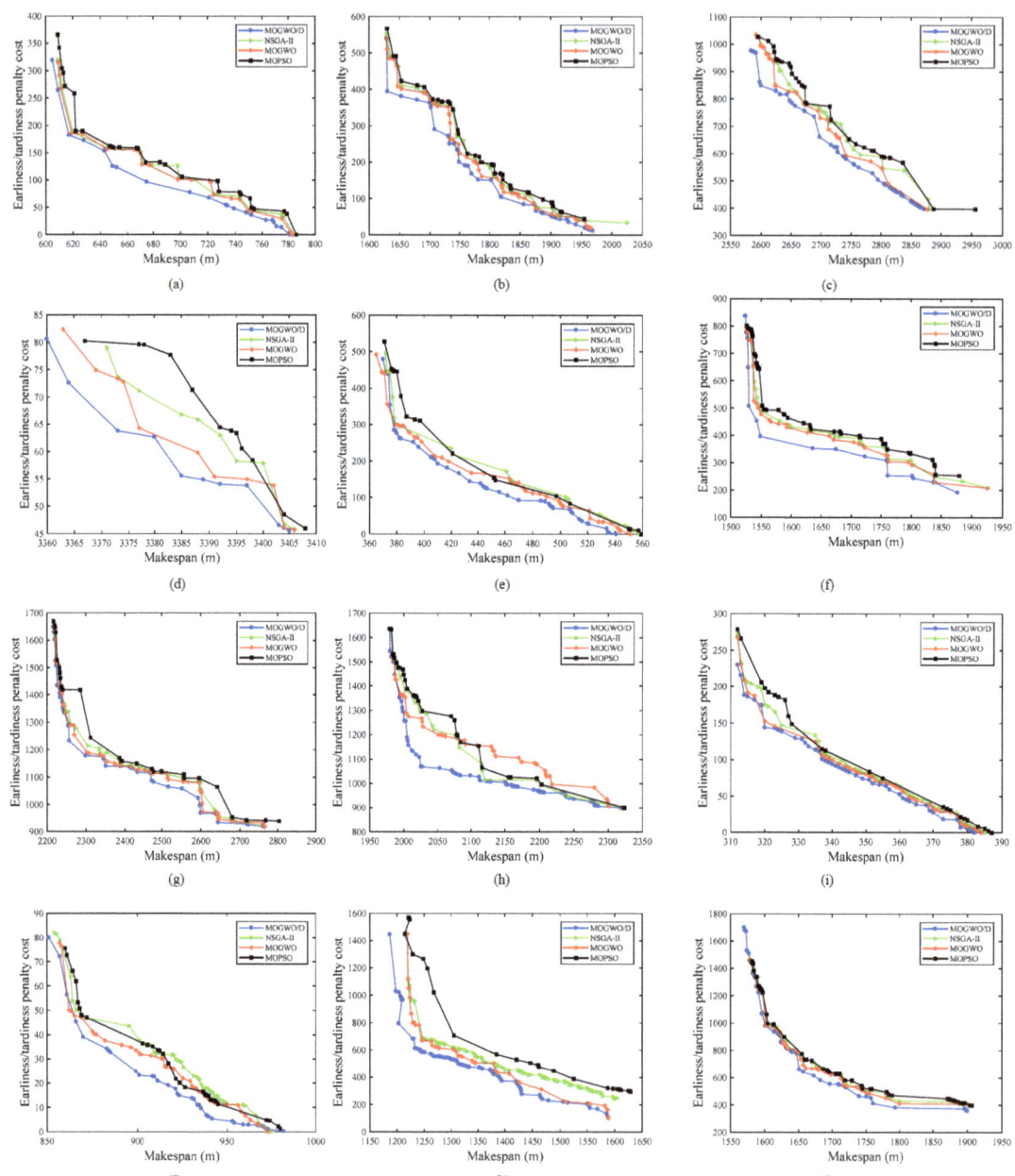

Figure 5. The convergence curves for the instance set with $z = 20$. (**a**) 1_2_20; (**b**) 1_4_20; (**c**) 1_6_20; (**d**) 1_8_20; (**e**) 2_2_20; (**f**) 2_4_20; (**g**) 2_6_20; (**h**) 2_8_20; (**i**) 3_2_20; (**j**) 3_4_20; (**k**) 3_6_20; (**l**) 3_8_20.

6. Conclusions

In this paper, a multi-objective NPPLS-JP derived from the real-life axle housing machining workshop of an axle manufacturer is studied. In the established NPPLS-JP model, the structures of all parallel lines are symmetrical. However, because of the demands of multivariety mixed production, the process capabilities and production capacities of these parallel production lines are asymmetric, and some types of job operations must be processed on the specific lines. This situation greatly affects the production efficiency of the production system and increases the difficulty of scheduling. To make multivariety mixed production more efficient and to maximize the utilization of the production capacity, a jumping process operation is introduced into the proposed model, which is the largest difference relative to the other general parallel production line scheduling problems. In the NPPLS-JP model, the multiline scheduling, multivariety mixed production, machine eligibility constraints, and MOPs are involved, so it is an NP-hard scheduling problem. In view of this model, we propose a hybrid multi-objective optimization algorithm that incorporates the single-objective GWO into the MOEA/D. The basic idea is to compensate for the shortcomings of the original algorithms by the reasonable mixing of several algorithms to balance their exploration and exploitation of abilities. To verify the effectiveness of the proposed algorithm, a set of instances is designed, and comparative experiments are conducted using the MOGWO/D algorithm as well as three other famous multi-objective optimization algorithms. The experimental results demonstrate that the proposed algorithm is superior to the compared algorithms for solving the NPPLS-JP problem. Furthermore, the experiment also proves that algorithm mixing can improve the performance and expand the application field of the constitutive algorithms.

In future research, we could solve the NPPLS-JP problem under the condition of considering the sequence-dependent setup times, or we could design new metaheuristics to solve the NPPLS-JP problem. Another interesting research direction is to explore new problem-specific rules to improve the performance of the MOGWO/D algorithm in terms of solving the NPPLS-JP. We can also focus on utilizing the MOGWO/D algorithm to solve other workshop scheduling problems, such as job shop scheduling problems and regular flow shop scheduling problems.

Author Contributions: Conceptualization, L.Y.; Formal analysis, G.X. and J.M.; Investigation, G.X. and J.L.; Methodology, G.X., L.Y. and J.M.; Project administration, Z.G.; Resources, L.Y.; Software, G.X. and J.L.; Writing—original draft, G.X. and L.Y.; Writing—review & editing, L.Y. and J.M.; funding acquisition, L.Y. All authors have read and agreed to the published version of the manuscript.

Funding: This research was funded by the Youth Program of National Natural Science Foundation of China (Grant No. 51905196) and the National Key R&D Program of China (Grant No. 2018YFB1702700).

Institutional Review Board Statement: Not applicable.

Informed Consent Statement: Not applicable.

Data Availability Statement: Not applicable.

Conflicts of Interest: The authors declare that they have no competing interest.

Abbreviations

The mathematical modeling notations are listed as follows.

o	the index of orders, $o = (1, 2, \cdots, O)$
O	the number of orders
N_o	the number of jobs in order o
N_J	the total number of jobs, $N' = \sum_{o=1}^{O} n_o$
J	a set of jobs, $J = \{1, 2, \cdots, N_J\}$
i, i'	the index of jobs, $i, i' \in J$
N_L	the total number of production lines
L	a set of production lines, $L = \{1, 2, \cdots, N_L\}$
l, l'	the index of production lines, $l, l' \in L$

N_S	the number of operations, which is equals to the number of stages
S	a set of operations, $S = \{1, 2, \cdots, N_S\}$
s	the index of operations, which is also the index of stages, $s \in S$
s_l	the jumping process point of production line l, $s_l = (1, 2, \cdots, S-1)$
t	the index of job types
N_t	the number of job types
k	the index of machines
$M_{l,s}$	the number of machines at stage s of production line l
$pt_{k,t,s}$	the processing time of operation s for a job of type t on machine k
α_o	the earliness penalty cost coefficient of order o
β_o	the tardiness penalty cost coefficient of order o
M	a sufficiently large positive number
$x_{t,s,l}$	takes a value of 1 if stage s of type t can be processed on production line l and 0 otherwise
$x_{i,o}$	takes a value of 1 if job i is included in order o and 0 otherwise
$x_{o,t}$	takes a value of 1 if the type of jobs in order o is t and 0 otherwise
$(d_o^1, d_o^2, d_o^3, d_o^4)$	the trapezoidal fuzzy number for trapezoidal fuzzy due date of order o, where $d_o^1 \leq \leq d_o^2 \leq d_o^3 \leq d_o^4$

Decision variables

$X_{i,s,l}$	binary variable, taking a value of 1 if operation s of job i is processed on production line l and 0 otherwise
$X_{k,s,i,l}$	binary variable, taking a value of 1 if operation s and s' of job i are both processed on production line l and 0 otherwise
$Y_{i,s,s'}$	binary variable, taking a value of 1 if operation s of job i is processed before job i' on machine k of production line l and 0 otherwise
$Z_{k,i,i',s,l}$	binary variable, taking a value of 1 if operation s of job i is processed before job i' on machine k of production line l and 0 otherwise
$C_{k,s,i,l}$	the completion time of operation s of job i on machine k of production line l
C_o	the completion time for order o
P_o	the fuzzy due date earliness/tardiness penalty cost of order o

References

1. Mohammadi, M.; Ghomi, S.; Jafari, N. A genetic algorithm for simultaneous lotsizing and sequencing of the permutation flow shops with sequence-dependent setups. *Int. J. Comput. Integr. Manuf.* **2011**, *24*, 87–93. [CrossRef]
2. Varmazyar, M.; Salmasi, N. Sequence-dependent flow shop scheduling problem minimising the number of tardy jobs. *Int. J. Prod. Res.* **2012**, *50*, 5843–5858. [CrossRef]
3. Yue, L.; Guan, Z.; Zhang, L.; Ullah, S.; Cui, Y. Multi objective lotsizing and scheduling with material constraints in flexible parallel lines using a Pareto based guided artificial bee colony algorithm. *Comput. Ind. Eng.* **2019**, *128*, 659–680. [CrossRef]
4. Jungwattanakit, J.; Reo De Cha, M.; Chaovalitwongse, P.; Werner, F. A comparison of scheduling algorithms for flexible flow shop problems with unrelated parallel machines, setup times, and dual criteria. *Comput. Oper. Res.* **2009**, *36*, 358–378. [CrossRef]
5. Low, C. Simulated annealing heuristic for flow shop scheduling problems with unrelated parallel machines. *Comput. Oper. Res.* **2005**, *32*, 2013–2025. [CrossRef]
6. Soltani, S.A.; Karimi, B. Cyclic hybrid flow shop scheduling problem with limited buffers and machine eligibility constraints. *Int. J. Adv. Manuf. Technol.* **2014**, *76*, 1739–1755. [CrossRef]
7. Tadayon, B.; Salmasi, N. A two-criteria objective function flexible flowshop scheduling problem with machine eligibility constraint. *Int. J. Adv. Manuf. Technol.* **2013**, *64*, 1001–1015. [CrossRef]
8. Ruiz, R.; Maroto, C. A genetic algorithm for hybrid flowshops with sequence dependent setup times and machine eligibility. *Eur. J. Oper. Res.* **2007**, *169*, 781–800. [CrossRef]
9. Zhang, X.Y.; Chen, L. A re-entrant hybrid flow shop scheduling problem with machine eligibility constraints. *Int. J. Prod. Res.* **2018**, *56*, 5293–5305. [CrossRef]
10. Oddi, A.; Rasconi, R.; Cortellessa, G.; Magazzeni, D.; Maratea, M.; Serina, I. Leveraging constraint-based approaches formulti-objective flexible flow-shop scheduling with energy costs. *Intell. Artif.* **2016**, *10*, 147–160.
11. Méndez, C.A.; Henning, G.P.; Cerdá, J. An MILP continuous-time approach to short-term scheduling of resource-constrained multistage flowshop batch facilities. *Comput. Chem. Eng.* **2001**, *25*, 701–711. [CrossRef]
12. Malik, A.I.; Kim, B.S. A multi-constrained supply chain model with optimal production rate in relation to quality of products under stochastic fuzzy demand. *Comput. Ind. Eng.* **2020**, *149*, 106814. [CrossRef]

13. Malik, A.I.; Sarkar, B. Coordinating supply-chain management under stochastic fuzzy environment and lead-time reduction. *Mathematics* **2019**, *7*, 480. [CrossRef]
14. Wu, H.C. Solving the fuzzy earliness and tardiness in scheduling problems by using genetic algorithms. *Expert Syst. Appl.* **2010**, *37*, 4860–4866. [CrossRef]
15. Bukchin, J.; Dar-El, E.M.; Rubinovitz, J. Mixed model assembly line design in a make-to-order environment. *Comput. Ind. Eng.* **2002**, *41*, 405–421. [CrossRef]
16. Caridi, M.; Sianesi, A. Multi-Agent Systems in production planning and control: An application to the scheduling of mixed-model assembly lines. *Int. J. Prod. Econ.* **2000**, *68*, 29–42. [CrossRef]
17. Askin, R.G.; Zhou, M. A parallel station heuristic for the mixed-model production line balancing problem. *Int. J. Prod. Res.* **1997**, *35*, 3095–3106. [CrossRef]
18. Emde, S.; Boysen, N. Optimally locating in-house logistics areas to facilitate JIT-supply of mixed-model assembly lines. *Int. J. Prod. Econ.* **2010**, *135*, 393–402. [CrossRef]
19. Lopes, T.C.; Michels, A.S.; Sikora, C.; Magatão, L. Balancing and cyclical scheduling of asynchronous mixed-model assembly lines with parallel stations. *J. Manuf. Syst.* **2019**, *50*, 193–200. [CrossRef]
20. Zhao, X.; Liu, J.; Ohno, K.; Kotani, S. Modeling and analysis of a mixed-model assembly line with stochastic operation times. *Nav. Res. Logist.* **2010**, *54*, 681–691. [CrossRef]
21. Khalid, Q.S.; Arshad, M.; Maqsood, S.; Kim, S. Hybrid particle swarm algorithm for products' scheduling problem in cellular manufacturing system. *Symmetry* **2019**, *11*, 729. [CrossRef]
22. Mcmullen, P.R.; Tarasewich, P. A beam search heuristic method for mixed-model scheduling with setups. *Int. J. Prod. Econ.* **2005**, *96*, 273–283. [CrossRef]
23. Leu, S.S.; Hwang, S.T. GA-based resource-constrained flow-shop scheduling model for mixed precast production. *Autom. Constr.* **2002**, *11*, 439–452. [CrossRef]
24. Wang, B.; Guan, Z.; Ullah, S.; Xu, X.; He, Z. Simultaneous order scheduling and mixed-model sequencing in assemble-to-order production environment: A multi-objective hybrid artificial bee colony algorithm. *J. Intell. Manuf.* **2017**, *28*, 419–436. [CrossRef]
25. Bahman, N.; Ahmed, A.; Katayoun, B. A realistic multi-manned five-sided mixed-model assembly line balancing and scheduling problem with moving workers and limited workspace. *Int. J. Prod. Res.* **2019**, *57*, 643–661.
26. Alghazi, A.; Kurz, M.E. Mixed model line balancing with parallel stations, zoning constraints, and ergonomics. *Constraints* **2018**, *23*, 123–153. [CrossRef]
27. Rajeswari, N.; Shahabudeen, P. Bicriteria parallel flow line scheduling using hybrid population-based heuristics. *Int. J. Adv. Manuf. Technol.* **2009**, *43*, 799–804. [CrossRef]
28. Haq, A.N.; Balasubramanian, K.; Sashidharan, B.; Karthick, R.B. Parallel line job shop scheduling using genetic algorithm. *Int. J. Adv. Manuf. Technol.* **2008**, *35*, 1047–1052.
29. Meyr, H.; Mann, M. A decomposition approach for the general lotsizing and scheduling problem for parallel production lines. *Eur. J. Oper. Res.* **2013**, *229*, 718–731. [CrossRef]
30. Mumtaz, J.; Guan, Z.; Yue, L.; Zhang, L.; He, C. Hybrid spider monkey optimisation algorithm for multi-level planning and scheduling problems of assembly lines. *Int. J. Prod. Res.* **2020**, *58*, 6252–6267. [CrossRef]
31. Ebrahimipour, V.; Najjarbashi, A.; Sheikhalishahi, M. Multi-objective modeling for preventive maintenance scheduling in a multiple production line. *J. Intell. Manuf.* **2013**, *26*, 1–12. [CrossRef]
32. Mumtaz, J.; Guan, Z.; Yue, L.; Wang, Z.; Rauf, M. Multi-level planning and scheduling for parallel pcb assembly lines using hybrid spider monkey optimization approach. *IEEE Access* **2019**, *7*, 2169–3536. [CrossRef]
33. Liu, Z.Z.; Wang, Y.; Huang, P.Q. A many-objective evolutionary algorithm with angle-based selection and shift-based density estimation. *Inf. Sci.* **2017**, *509*, 400–419. [CrossRef]
34. Goldberg, D.E.; Korb, B.; Deb, K. Messy genetic algorithms: Motivation, analysis, and first results. *Complex Syst.* **1989**, *3*, 493–530.
35. Zitzler, E.; Thiele, L. Multiobjective evolutionary algorithms: A comparative case study and the strength Pareto approach. *IEEE Trans. Evol. Comput.* **1999**, *3*, 257–271. [CrossRef]
36. Zitzler, E.; Laumanns, M.; Thiele, L. *SPEA2: Improving the Performance of the Strength Areto Evolutionary Algorithm*; Evolutionary Methods for Design, Optimization and Control with Applications to Industrial Problems (EUROGEN 2001), Athens, Greece, September; Giannakoglou, K.C., Tsahalis, D.T., Périaux, J., Papailiou, K.D., Fogarty, T., Eds.; International Center for Numerical Methods in Engineering (CIMNE): Barcelona, Spain, 2002; pp. 95–100.
37. Deb, K.; Pratap, A.; Agarwal, S.; Meyarivan, T. A fast and elitist multiobjective genetic algorithm: NSGA-II. *IEEE Trans. Evol. Comput.* **2002**, *6*, 182–197. [CrossRef]
38. Beume, N.; Naujoks, B.; Emmerich, M. Sms-emoa: Multiobjective selection based on dominated hypervolume. *Eur. J. Oper. Res.* **2007**, *181*, 1653–1669. [CrossRef]
39. Bader, J.; Zitzler, E. Hype: An algorithm for fast hypervolume-based many-objective optimization. *Evol. Comput.* **2011**, *19*, 45–76. [CrossRef] [PubMed]
40. Zhang, Q.; Hui, L. MOEA/D: A multiobjective evolutionary algorithm based on decomposition. *IEEE Trans. Evol. Comput.* **2007**, *11*, 712–731. [CrossRef]
41. Li, H.; Landa-Silva, D. An adaptive evolutionary multi-objective approach based on simulated annealing. *Evol. Comput.* **2014**, *19*, 561–595. [CrossRef] [PubMed]

42. Tan, Y.Y.; Jiao, Y.C.; Hong, L.; Wang, X.K. MOEA/D-SQA: A multi-objective memetic algorithm based on decomposition. *Eng. Optim.* **2012**, *44*, 1–21. [CrossRef]
43. Cai, D.; Yuping, W.; Miao, Y. A new multi-objective particle swarm optimization algorithm based on decomposition. *Inf. Sci.* **2015**, *325*, 541–557.
44. Ke, L.; Zhang, Q.; Battiti, R. MOEA/D-ACO: A Multiobjective Evolutionary Algorithm Using Decomposition and Antcolony. *IEEE Trans. Cybern.* **2013**, *43*, 1845–1859. [CrossRef] [PubMed]
45. Alhindi, A.; Alhindi, A.; Alhejali, A.; Alsheddy, A.; Tairan, N.; Alhakami, H. MOEA/D-GLS: A multiobjective memetic algorithm using decomposition and guided local search. *Soft Comput.* **2019**, *23*, 9605–9615. [CrossRef]
46. Zhang, Q.; Liu, W.; Tsang, E.; Virginas, B. Expensive multiobjective optimization by MOEA/D with gaussian process model. *IEEE Trans. Evol. Comput.* **2010**, *14*, 456–474. [CrossRef]
47. Wang, Z.; Zhang, Q.; Zhou, A.; Gong, M.; Jiao, L. Adaptive replacement strategies for MOEA/D. *IEEE Trans. Cybern.* **2017**, *46*, 474–486. [CrossRef]
48. Ho, Y.C.; Pepyne, D.L. Simple explanation of the no-free-lunch theorem and its implications. *J. Optim. Theory Appl.* **2002**, *115*, 549–570. [CrossRef]
49. Murata, T.; Gen, M.; Ishibuchi, H. Multi-objective scheduling with fuzzy due-date. *Comput. Ind. Eng.* **1998**, *35*, 439–442. [CrossRef]
50. Vela, C.R.; Afsar, S.; Palacios, J.J.; González-Rodríguez, I.; Puente, J. Evolutionary tabu search for flexible due-date satisfaction in fuzzy job shop scheduling. *Comput. Oper. Res.* **2020**, *119*, 104931. [CrossRef]
51. Wen, X.; Li, X.; Gao, L.; Wang, K.; Li, H. Modified honey bees mating optimization algorithm for multi-objective uncertain integrated process planning and scheduling problem. *Int. J. Adv. Robot. Syst.* **2020**, *17*, 172988142092523. [CrossRef]
52. Mirjalili, S.; Mirjalili, S.M.; Lewis, A. Grey Wolf Optimizer. *Adv. Eng. Softw.* **2014**, *69*, 46–61. [CrossRef]
53. Scheffé, H. Experiments with Mixtures. *J. Roy. Statist. Soc.* **1958**, *20*, 344–360. [CrossRef]
54. Li, B.B.; Wang, L.; Liu, B. An effective PSO-based hybrid algorithm for multiobjective permutation flow shop scheduling. *IEEE Trans. Syst. Man Cybern. Paart A Syst. Hum.* **2008**, *38*, 818–831. [CrossRef]
55. Mirjalili, S.; Saremi, S.; Mirjalili, S.M.; Coelho, L. Multi-objective grey wolf optimizer: A novel algorithm for multi-criterion optimization. *Expert Syst. Appl.* **2015**, *47*, 106–119. [CrossRef]
56. Coello, C.A.C.; Pulido, G.T.; Lechuga, M.S. Handling multiple objectives with particle swarm optimization. *IEEE Trans. Evol. Comput.* **2004**, *8*, 256–279. [CrossRef]
57. Zitzler, E.; Thiele, L.; Laumanns, M.; Fonseca, C.M.; Fonseca, V. Performance assessment of multiobjective optimizers: An analysis and review. *IEEE Trans. Evol. Comput.* **2003**, *7*, 117–132. [CrossRef]
58. Jiang, S.; Ong, Y.; Zhang, J.; Feng, L. Consistencies and contradictions of performance metrics in multiobjective optimization. *IEEE Trans. Cybern.* **2014**, *44*, 2391–2404. [CrossRef] [PubMed]
59. Oguz, C.; Ercan, M.F. A genetic algorithm for hybrid flow-shop scheduling with multiprocessor tasks. *Complex Syst.* **2005**, *8*, 323–351.

Article
Total Domination in Rooted Product Graphs

Abel Cabrera Martínez and **Juan A. Rodríguez-Velázquez** *

Departament d'Enginyeria Informàtica i Matemàtiques, Universitat Rovira i Virgili, Av. Països Catalans 26, 43007 Tarragona, Spain; abel.cabrera@urv.cat
* Correspondence: juanalberto.rodriguez@urv.cat

Received: 2 October 2020; Accepted: 20 November 2020; Published: 23 November 2020

Abstract: During the last few decades, domination theory has been one of the most active areas of research within graph theory. Currently, there are more than 4400 published papers on domination and related parameters. In the case of total domination, there are over 580 published papers, and 50 of them concern the case of product graphs. However, none of these papers discusses the case of rooted product graphs. Precisely, the present paper covers this gap in the theory. Our goal is to provide closed formulas for the total domination number of rooted product graphs. In particular, we show that there are four possible expressions for the total domination number of a rooted product graph, and we characterize the graphs reaching these expressions.

Keywords: total domination; domination; rooted product graph

Let G be a graph. The open neighborhood of a vertex $v \in V(G)$ is defined to be $N(v) = \{u \in V(G) : u \text{ is adjacent to } v\}$. A set $S \subseteq V(G)$ is a dominating set of G if $N(v) \cap S \neq \emptyset$ for every vertex $v \in V(G) \setminus S$. Let $\mathcal{D}(G)$ be the set of dominating sets of G. The domination number of G is defined to be,

$$\gamma(G) = \min\{|S| : S \in \mathcal{D}(G)\}.$$

A set $S \subseteq V(G)$ is a total dominating set, TDS, of a graph G without isolated vertices if every vertex $v \in V(G)$ is adjacent to at least one vertex in S. Let $\mathcal{D}_t(G)$ be the set of total dominating sets of G.

The total domination number of G is defined to be,

$$\gamma_t(G) = \min\{|S| : S \in \mathcal{D}_t(G)\}.$$

By definition, $\mathcal{D}_t(G) \subseteq \mathcal{D}(G)$, so that $\gamma(G) \leq \gamma_t(G)$.

We define a $\gamma_t(G)$-set as a set $S \in \mathcal{D}_t(G)$ with $|S| = \gamma_t(G)$. The same agreement will be assumed for optimal parameters associated with other characteristic sets defined in the paper. For instance, a $\gamma(G)$-set will be a set $S \in \mathcal{D}(G)$ with $|S| = \gamma(G)$.

The theory of domination in graphs has been extensively studied. For instance, there are more than 4400 papers already published on domination and related parameters. In particular, we cite the following books [1,2]. In the case of total domination, there are over 580 published papers and one book [3]. Among these papers on total domination in graphs, there are over 50 which concern the case of product graphs. Surprisingly, none of these papers discusses the case of rooted product graphs. The present paper covers that gap in the theory.

In order to present our results, we need to introduce some additional notation and terminology. The closed neighborhood of $v \in V(G)$ is defined to be $N[v] = N(v) \cup \{v\}$. A vertex $v \in V(G)$ is universal if $N[v] = V(G)$, while it is a leaf if $|N(v)| = 1$. The set of leaves of G will be denoted by $\mathcal{L}(G)$. A support vertex is a vertex v with $N(v) \cap \mathcal{L}(G) \neq \emptyset$. The set of support vertices of G will be denoted by $\mathcal{S}(G)$. If v is a vertex of a graph G, then the vertex-deletion subgraph $G - \{v\}$ is the subgraph of G induced by $V(G) \setminus \{v\}$. By analogy, we define the subgraph $G - S$ for an arbitrary subset $S \subseteq V(G)$.

The concept of rooted product graph was introduced in 1978 by Godsil and McKay [4]. Given a graph G of order $n(G)$ and a graph H with root vertex v, the rooted product graph $G \circ_v H$ is defined as the graph obtained from G and H by taking one copy of G and $n(G)$ copies of H and identifying the ith vertex of G with the root vertex v in the ith copy of H for every $i \in \{1, 2, \ldots, n(G)\}$. If H or G is a trivial graph, then $G \circ_v H$ is equal to G or H, respectively. In this sense, hereafter we will only consider graphs G and H with no isolated vertex.

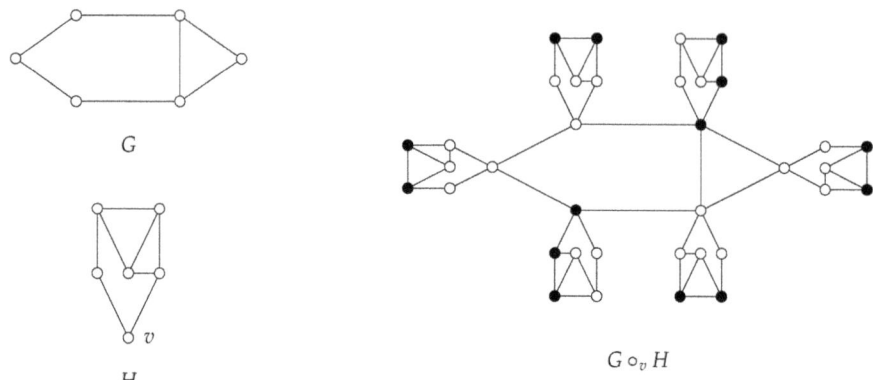

Figure 1. The set of black-coloured vertices forms a $\gamma_t(G \circ_v H)$-set.

Figure 1 shows an example of a rooted product graph. In this case, the set of black-coloured vertices forms a TDS of $G \circ_v H$ and $\gamma_t(G \circ_v H) = 14 = \gamma(G) + n(G)(\gamma_t(H) - 1)$.

For every $x \in V(G)$, $H_x \cong H$ will denote the copy of H in $G \circ_v H$ containing x. The restriction of any set $S \subseteq V(G \circ_v H)$ to $V(H_x)$ will be denoted by S_x, and the restriction to $V(H_x - \{x\})$ will be denoted by S_x^-; i.e., $S_x = S \cap V(H_x)$ and $S_x^- = S_x \setminus \{x\}$. In some cases, we will need to define $S \subseteq V(G \circ_v H)$ from the sets $S_x \subseteq V(H_x)$ as $S = \bigcup_{x \in V(G)} S_x$.

Since $V(G \circ_v H) = \bigcup_{x \in V(G)} V(H_x)$, we have that for every set $S \subseteq V(G \circ_v H)$,

$$|S| = \sum_{x \in V(G)} |S_x| = \sum_{x \in V(G)} |S_x^-| + |S \cap V(G)|. \quad (1)$$

A basic problem in the study of product graphs consists of finding closed formulas or sharp bounds for specific invariants of the product of two graphs and expressing these in terms of parameters of the graphs involved in the product. In this sense, for recent results on rooted product graphs, we cite the following works [5–19]. As we can expect, the products of graphs are not alien to applications in other fields. In particular, in [5] the authors show that several important classes of chemical graphs can be expressed as rooted product graphs, and as described in [20], there exist a number of molecular graphs of high-tech interest that can be generated using the rooted product of graphs.

1. Closed Formulas for the Total Domination Number

The following three lemmas will be the main tools to deduce our results.

Lemma 1. *Given a graph H with no isolated vertex and any $v \in V(H) \setminus S(H)$, the following statements hold.*

(i) $\gamma_t(H - \{v\}) \geq \gamma_t(H) - 1$.

(ii) *If $\gamma_t(H - \{v\}) = \gamma_t(H) - 1$, then the following statements hold.*

(a) $N(v) \cap S = \emptyset$ *for every $\gamma_t(H - \{v\})$-set S.*

(b) There exists a $\gamma_t(H)$-set S such that $v \notin S$.

(iii) If $\gamma_t(H - \{v\}) > \gamma_t(H)$, then $v \in S$ for every $\gamma_t(H)$-set S.

Proof. Let $v \in V(H) \setminus S(H)$ and S a $\gamma_t(H - \{v\})$-set. For every $u \in N(v)$ we have that $S \cup \{u\}$ is a TDS of H, which implies that $\gamma_t(H) \leq |S \cup \{u\}| \leq \gamma_t(H - \{v\}) + 1$. Therefore, (i) follows.

Now, in order to prove (ii), we assume that $|S| = \gamma_t(H) - 1$. If there exists a vertex $y \in N(v) \cap S$, then S is also a TDS of H, which is a contradiction. Therefore, $N(v) \cap S = \emptyset$ and so (a) follows. In addition, for any $y \in N(v)$, the set $S \cup \{y\}$ is a $\gamma_t(H)$-set not containing v. Therefore, (b) also follows.

Finally, we proceed to prove (iii). If there exists a $\gamma_t(H)$-set D such that $v \notin D$, then D is also a TDS of $H - \{v\}$, and so $\gamma_t(H - \{v\}) \leq |D| = \gamma_t(H)$. Therefore, we conclude that if $\gamma_t(H - \{v\}) > \gamma_t(H)$, then $v \in D$ for every $\gamma_t(H)$-set D, which completes the proof. □

Lemma 2. *Let H be a graph and $v \in V(H)$. If v is not a universal vertex and $H - N[v]$ does not have isolated vertices, then*

$$\gamma_t(H - N[v]) \geq \gamma_t(H) - 2.$$

Furthermore, if $\gamma_t(H - \{v\}) = \gamma_t(H) - 1$, then

$$\gamma_t(H) - 2 \leq \gamma_t(H - N[v]) \leq \gamma_t(H) - 1.$$

Proof. Assume that v is not a universal vertex and $H - N[v]$ does not have isolated vertices. Let S be a $\gamma_t(H - N[v])$-set and $u \in N(v)$. Since $S \cup \{u,v\}$ is a TDS of H, we have that $\gamma_t(H) \leq |S \cup \{u,v\}| = \gamma_t(H - N[v]) + 2$, as required.

Now, assume $\gamma_t(H - \{v\}) = \gamma_t(H) - 1$. In this case, by Lemma 1 (ii) we have that $N(v) \cap D = \emptyset$ for every $\gamma_t(H - \{v\})$-set D, which implies that D is a TDS of $H - N[v]$, and so $\gamma_t(H - N[v]) \leq |D| = \gamma_t(H - \{v\}) = \gamma_t(H) - 1$. Therefore, the result follows. □

Lemma 3. *Given a $\gamma_t(G \circ_v H)$-set S and a vertex $x \in V(G)$, the following statements hold.*

(i) $|S_x| \geq \gamma_t(H) - 1$.
(ii) *If $|S_x| = \gamma_t(H) - 1$, then $N(x) \cap S_x = \emptyset$.*

Proof. Let $x \in V(G)$. Notice that every vertex in $V(H_x) \setminus \{x\}$ is adjacent to some vertex in S_x. For any $y \in N(x) \cap V(H_x)$, the set $S_x \cup \{y\}$ is a TDS of H_x, and so $\gamma_t(H) = \gamma_t(H_x) \leq |S_x \cup \{y\}| = |S_x| + 1$. Therefore, (i) follows.

Finally, assume that $|S_x| = \gamma_t(H) - 1$. If there exists a vertex $y \in N(x) \cap S_x$, then S_x is a TDS of H_x, which is a contradiction. Therefore, $N(x) \cap S_x = \emptyset$, and so (ii) follows. □

Given a $\gamma_t(G \circ_v H)$-set S, we define the following subsets of $V(G)$ associated with S.

$$\mathcal{A}_S = \{x \in V(G) : |S_x| \geq \gamma_t(H)\} \text{ and } \mathcal{B}_S = \{x \in V(G) : |S_x| = \gamma_t(H) - 1\}.$$

These sets will play an important role in the inference results. By Lemma 3, $V(G) = \mathcal{A}_S \cup \mathcal{B}_S$. In particular, if $\mathcal{A}_S = \emptyset$, then $\gamma_t(G \circ_v H) = n(G)(\gamma_t(H) - 1)$, and as we will show in the proof of Theorem 2, if $\mathcal{B}_S = \emptyset$, then $\gamma_t(G \circ_v H) = n(G)\gamma_t(H)$. As we can expect, these are the extreme values of $\gamma_t(G \circ_v H)$.

Theorem 1. *For any graphs G and H with no isolated vertex and any $v \in V(H)$,*

$$n(G)(\gamma_t(H) - 1) \leq \gamma_t(G \circ_v H) \leq n(G)\gamma_t(H).$$

Furthermore, if $\gamma_t(H - \{v\}) = \gamma_t(H) - 1$, then

$$\gamma_t(G \circ_v H) \leq \gamma_t(G) + n(G)(\gamma_t(H) - 1).$$

Proof. The lower bound follows from Lemma 3, as for any $\gamma_t(G \circ_v H)$-set S,

$$\gamma_t(G \circ_v H) = |S| = \sum_{x \in V(G)} |S_x| \geq n(G)(\gamma_t(H) - 1).$$

Now, we proceed to prove the upper bound. Let $D \subseteq V(G \circ_v H)$ such that D_x is a $\gamma_t(H_x)$-set for every $x \in V(G)$. It is readily seen that D is a TDS of $G \circ_v H$. Hence,

$$\gamma_t(G \circ_v H) \leq |D| = \sum_{x \in V(G)} |D_x| = \sum_{x \in V(G)} \gamma_t(H_x) = n(G)\gamma_t(H).$$

From now on, assume $\gamma_t(H - \{v\}) = \gamma_t(H) - 1$. Notice that, by assumption, $H - \{v\}$ does not have isolated vertices.

Let $W \subseteq V(G \circ_v H)$ such that $W_x^- = W_x \setminus \{x\}$ is a $\gamma_t(H_x - \{x\})$-set for every $x \in V(G)$ and $W \cap V(G)$ is a $\gamma_t(G)$-set. Clearly, W is a TDS of $G \circ_v H$, which implies that

$$\gamma_t(G \circ_v H) \leq |W \cap V(G)| + \sum_{x \in V(G)} |W_x^-| = \gamma_t(G) + \sum_{x \in V(G)} \gamma_t(H_x - \{x\}) = \gamma_t(G) + n(G)(\gamma_t(H) - 1).$$

Therefore, the result follows. □

The following lemma is another important tool for determining all possible values of $\gamma_t(G \circ_v H)$.

Lemma 4. *Given a $\gamma_t(G \circ_v H)$-set S with $\mathcal{B}_S \neq \emptyset$, the following statements hold.*

(i) *If $\mathcal{B}_S \cap S \neq \emptyset$, then $\gamma_t(G \circ_v H) = n(G)(\gamma_t(H) - 1)$.*
(ii) *If $\mathcal{B}_S \cap S = \emptyset$, then $\gamma_t(H - \{v\}) = \gamma_t(H) - 1$, and as a consequence,*

$$\gamma(G) + n(G)(\gamma_t(H) - 1) \leq \gamma_t(G \circ_v H) \leq \gamma_t(G) + n(G)(\gamma_t(H) - 1).$$

Proof. First, we proceed to prove (i). Given a fixed $x' \in \mathcal{B}_S \cap S$, let $D \subseteq V(G \circ_v H)$ such that for every $x \in V(G)$ the set D_x is induced by $S_{x'}$. Obviously, D is a TDS of $G \circ_v H$. Hence, $\gamma_t(G \circ_v H) \leq |D| = \sum_{x \in V(G)} |D_x| = n(G)|S_{x'}| = n(G)(\gamma_t(H) - 1)$. Therefore, Theorem 1 leads to $\gamma_t(G \circ_v H) = n(G)(\gamma_t(H) - 1)$.

In order to prove (ii), assume that $\mathcal{B}_S \cap S = \emptyset$, and let $x \in \mathcal{B}_S$. By Lemma 3 we have that $N[x] \cap S_x = \emptyset$. So, $x \notin S(H_x)$ and S_x is a TDS of $H_x - \{x\}$. Hence, $\gamma_t(H - \{v\}) = \gamma_t(H_x - \{x\}) \leq |S_x| = \gamma_t(H) - 1$, and so Lemma 1 leads to $\gamma_t(H - \{v\}) = \gamma_t(H) - 1$. Therefore, by Theorem 1 we have that $\gamma_t(G \circ_v H) \leq \gamma_t(G) + n(G)(\gamma_t(H) - 1)$.

Moreover, since $N[x] \cap S_x = \emptyset$ for every $x \in \mathcal{B}_S$, we have that $\mathcal{A}_S$ is a dominating set of G. Hence,

$$\begin{aligned}
\gamma_t(G \circ_v H) &= \sum_{x \in \mathcal{A}_S} |S_x| + \sum_{x \in \mathcal{B}_S} |S_x| \\
&\geq |\mathcal{A}_S|\gamma_t(H) + |\mathcal{B}_S|(\gamma_t(H) - 1) \\
&\geq |\mathcal{A}_S| + n(G)(\gamma_t(H) - 1) \\
&\geq \gamma(G) + n(G)(\gamma_t(H) - 1).
\end{aligned}$$

Therefore, the result follows. □

Next we give one of the main results of this section, which states the four possible values of $\gamma_t(G \circ_v H)$.

Theorem 2. *Let G and H be two graphs with no isolated vertex. For any $v \in V(H)$,*

$$\gamma_t(G \circ_v H) \in \{n(G)(\gamma_t(H) - 1), \gamma(G) + n(G)(\gamma_t(H) - 1), \gamma_t(G) + n(G)(\gamma_t(H) - 1), n(G)\gamma_t(H)\}.$$

Proof. Let S be a $\gamma_t(G \circ_v H)$-set and consider the subsets $\mathcal{A}_S, \mathcal{B}_S \subseteq V(G)$ associated with S. We distinguish the following cases.

Case 1. $\mathcal{B}_S = \emptyset$. In this case, for any $x \in V(G)$ we have that $|S_x| \geq \gamma_t(H)$, and as a consequence, $\gamma_t(G \circ_v H) = \sum_{x \in V(G)} |S_x| \geq n(G) \gamma_t(H)$. Thus, Theorem 1 leads to the equality $\gamma_t(G \circ_v H) = n(G) \gamma_t(H)$.

Case 2. $\mathcal{B}_S \neq \emptyset$. If $\mathcal{B}_S \cap S \neq \emptyset$, then from Lemma 4 (i) we have that $\gamma_t(G \circ_v H) = n(G)(\gamma_t(H) - 1)$. From now on we assume that $\mathcal{B}_S \cap S = \emptyset$. Hence, Lemma 4 (ii) leads to

$$\gamma(G) + n(G)(\gamma_t(H) - 1) \leq \gamma_t(G \circ_v H) \leq \gamma_t(G) + n(G)(\gamma_t(H) - 1).$$

We only need to prove that $\gamma_t(G \circ_v H)$ only can take the extreme values. To this end, we shall need to introduce the following notation. Let $\mathcal{A}'_S = \{x \in \mathcal{A}_S : |S_x| = \gamma_t(H)\}$ and $\mathcal{A}''_S = \mathcal{A}_S \setminus \mathcal{A}'_S$.

Subcase 2.1. There exists $x' \in \mathcal{A}'_S$ such that $S_{x'}$ is a $\gamma_t(H_{x'})$-set containing x'. From a fixed vertex $y \in \mathcal{B}_S$ and any $\gamma(G)$-set D, we can construct a set $W \subseteq V(G \circ_v H)$ as follows. If $x \in D$, then W_x is induced by $S_{x'}$, while if $x \in V(G) \setminus D$, then W_x is induced by S_y. Notice that W is a TDS of $G \circ_v H$, which implies that $\gamma_t(G \circ_v H) \leq |W| = \gamma(G) + n(G)(\gamma_t(H) - 1)$. Therefore, $\gamma_t(G \circ_v H) = \gamma(G) + n(G)(\gamma_t(H) - 1)$.

Subcase 2.2. $\mathcal{A}'_S = \emptyset$ or for any $x \in \mathcal{A}'_S$, either S_x is not a $\gamma_t(H_x)$-set or $x \notin S_x$. If $\mathcal{A}'_S \neq \emptyset$, then every vertex $x \in \mathcal{A}'_S$ satisfies one of the following conditions.

(a) S_x is a $\gamma_t(H_x)$-set such that $x \notin S_x$.
(b) S_x is not a TDS of H_x and $x \in S_x$.

Notice that we do not consider the case where S_x is not a TDS of H_x and $x \notin S_x$, as in this case we can replace S with the $\gamma_t(G \circ_v H)$-set $(S \setminus S_x) \cup S'_x$ for some $\gamma_t(H_x)$-set S'_x. In such a case, if $x \in S'_x$, then we proceed as in Subcase 2.1, while if $x \notin S'_x$, then x satisfies (a).

Let us construct a TDS X of G as follows.

- $\mathcal{A}_S \subseteq X$.
- For any $x \in \mathcal{A}'_S$ which satisfies condition (a) and $N(x) \cap S \cap V(G) = \emptyset$, we choose one vertex $y \in N(x) \cap V(G)$ and set $y \in X$.
- For any $x \in \mathcal{A}''_S$ with $N(x) \cap S \cap V(G) = \emptyset$, we choose one vertex $y \in N(x) \cap V(G)$ and set $y \in X$.

We proceed to show that X is a TDS of G. If $x \in V(G) \setminus X$, then either $x \in \mathcal{B}_S$ or $x \in \mathcal{A}'_S \setminus S$. If $x \in \mathcal{B}_S$, then $N(x) \cap S \cap \mathcal{A}_S \neq \emptyset$, which implies that $N(x) \cap X \neq \emptyset$. Obviously, if $x \in \mathcal{A}'_S \setminus S$, then $N(x) \cap X \neq \emptyset$, by definition of X. Now, let $x \in X$. If $x \in \mathcal{A}''_S \cup (\mathcal{A}'_S \setminus S)$, then $N(x) \cap X \neq \emptyset$ by definition. If $x \in \mathcal{A}'_S \cap S$, then x satisfies condition (b). This implies that $N(x) \cap S_x = \emptyset$. Hence, there exists a vertex $y \in N(x) \cap V(G) \cap S \subseteq X$, as desired.

Therefore, X is a TDS of G, which implies that $\gamma_t(G) \leq |X| \leq 2|\mathcal{A}''_S| + |\mathcal{A}'_S|$. Thus,

$$\begin{aligned}
\gamma_t(G \circ_v H) &\geq \sum_{x \in \mathcal{A}''_S} |S_x| + \sum_{x \in \mathcal{A}'_S} |S_x| + \sum_{x \in \mathcal{B}_S} |S_x| \\
&\geq |\mathcal{A}''_S|(\gamma_t(H) + 1) + |\mathcal{A}'_S|\gamma_t(H) + |\mathcal{B}_S|(\gamma_t(H) - 1) \\
&\geq (2|\mathcal{A}''_S| + |\mathcal{A}'_S|) + n(G)(\gamma_t(H) - 1) \\
&\geq \gamma_t(G) + n(G)(\gamma_t(H) - 1),
\end{aligned}$$

which completes the proof. □

Later on, we will characterize the graphs that reach each of the previous expressions. However, we have to admit that when applying some of these characterizations we will need to calculate the total domination number of $H - \{v\}$ or $H - N[v]$ which may not be easy. Before giving the above mentioned characterizations, we shall show a simple example in which we can observe that these expressions of $\gamma_t(G \circ_v H)$ are realizable.

Example 1. *Let G be a graph with no isolated vertex. If H is one of the graphs shown in Figure 2, then the resulting values of $\gamma_t(G \circ_v H)$ for some specific roots are described below.*

- $\gamma_t(G \circ_{v'} H_2) = 3n(G) = n(G)(\gamma_t(H_2) - 1)$.
- $\gamma_t(G \circ_v H_2) = \gamma(G) + 3n(G) = \gamma(G) + n(G)(\gamma_t(H_2) - 1)$.
- $\gamma_t(G \circ_v H_1) = \gamma_t(G) + 2n(G) = \gamma_t(G) + n(G)(\gamma_t(H_1) - 1)$.
- $\gamma_t(G \circ_{v'} H_1) = \gamma_t(G \circ_{v''} H_1) = 3n(G) = n(G)\gamma_t(H_1)$.

For these cases, it is not difficult to construct a $\gamma_t(G \circ_v H)$-set. For instance, a $\gamma_t(G \circ_v H_2)$-set S can be formed as follows. Given a fixed $\gamma(G)$-set X, we take S in such a way that the set S_x is induced by $\{a, b, v', v\}$ for every $x \in X$, and induced by $\{a, b, c\}$ for every $x \in V(G) \setminus X$.

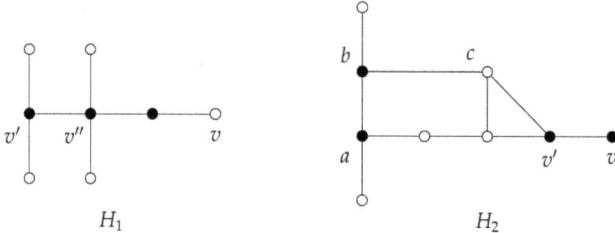

Figure 2. The set of black-coloured vertices forms a $\gamma_t(H_i)$-set for $i \in \{1, 2\}$. The set $\{v', v''\}$ forms a $\gamma_t(H_1 - \{v\})$-set, while $\{a, b, c\}$ forms a $\gamma_t(H_2 - \{v\})$-set.

As we have observed in Lemma 2, if $v \in V(H)$ is not a universal vertex and $H - N[v]$ does not have isolated vertices, then $\gamma_t(H - N[v]) \geq \gamma_t(H) - 2$. Next we show that the extreme case $\gamma_t(H - N[v]) = \gamma_t(H) - 2$ characterizes the graphs with $\gamma_t(G \circ_v H) = n(G)(\gamma_t(H) - 1)$.

Theorem 3. *Given two graphs G and H with no isolated vertex and $v \in V(H)$, the following statements are equivalent.*

(i) $\gamma_t(G \circ_v H) = n(G)(\gamma_t(H) - 1)$.
(ii) *v is a universal vertex of H or $\gamma_t(H - N[v]) = \gamma_t(H) - 2$.*

Proof. First, assume that (i) holds. Let S be a $\gamma_t(G \circ_v H)$-set. If v is a universal vertex of H, then we are done. Assume that $v \in V(H)$ is not a universal vertex. In this case, Lemma 3 leads to $\mathcal{B}_S = V(G)$ and $N(x) \cap S_x = \emptyset$ for every $x \in \mathcal{B}_S$. Thus, $\mathcal{B}_S \cap S$ is a dominating set of G and for any $x \in \mathcal{B}_S \cap S$ we have that $H_x - N[x]$ does not have isolated vertices and $S_x \setminus \{x\}$ is a TDS of $H_x - N[x]$, which implies that $\gamma_t(H - N[v]) = \gamma_t(H_x - N[x]) \leq |S_x \setminus \{x\}| = \gamma_t(H) - 2$. Hence, Lemma 2 leads to $\gamma_t(H - N[v]) = \gamma_t(H) - 2$. Therefore, (ii) follows.

Conversely, assume that (ii) holds. If v is a universal vertex of H, then $V(G)$ is a TDS of $G \circ_v H$, which implies that $\gamma_t(G \circ_v H) \leq |V(G)| = n(G) = n(G)(\gamma_t(H) - 1)$. Thus, by Theorem 1 we conclude that $\gamma_t(G \circ_v H) = n(G)(\gamma_t(H) - 1)$.

From now on, we assume that v is not a universal vertex. For any $x \in V(G)$, let D'_x be a $\gamma_t(H_x - N[x])$-set and $D_x = D'_x \cup \{x\}$. Observe that $D = \bigcup_{x \in V(G)} D_x$ is a TDS of $G \circ_v H$, which implies that $\gamma_t(G \circ_v H) \leq |D| = n(G)(\gamma_t(H - N[v]) + 1) = n(G)(\gamma_t(H) - 1)$. By Theorem 1 we conclude that $\gamma_t(G \circ_v H) = n(G)(\gamma_t(H) - 1)$, which completes the proof. □

Lemma 5. *Let G and H be two graphs with no isolated vertex and $v \in V(H) \setminus \mathcal{S}(H)$. If $\gamma_t(H - \{v\}) \geq \gamma_t(H)$, then*

$$\gamma_t(G \circ_v H) \in \{n(G)\gamma_t(H), n(G)(\gamma_t(H) - 1)\}.$$

Proof. By Theorem 1 we have that $\gamma_t(G \circ_v H) \leq n(G)\gamma_t(H)$. Let S be a $\gamma_t(G \circ_v H)$-set. If $|S| = n(G)\gamma_t(H)$, then we are done. Suppose that $|S| < n(G)\gamma_t(H)$. Hence, there exists $x \in V(G)$ such that $|S_x| < \gamma_t(H)$, which implies that $x \in \mathcal{B}_S$ by Lemma 3. Since $\gamma_t(H - \{v\}) \geq \gamma_t(H)$, Lemma 4 (ii) leads to $x \in S$, and by Lemma 4 (i) we deduce that $\gamma_t(G \circ_v H) = n(G)(\gamma_t(H) - 1)$. □

Lemma 6. *Let G and H be two graphs with no isolated vertex and $v \in V(H)$. If v belongs to every $\gamma_t(H)$-set, then*
$$\gamma_t(G \circ_v H) \in \{n(G)\gamma_t(H), n(G)(\gamma_t(H) - 1)\}.$$

Proof. We first consider the case where $v \in V(H) \setminus \mathcal{S}(H)$. By Lemma 1 we deduce that $\gamma_t(H - \{v\}) \geq \gamma_t(H)$, and so Lemma 5 leads to the result. Now, assume that $v \in \mathcal{S}(H)$ and let S be a $\gamma_t(G \circ H)$-set. If $\gamma_t(G \circ_v H) = n(G)\gamma_t(H)$, then we are done. Thus, we assume that $\gamma_t(G \circ_v H) < n(G)\gamma_t(H)$. In such a case, there exists $x \in \mathcal{B}_S$, and since $x \in \mathcal{S}(H_x)$, it follows that $x \in \mathcal{S}(G \circ H)$. Therefore, $x \in S$, and by Lemma 4 (i) we deduce that $\gamma_t(G \circ_v H) = n(G)(\gamma_t(H) - 1)$, which completes the proof. □

We are now ready to characterize the graphs with $\gamma_t(G \circ_v H) = \gamma(G) + n(G)(\gamma_t(H) - 1)$.

Theorem 4. *Let G and H be two graphs with no isolated vertex and $v \in V(H)$. The following statements are equivalent.*

(i) $\gamma_t(G \circ_v H) = \gamma(G) + n(G)(\gamma_t(H) - 1)$.
(ii) $\gamma_t(H - N[v]) = \gamma_t(H - \{v\}) = \gamma_t(H) - 1$, *and in addition,* $\gamma_t(G) = \gamma(G)$ *or there exists a $\gamma_t(H)$-set D such that $v \in D$.*

Proof. First, assume that (i) holds. Since $1 \leq \gamma(G) < n(G)$, by Lemma 6, $v \notin \mathcal{S}(H)$, so that from Lemma 5 we deduce that $\gamma_t(H - \{v\}) \leq \gamma_t(H) - 1$ and Lemma 1 leads to $\gamma_t(H - \{v\}) = \gamma_t(H) - 1$. Hence, by Lemma 2 it follows that $\gamma_t(H - N[v]) \in \{\gamma_t(H) - 2, \gamma_t(H) - 1\}$ and by Theorem 3 we obtain that $\gamma_t(H - N[v]) = \gamma_t(H) - 1$.

Now, let S be a $\gamma_t(G \circ_v H)$-set. Since $1 \leq \gamma(G) < n(G)$, Lemma 3 leads to $\mathcal{A}_S \neq \emptyset$ and $\mathcal{B}_S \neq \emptyset$. Additionally, by Lemma 4 we deduce that $\mathcal{B}_S \cap S = \emptyset$, and by Lemma 3 we have that $N(x) \cap S_x = \emptyset$ for every $x \in \mathcal{B}_S$. Hence, $\mathcal{A}_S$ is a dominating set of G and $\mathcal{A}_S \cap S \neq \emptyset$. Thus, $\gamma_t(G \circ_v H) \geq |\mathcal{A}_S| + n(G)(\gamma_t(H) - 1) \geq \gamma(G) + n(G)(\gamma_t(H) - 1) = \gamma_t(G \circ_v H)$, which implies that $\mathcal{A}_S$ is a $\gamma(G)$-set and for every $x \in \mathcal{A}_S \cap S$ we have that $|S_x| = \gamma_t(H)$. Therefore, there exists $x \in \mathcal{A}_S \cap S$ such that S_x is a $\gamma_t(H_x)$-set or $\mathcal{A}_S$ is a $\gamma_t(G)$-set, which implies that (ii) holds.

Conversely, assume that (ii) holds. As above, let S be a $\gamma_t(G \circ_v H)$-set. Since $\gamma_t(H - \{v\}) = \gamma_t(H) - 1$, by Theorem 1, $\gamma_t(G \circ_v H) \leq \gamma_t(G) + n(G)(\gamma_t(H) - 1)$.

Suppose that $\mathcal{B}_S = \emptyset$. In such a case, $\gamma_t(G \circ_v H) = n(G)\gamma_t(H)$, which implies that $\gamma(G) < \gamma_t(G) = n(G)$, and so $G \cong \cup K_2$. Let $A \cup B = V(G)$ be the bipartition of the vertex set of G, i.e., every edge has one endpoint in A and the other one in B. Thus, for every $x \in V(G)$ we define a subset $Y_x \subseteq V(H_x)$ as follows. If $x \in A$, then Y_x is a $\gamma_t(H_x)$-set which contains x, while if $x \in B$, then Y_x is a $\gamma_t(H_x - \{x\})$-set. Hence, $Y = \cup_{x \in V(G)} Y_x$ is a TDS of $G \circ_v H$ and so $\gamma_t(G \circ_v H) \leq |Y| = n(G)\gamma_t(H) - \frac{n(G)}{2} < n(G)\gamma_t(H)$, which is a contradiction. From now on we assume that $\mathcal{B}_S \neq \emptyset$.

If there exists a vertex $x \in \mathcal{B}_S \cap S$, then by Lemma 3 we have that $N(x) \cap S_x = \emptyset$, which implies that $S_x \setminus \{x\}$ is a TDS of $H_x - N[x]$. Hence, $\gamma_t(H - N[v]) = \gamma_t(H_x - N[x]) \leq |S_x \setminus \{x\}| = \gamma_t(H) - 2$, which is a contradiction with the assumption $\gamma_t(H - N[v]) = \gamma_t(H) - 1$. Therefore, $\mathcal{B}_S \cap S = \emptyset$, and by Lemma 4 we deduce that $\gamma_t(G \circ_v H) \geq \gamma(G) + n(G)(\gamma_t(H) - 1)$.

It is still necessary to prove that $\gamma_t(G \circ_v H) \leq \gamma(G) + n(G)(\gamma_t(H) - 1)$. If $\gamma(G) = \gamma_t(G)$, then we are done. Assume $\gamma(G) < \gamma_t(G)$. Now we take a $\gamma(G)$-set X and for every $x \in V(G)$ we define a set $Z_x \subseteq V(H_x)$ as follows. If $x \in X$, then Z_x is a $\gamma_t(H_x)$-set such that $x \in Z_x$, while if $x \in V(G) \setminus X$, then Z_x is a $\gamma_t(H_x - \{x\})$-set. Notice that $Z = \cup_{x \in V(G)} Z_x$ is a TDS of $G \circ_v H$. Therefore, $\gamma_t(G \circ_v H) \leq |Z| = \gamma(G) + n(G)(\gamma_t(H) - 1)$, as required. □

Next we proceed to characterize the graphs with $\gamma_t(G \circ_v H) = \gamma_t(G) + n(G)(\gamma_t(H) - 1)$. Notice that it is excluded the case $G \cong \cup K_2$. In such a case, $\gamma_t(G) = n(G)$, and so $\gamma_t(G) + n(G)(\gamma_t(H) - 1) = n(G)\gamma_t(H)$, which implies that the characterization of this particular case can be derived by elimination from Theorems 3 and 4. Analogously, the case $\gamma(G) = \gamma_t(G)$ is excluded, as it was discusses in Theorem 4.

Theorem 5. *Let $G \not\cong \cup K_2$ and H be two graphs with no isolated vertex such that $\gamma(G) < \gamma_t(G)$, and let $v \in V(H)$. The following statements are equivalent.*

(i) $\gamma_t(G \circ_v H) = \gamma_t(G) + n(G)(\gamma_t(H) - 1)$.
(ii) $\gamma_t(H - \{v\}) = \gamma_t(H) - 1$ and $v \notin D$ for every $\gamma_t(H)$-set D.

Proof. First, assume that (i) holds. Since, $G \not\cong \cup K_2$, we have that $\gamma_t(G) < n(G)$. Thus, by Lemma 6, $v \notin \mathcal{S}(H)$ and then by Lemma 5 we deduce that $\gamma_t(H - \{v\}) \leq \gamma_t(H) - 1$ and Lemma 1 leads to $\gamma_t(H - \{v\}) = \gamma_t(H) - 1$.

Suppose that there exists a $\gamma_t(H)$-set containing v. Let X be a $\gamma(G)$-set. For every $x \in V(G)$ we define a set $Z_x \subseteq V(H_x)$ as follows. If $x \in X$, then Z_x is a $\gamma_t(H_x)$-set such that $x \in Z_x$, while if $x \in V(G) \setminus X$, then Z_x is a $\gamma_t(H_x - \{x\})$-set. Notice that $Z = \cup_{x \in V(G)} Z_x$ is a TDS of $G \circ_v H$. Therefore, $\gamma_t(G \circ_v H) \leq |Z| = \gamma(G) + n(G)(\gamma_t(H) - 1)$, which is a contradiction, as $\gamma_t(G) > \gamma(G)$. Therefore, $v \notin D$ for every $\gamma_t(H)$-set D, which implies that (ii) follows.

Conversely, assume that (ii) holds. Since $\gamma_t(H - \{v\}) = \gamma_t(H) - 1$, by Theorem 1 we have that $\gamma_t(G \circ_v H) \leq \gamma_t(G) + n(G)(\gamma_t(H) - 1)$. Let S be a $\gamma_t(G \circ_v H)$-set. If $\mathcal{B}_S = \emptyset$, then $\gamma_t(G \circ_v H) = n(G)\gamma_t(H)$, and so $\gamma_t(G) = n(G)$, which is a contradiction, as $G \not\cong \cup K_2$. Hence, from now on we assume that $\mathcal{B}_S \neq \emptyset$.

If there exists a vertex $x \in \mathcal{B}_S \cap S$, then for any vertex $y \in N(v) \cap V(H_x)$, the set $S_x \cup \{y\}$ is a $\gamma_t(H_x)$-set, which is a contradiction. Thus, $\mathcal{B}_S \cap S = \emptyset$, and so by Lemma 3, $\mathcal{A}_S$ is a dominating set of G. Moreover, by Lemma 4 and Theorem 2 we deduce that either $\gamma_t(G \circ_v H) = \gamma(G) + n(G)(\gamma_t(H) - 1)$ or $\gamma_t(G \circ_v H) = \gamma_t(G) + n(G)(\gamma_t(H) - 1)$. Now, let $\mathcal{A}_S = A^- \cup A^+$ where $x \in A^-$ if $x \in \mathcal{A}_S$ and $N(x) \cap \mathcal{A}_S = \emptyset$. Let $B \subseteq \mathcal{B}_S$ such that $|B| \leq |A^-|$ and $N(x) \cap B \neq \emptyset$ for every $x \in A^-$. Obviously, $B \cup A^+$ is a total dominating set of G, and so $\gamma_t(G) + n(G)(\gamma_t(H) - 1) \leq |B \cup A^+| + n(G)(\gamma_t(H) - 1) \leq |\mathcal{A}_S| + n(G)(\gamma_t(H) - 1) \leq \gamma_t(G \circ_v H)$. Therefore, the result follows. □

From Theorem 2 we learned that there are four possible expressions for $\gamma_t(G \circ_v H)$. In the case of the first three expressions, the graphs (and the root) reaching the equality were characterized in Theorems 3–5. In the case of the expression $\gamma_t(G \circ_v H) = n(G)\gamma_t(H)$, the corresponding characterization can be derived by elimination from the previous results, although it must be recognized that the formulation of such a characterization is somewhat cumbersome. To conclude this section, we will just give a couple of examples where this expression is obtained.

The following result shows an example where $\gamma_t(G \circ_v H) = n(G)\gamma_t(H)$, which covers the cases in which v is a neighbor of a support vertex, excluding the case where v is the only leaf adjacent to its support.

Proposition 1. *Let G and H be two graphs with no isolated vertex and $v \in V(H)$. If there exists $u \in N(v)$ such that $N(u) \cap (\mathcal{L}(H) \setminus \{v\}) \neq \emptyset$, then*

$$\gamma_t(G \circ_v H) = n(G)\gamma_t(H).$$

Proof. Assume first that $v \notin \mathcal{S}(H)$. Let D be a $\gamma_t(H - \{v\})$-set. Since $u \in \mathcal{S}(H - \{v\})$, we have that $u \in D$. Hence, D is a TDS of H, and so $\gamma_t(H - \{v\}) = |D| \geq \gamma_t(H)$. Therefore, Lemma 5 leads to $\gamma_t(G \circ_v H) = n(G)\gamma_t(H)$ or $\gamma_t(G \circ_v H) = n(G)(\gamma_t(H) - 1)$. Now, suppose that $\gamma_t(G \circ_v H) = n(G)(\gamma_t(H) - 1)$. Let S be a $\gamma_t(G \circ_v H)$-set. By Lemma 3, $\mathcal{B}_S = V(G)$ and $N(x) \cap S_x = \emptyset$

for every $x \in \mathcal{B}_S$, which is a contradiction, as $N(x) \cap \mathcal{S}(H_x) \neq \emptyset$ and $\mathcal{S}(H_x) \subseteq S_x$. Therefore, $\gamma_t(G \circ_v H) = n(G)\gamma_t(H)$.

Now, if $v \in \mathcal{S}(H)$, then $u, v \in \mathcal{S}(G \circ_v H)$. Hence, for every $\gamma_t(G \circ_v H)$-set S and every vertex $x \in V(G)$, we have that S_x is a TDS of H_x. Thus, $\mathcal{B}_S = \emptyset$, which implies that $\gamma_t(G \circ_v H) = n(G)\gamma_t(H)$, as required. □

We next consider another example where $\gamma_t(G \circ_v H) = n(G)\gamma_t(H)$.

Proposition 2. *Let G and H be two graphs with no isolated vertex and $v \in V(H) \setminus \mathcal{S}(H)$. If $\gamma_t(H - \{v\}) \geq \gamma_t(H)$ and v does not belong to any $\gamma_t(H)$-set, then*

$$\gamma_t(G \circ_v H) = n(G)\gamma_t(H).$$

Proof. If $\gamma_t(H - \{v\}) \geq \gamma_t(H)$, then by Lemma 5 we have that $\gamma_t(G \circ_v H) = n(G)\gamma_t(H)$ or $\gamma_t(G \circ_v H) = n(G)(\gamma_t(H) - 1)$. Now, assume that v does not belong to any $\gamma_t(H)$-set. If $\gamma_t(G \circ_v H) = n(G)(\gamma_t(H) - 1)$, then $\mathcal{B}_S = V(G)$. Hence, by Lemma 4 (ii) there exists $x \in \mathcal{B}_S \cap S$, which is a contradiction as from any $x' \in N(x) \cap V(H_x)$ the set $S_x \cup \{x'\}$ is a $\gamma_t(H_x)$-set containing x. Therefore, $\gamma_t(G \circ_v H) = n(G)\gamma_t(H)$. □

2. An Observation on the Domination Number

It was shown in [15] that there are two possibilities for the domination number of a rooted product graph. Since the graphs reaching these expressions have not been characterized, we consider that it is appropriate to derive a result in this direction. Specifically, we will provide a characterization in Theorem 7.

Theorem 6. [15] *For any nontrivial graphs G and H and any $v \in V(H)$,*

$$\gamma(G \circ_v H) \in \{n(G)\gamma(H), \gamma(G) + n(G)(\gamma(H) - 1)\}.$$

In order to derive our result, we need to introduce the following two lemmas.

Lemma 7. [21] *Let H be a graph. For any vertex $v \in V(H)$,*

$$\gamma(H - \{v\}) \geq \gamma(H) - 1.$$

Lemma 8. *For any $\gamma(G \circ_v H)$-set D and any vertex $x \in V(G)$,*

$$|D_x| \geq \gamma(H) - 1.$$

Furthermore, if $|D_x| = \gamma(H) - 1$, then $N[x] \cap D_x = \emptyset$.

Proof. Let $x \in V(G)$. Notice that every vertex in $V(H_x) \setminus \{x\}$ is adjacent to some vertex in D_x. Since $D_x \cup \{x\}$ is a dominating set of H_x, we have that $\gamma(H) = \gamma(H_x) \leq |D_x \cup \{x\}| \leq |D_x| + 1$, as required.

Now, assume that $|D_x| = \gamma(H) - 1$. If $N[x] \cap D_x \neq \emptyset$, then D_x is a dominating set of H_x, which is a contradiction as $|D_x| = \gamma(H_x) - 1$. Therefore, the result follows. □

Theorem 7. *For any pair of nontrivial graphs G and H, and any $v \in V(H)$,*

$$\gamma(G \circ_v H) = \begin{cases} \gamma(G) + n(G)(\gamma(H) - 1) & \text{if } \gamma(H - \{v\}) = \gamma(H) - 1, \\ n(G)\gamma(H) & \text{otherwise.} \end{cases}$$

Proof. By Theorem 6 we only need to prove that $\gamma(G \circ_v H) = \gamma(G) + n(G)(\gamma(H) - 1)$ if and only if $\gamma(H - \{v\}) = \gamma(H) - 1$.

We first assume $\gamma(H - \{v\}) = \gamma(H) - 1$. Let $D \subseteq V(G \circ_v H)$ such that $D_x^- = D_x \setminus \{x\}$ is a $\gamma(H_x - \{x\})$-set for every $x \in V(G)$, and $D \cap V(G)$ is a $\gamma(G)$-set. It is readily seen that D is a dominating set of $G \circ_v H$, which implies that $\gamma(G \circ_v H) \leq |D| = \gamma(G) + \sum_{x \in V(G)} |D_x^-| = \gamma(G) + n(G)(\gamma(H) - 1)$, and by Theorem 6 we conclude that the equality holds.

Conversely, assume $\gamma(G \circ_v H) = \gamma(G) + n(G)(\gamma(H) - 1)$. Let S be a $\gamma(G \circ_v H)$-set. Since $|S| < n(G)\gamma(H)$, there exists $x \in V(G)$ such that $|S_x| < \gamma(H)$. Hence, by Lemma 8, $|S_x| = \gamma(H) - 1$ and $N[x] \cap S_x = \varnothing$. This implies that S_x is a dominating set of $H_x - \{x\}$, and so $\gamma(H - \{v\}) = \gamma(H_x - \{x\}) \leq |S_x| = \gamma(H) - 1$. By Lemma 7 we conclude that $\gamma(H - \{v\}) = \gamma(H) - 1$, which completes the proof. □

Author Contributions: All authors contributed equally to this work. All authors have read and agreed to the published version of the manuscript

Funding: This research received no external funding.

Conflicts of Interest: The authors declare no conflict of interest.

References

1. Haynes, T.W.; Hedetniemi, S.T.; Slater, P.J. *Domination in Graphs: Advanced Topics.*; Chapman and Hall/CRC Pure and Applied Mathematics Series; Marcel Dekker, Inc.: New York, NY, USA, 1998.
2. Haynes, T.W.; Hedetniemi, S.T.; Slater, P.J. *Fundamentals of Domination in Graphs*; Chapman and Hall/CRC Pure and Applied Mathematics Series; Marcel Dekker, Inc.: New York, NY, USA, 1998.
3. Henning, M.; Yeo, A. *Total Domination in Graphs. Springer Monographs in Mathematics*; Springer: New York, NY, USA, 2013.
4. Godsil, C.D.; McKay, B.D. A new graph product and its spectrum *Bull. Austral. Math. Soc.* **1978**, *18*, 21–28. [CrossRef]
5. Azari, M.; Iranmanesh, A. Chemical Graphs Constructed from Rooted Product and Their Zagreb Indices. *MATCH Commun. Math. Comput. Chem.* **2013**, *70*, 901–919.
6. Cabrera Martínez, A. Double outer-independent domination number of graphs. *Quaest. Math.* **2020**. [CrossRef]
7. Cabrera Martínez, A.; Cabrera García, S.; Carrión García, A.; Hernández Mira, F.A. Total Roman domination number of rooted product graphs. *Mathematics* **2020**, *8*, 1850. [CrossRef]
8. Cabrera Martínez, A.; Cabrera García, S.; Carrión García, A.; Grisales del Rio, A.M. On the outer-independent Roman domination in graphs. *Symmetry* **2020**, *12*, 1846. [CrossRef]
9. Cabrera Martínez, A.; Estrada-Moreno, A.; Rodríguez-Velázquez, J.A. Secure total domination in rooted product graphs. *Mathematics* **2020**, *8*, 600. [CrossRef]
10. Cabrera Martínez, A.; Montejano, L.P.; Rodríguez-Velázquez, J.A. Total weak Roman domination in graphs. *Symmetry* **2019**, *11*, 831. [CrossRef]
11. Chris Monica, M.; Santhakumar, S. Partition dimension of rooted product graphs. *Discrete Appl. Math.* **2019**, *262*, 138–147. [CrossRef]
12. Hernández-Ortiz, R.; Montejano, L.P.; Rodríguez-Velázquez, J.A. Italian domination in rooted product graphs. *Bull. Malays. Math. Sci. Soc.* **2020**. [CrossRef]
13. Hernández-Ortiz, R.; Montejano, L.P.; Rodríguez-Velázquez, J.A. Secure domination in rooted product graphs. *J. Comb. Optim.* to appear.
14. Klavžar, S.; Yero, I.G. The general position problem and strong resolving graphs. *Open Math.* **2019**, *17*, 1126–1135. [CrossRef]
15. Kuziak, D.; Lemanska, M.; Yero, I.G. Domination-Related parameters in rooted product graphs. *Bull. Malays. Math. Sci. Soc.* **2019**, *39*, 199–217. [CrossRef]
16. Kuziak, D.; Yero, I.G.; Rodríguez-Velázquez, J.A. Strong metric dimension of rooted product graphs. *Int. J. Comput. Math.* **2016**, *93*, 1265–1280. [CrossRef]

17. Jakovac, M. The k-path vertex cover of rooted product graphs. *Discrete Appl. Math.* **2015**, *187*, 111–119. [CrossRef]
18. Lou, Z.; Huang, Q.; Huang, X. On the construction of Q-controllable graphs. *Electron. J. Linear Algebra* **2017**, *32*, 365–379. [CrossRef]
19. Yang, Y.; Klein, D.J. Resistance distances in composite graphs. *J. Phys. A* **2014**, *47*, 375203. [CrossRef]
20. Farrell, E.J.; Rosenfeld, V.R. Block and articulation node polynomials of the generalized rooted product of graphs. *J. Math. Sci. India* **2000**, *11*, 35–47.
21. Sampathkumar, E.; Neeralagi, P.S. Domination and neighbourhood critical, fixed, free and totally free points. *Sankhya* **1992**, *54*, 403–407.

Publisher's Note: MDPI stays neutral with regard to jurisdictional claims in published maps and institutional affiliations.

© 2020 by the authors. Licensee MDPI, Basel, Switzerland. This article is an open access article distributed under the terms and conditions of the Creative Commons Attribution (CC BY) license (http://creativecommons.org/licenses/by/4.0/).

Article

Determination of a Good Indicator for Estimated Prime Factor and Its Modification in Fermat's Factoring Algorithm

Rasyid Redha Mohd Tahir [1], Muhammad Asyraf Asbullah [1,2,*], and Muhammad Rezal Kamel Ariffin [1,3] and Zahari Mahad [1]

1. Laboratory of Cryptography, Structure and Analysis, Institute for Mathematical Research (INSPEM), Universiti Putra Malaysia, Serdang 43400, Malaysia; gs50509@student.upm.edu.my (R.R.M.T.); rezal@upm.edu.my (M.R.K.A.); zaharimahad@upm.edu.my (Z.M.)
2. Centre of Foundation Study for Agricultural Science, Universiti Putra Malaysia, Serdang 43400, Malaysia
3. Mathematic Department, Faculty of Science, Universiti Putra Malaysia, Serdang 43400, Malaysia
* Correspondence: ma_asyraf@upm.edu.my

Abstract: Fermat's Factoring Algorithm (FFA) is an integer factorisation methods factoring the modulus N using exhaustive search. The appearance of the Estimated Prime Factor (EPF) method reduces the cost of FFA's loop count. However, the EPF does not work for balanced primes. This paper proposed the modified Fermat's Factoring Algorithm 1-Estimated Prime Factor (mFFA1-EPF) that improves the EPF method. The algorithm works for factoring a modulus with unbalanced and balanced primes, respectively. The main results of mFFA1-EPF focused on three criteria: (i) the approach to select good candidates from a list of convergent continued fraction, (ii) the establishment of new potential initial values based on EPF, and (iii) the application of the above modification upon FFA. The resulting study shows the significant improvement that reduces the loop count of FFA1 via (improved) EPF compared to existing methods. The proposed algorithm can be executed without failure and caters for both the modulus N with unbalanced and balanced primes factor. The algorithm works for factoring a modulus with unbalanced and balanced primes.

Keywords: estimated prime factor; integer factorisation problem; continued fraction; Fermat's Factoring Algorithm

Citation: Tahir, R.R.M.; Asbullah, M.A.; Ariffin, M.R.K.; Mahad, Z. Determination of a Good Indicator for Estimated Prime Factor and Its Modification in Fermat's Factoring Algorithm. *Symmetry* **2021**, *13*, 735. https://doi.org/10.3390/sym13050735

Academic Editor: Juan Alberto Rodríguez Velázquez

Received: 11 February 2021
Accepted: 13 March 2021
Published: 21 April 2021

Publisher's Note: MDPI stays neutral with regard to jurisdictional claims in published maps and institutional affiliations.

Copyright: © 2021 by the authors. Licensee MDPI, Basel, Switzerland. This article is an open access article distributed under the terms and conditions of the Creative Commons Attribution (CC BY) license (https://creativecommons.org/licenses/by/4.0/).

1. Introduction

Cryptography has its crucial parts in Industrial Revolution 4 (IR4) where technology is embedded in artificial intelligent to maintain the secureness of the information data. Regarding cryptography, there are two types of cryptography: symmetric and asymmetric cryptography. Symmetric cryptography uses the same key for the encryption and decryption process while asymmetric cryptography uses different keys for each encryption and decryption process. A lot of asymmetric cryptography strength relies on the Integer Factorisation Problem (IFP). IFP is one of the oldest hard mathematical problems in history. IFP is defined as finding the two distinct primes, p and q, for a given integer (a modulus) $N = pq$, which is the multiplication of those two primes. We et al. [1] mentioned that from the existing classical sense of computation, a modulus with a minimum 1024-bit length is still very hard to be factorised. There are several general purpose algorithms to solve the IFP, such as Pollard's $p - 1$, General Number Field Sieve, Quadratic Sieve, Elliptic Curve Factoring, and Fermat's Factoring Algorithm [2].

Pierre de Fermat explored Fermat's Factoring Algorithm (FFA) as one of the IFP methods that is used to factor the modulus N with balanced primes (Ambedkar et al. [3]). According to Somsuk and Tientanopajai [4], the modulus N in FFA is written as $N = \left(\frac{p+q}{2}\right)^2 - \left(\frac{p-q}{2}\right)^2$. In this work, the FFA is categorised into Fermat's Factoring Algorithm 1 (FFA1) and Fermat's Factoring Algorithm 2 (FFA2). FFA1 uses a square root, while FFA2 uses multiplication as their main processes that lead to factorization of N. Both methods are having their advantages and disadvantages in terms of the number of loop count (iteration) and computational time to complete the factoring process. Many studies have introduced to improve the FFA, to make the algorithm efficient for factoring the modulus N [5–8]. The main purpose is to speed up the algorithms: either to reduce the loop count or to improve the algorithm's computational time for exhaustive search or both [9].

The EPF is introduced by Wu et al. [1]. Wu et al's study enhances the efficiency of FFA2 by shortening the search for the target value of $p + q$ and $p - q$. The EPF method was adopted as a mechanism to reset the initial values of FFA (in this case is FFA2), which results in reducing the loop count for the FFA to complete the search and successfully factor the modulus N. The authors of [1,9] use the continued fraction of $\frac{1}{\sqrt{N}}$ to produce a list of convergent and create an additional extension for the initial values. Potentially, EPF is considered as a good "device" to increase the efficiency of FFA.

However, as reported in [1], the absence of a deterministic approach to select the required parameter in EPF is the limitations of such an approach, and most of the cases cannot work on balanced primes. Somsuk [5] agreed that EPF works perfectly only on unbalanced primes. By observation and empirical evidence, the selected values on the list of convergent certainly cannot be used for the initial value because it may cause the FFA2 algorithm to fail. On the other hand, the authors of [1] overlook a convergent-selecting case that affects the effectiveness of EPF. In finding a solution regarding EPF, FFA1 is chosen to be the main integer factorisation method in this study. This is because FFA1 potentially avoids the failure of running algorithm via EPF and reduce the exhaustive search as it uses a single loop run the algorithm. The resulting study shows a significant improvement that reduces the loop count of FFA1 via (improved) EPF compared to previous methods (FFA1, FFA2, FFA2-EPF, and FFA-Euler).

The rest of this paper is sorted as follows. Section 2 introduces the background of the FFA1, FFA2, and EPF. The definition of a modulus N is provided considering unbalanced and balanced primes. Section 3 discusses the methodology that will support the finding of this work. Section 4 presents the mFFA1-EPF, which works on factoring a modulus N for both unbalanced and balanced primes. The results of numerical examples are shown and compared with other existing FFA-based method. The conclusion is drawn in Section 5.

2. Preliminaries

Some fundamental information for the study, about FFAs and EPF, and some definitions are provided.

2.1. Balanced and Unbalanced Primes

This section provided the definitions of balanced and unbalanced primes, which restructure from [10,11], respectively. The definition of balanced prime is as follows.

Definition 1. *Let $N = pq$ be a number with a multiplication of two primes p and q. The number N is defined to have balanced primes where p and q have the same bit-size and satisfy the relation $q < p < 2q$.*

The term of unbalanced primes is defined, as follows.

Definition 2. *Let $N = pq$ be a number with a multiplication of two primes p and q. The number N is defined to have unbalanced primes where p and q have the different bit-size and satisfy the relation $q < p < \alpha q$ where $\alpha > 2$.*

2.2. Fermat's Factoring Algorithm (FFA)

De Weger [12] studied FFA1 as an approach of IFP to factor the modulus N by searching the value of $p + q$ and $p - q$. There is modulus N written as the difference of square, $N = \left(\frac{p+q}{2}\right)^2 - \left(\frac{p-q}{2}\right)^2$. As the value of $p + q$ and $p - q$ are unknown, we need to find the closest of those values. According to Asbullah and Ariffin [10], the smallest value of $p+q$, based on balanced prime, is $2\sqrt{N}$. We start the initial value $x = 2\lceil\sqrt{N}\rceil$ and then, compute $y = \sqrt{x^2 - N}$. If y is an integer then we accept the pair (x,y). Otherwise, the algorithm is ran by increasing the value of x by 1. If there is a pair of integers (x,y), then compute the values $p = x+y$ and $q = x-y$. Note that FFA1 only run a loop on searching the integer value on $p+q$ via initial value of x.

Bressoud [13] introduced FFA2, which is uses two loops on searching $p+q$ and $p-q$ to reduce the running time. Wu et al. [1] reformulated the Bressoud's method. Suppose there is modulus $N = x^2 - y^2$ where $x = \frac{p+q}{2}$ and $y = \frac{p-q}{2}$. The modulus is derived into $4N = u^2 - v^2$ where $u = 2x$ and $v = 2y$. As the actual values of u and v are unknown, reset $u = 2\lceil\sqrt{N}\rceil$ and $v = 0$. Now compute $r = u^2 - v^2 - 4N$. If $r = 0$, thus the solution of (u,v) is found. There are two cases for value of r as $r \neq 0$:

- Case 1: When $r > 0$
  The value of v needs to set larger; $v \leftarrow v + 2$ and then $r \leftarrow r - (4v + 4)$
- Case 2: When $r < 0$
  The value of u needs to set larger; $u \leftarrow u + 2$ and then $r \leftarrow r + (4u + 4)$

When the initial value u and v are created, we need to check the value of r. If $r \neq 0$, there are two cases: $r > 0$ and $r < 0$. If $r > 0$, the new v is produced that is increased by 2 from the old v, then the new r is computed by $r - (4v + 4)$. Otherwise, if $r < 0$, the new u is produced that is increased by 2 from the old u, then the new r is computed from $r + (4u + 4)$. The iteration of both case will be run until $r = 0$. If (u,v) found, the factorisation of N occurs as $p = \frac{u+v}{2}$ and $q = \frac{u-v}{2}$.

Remark 1. *The sign $\leftarrow$ represents an assignment of changes on a same value. For example, $v \leftarrow v + 2$ means the new value v is computed from the old value v with increment by 2.*

2.3. Continued Fraction

The continued fraction is a non-integer expansion method that represents a decimal number into a list of integers. The list of the integer can establish a partial quotient that brings into a convergent list in term of a rational number. The continued fraction is suitable for the representation of the rational and irrational number like π, Euler's number, e, and $\sqrt{2}$. Chung et al. [14] mentioned that the continued fraction for a rational number may produce a finite yield of integer number while for an irrational number may produce an infinite yield an infinite list of an integer number.

Let r be a real number which has unique continued fraction expansion,

$$r = [m_0, m_1, m_2, \ldots, m_i, \ldots] = m_0 + \cfrac{1}{m_1 + \cfrac{1}{m_2 + \cfrac{1}{\ldots}}}$$

where $m_i \in \mathbb{Z}$ and $i \in \mathbb{N}$. The list of m_i is a list integer form of r, thus, let r_i with an amount of i represent the partial quotients as follows:

$$r_0 = m_0$$

$$r_1 = m_0 + \frac{1}{m_1}$$

$$r_2 = m_0 + \cfrac{1}{m_1 + \cfrac{1}{m_2}}$$

$$\vdots$$

$$r_i = m_0 + \cfrac{1}{m_1 + \cfrac{1}{m_2 + \cfrac{1}{\ldots + \frac{1}{m_i}}}}$$

The list of partial quotients, $[r_0, r_1, r_2, \ldots, r_i]$ is also known as a convergent list. The list of the convergence is significantly used for several purposes such as shortening the distance of the initial value in Fermat's Factoring Algorithm and creating an approximate value of a rational number. Wu et al. [1] purposed Estimated Prime Factor in which used in application such as shortening the searching distance for (FFA2). It may give a "hint" for a new position for initial values of FFA2. The EPF will be discussed in the next section.

2.4. Estimated Prime Factor (EPF)

Wu et al. [1,9] proposed an approach to estimate $p + q$ and $p - q$ using EPF. The authors of [1,9] mentioned that the continued fraction of $\frac{1}{\sqrt{N}}$ is used to give out the partial knowledge of $\frac{D_p - D_q}{D_p D_q}$ in which helps to find $p + q$ and $p - q$ as $D_p - D_q$ and $D_p D_q$ are unknown. From the list of convergent $\frac{1}{\sqrt{N}}$, $\frac{h_t}{k_t}$ is selected as the additional extension where $k_t \lesssim N$ and $h_t < D_p - D_q < h_{t+1}$. The in-depth discussion of convergent $\frac{h_t}{k_t}$ in EPF is provided in Appendix A.

Remark 2. *Let k_t be denominator of $\frac{h_t}{k_t}$ from convergent list pf $\frac{1}{\sqrt{N}}$. The value of k_t is approximately less than N, $k_t \lesssim N$, in which k_t need to be less and closer to the value of N. As the value of N is known, it is easy to select $\frac{h_t}{k_t}$ by k_t comparing with the value of N, and k_t could be good indicator to select a good convergent of $\frac{1}{\sqrt{N}}$ as there is i convergents on the list.*

We illustrate EPF process in Figure 1.

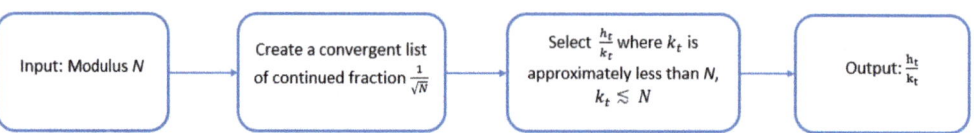

Figure 1. The process of Estimated Prime Factor (EPF).

3. Methodology

As early discussed in Section 1, Bressoud [13] claimed that the FFA2 has a better component for loop count without any multiplication or division and exhibit faster computational time. However, the FFA2 requires a huge number of cycles because it needs to search for the value of $p + q$ and $p - q$, separately. Compare to FFA2, FFA1 needs to search for the value of $p + q$ only. This eliminates the process of searching for the value of $p - q$, and thus reduces the number of loop count.

Table 1 shows the comparison of loop count between FFA1 and FFA2 based on three distinct moduli $N = pq$. The first modulus of $N = 1{,}783{,}647{,}329$ is taken from an example in [1]. Observed that the FFA1 dominates the smallest values of loop count (10,552) rather than the FFA2 (42,215) and the FFA2-EPF (11,455). For $N = 195{,}656{,}557$, again, the FFA1 dominates the smallest loop count compare to the other two methods. $N = 1{,}952{,}194{,}393$ is selected to illustrate the balanced prime situation. Similarly, the FFA1 recorded the smallest loop count, while the loop count for FFA2-EPF is not available as the initial value of u and v are larger than $p + q$ and $p - q$. Overall, the FFA1 requires a lesser loop count, therefore it can factor modulus N relatively faster than the FFA2 and the FFA2-EPF.

Table 1. The comparison on loop count between FFA1, FFA2, and FFA2-EPF based on 3 distinct modulus $N = pq$.

Modulus $N = pq$	Loop Count		
	FFA1	FFA2	FFA2-EPF
$1{,}783{,}647{,}329 = 84{,}449 \cdot 21{,}121$	10,552	42,215	11,455
$195{,}656{,}557 = 27{,}103 \cdot 7219$	3173	13,115	8131
$1{,}952{,}194{,}393 = 47{,}969 \cdot 40{,}697$	150	3785	N/A

Recall that the initial value of the FFA1 started from $\sqrt{N}$ and it increased by 1 until it reach the value $\frac{p+q}{2}$. Therefore, there exists a distance, denoted by d_0 from the initial value where it starts from $\sqrt{N}$ to $\frac{p+q}{2}$: $d_0 = \left|\frac{p+q}{2} - \sqrt{N}\right|$ (as shown in Figure 2). The methodology is to introduce a new parameter $\lambda \in \mathbb{N}$, such that $\sqrt{N} + \lambda$ will be serve as a new initial value for the FFA1. The reason is to establish a new distance, $d_{new} = \left|\frac{p+q}{2} - (\sqrt{N} + \lambda)\right|$, as shown in Figure 2, where $d_{new} < d_0$. Therefore, a good λ is needed toward initial value of FFA1 to obtain the d_{new}.

In this work, the EPF method is used to extend the initial value of FFA1. We show that the EPF approach is suitable to be used in FFA1 with following conditions. Suppose $x = \frac{p+q}{2}$ and $y = \frac{p-q}{2}$. Note that from Section 2.4, we have $p = \sqrt{N} + D_p$ and $q = \sqrt{N} - D_q$. Recall that modulus $N = x^2 - y^2 = \left(\frac{p+q}{2}\right)^2 - \left(\frac{p-q}{2}\right)^2$, therefore

$$N = \left(\frac{(D_p + \sqrt{N}) + (\sqrt{N} - D_q)}{2}\right)^2 - \left(\frac{(D_p + \sqrt{N}) - (\sqrt{N} - D_q)}{2}\right)^2$$

$$= \left(\frac{2\sqrt{N} + D_p - D_q}{2}\right)^2 - \left(\frac{D_p + D_q}{2}\right)^2$$

$$= \frac{4N + 4\sqrt{N}(D_p - D_q) + (D_p - D_q)^2}{4} - \frac{(D_p + D_q)^2}{4} \quad (1)$$

$$4N = 4N + 4\sqrt{N}(D_p - D_q) + (D_p - D_q)^2 - (D_p + D_q)^2$$

$$4\sqrt{N}(D_p - D_q) = 4 D_p D_q$$

$$\sqrt{N}(D_p - D_q) = D_p D_q$$

$$\frac{D_p - D_q}{D_p D_q} = \frac{1}{\sqrt{N}}$$

From Equation (3), the value $D_p - D_q$ is needed to improve FFA1, which obtained from the convergences of continued fraction of $\frac{1}{\sqrt{N}}$. Recall from Section 2.4, from the continued fraction of $\frac{1}{\sqrt{N}}$ will established a list of $\frac{h_i}{k_i}$. As a candidate $\frac{D_p - D_q}{D_p D_q}$, the $\frac{h_t}{k_t}$ is selected with $k_t \lesssim N$ and h_t be the additional extension λ to improve the initial value of FFA1.

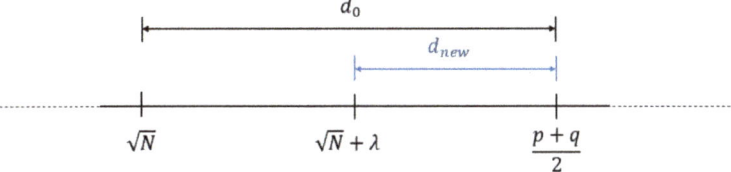

Figure 2. The position of $x = \sqrt{N}$ and $\frac{p+q}{2}$.

Now, the bound of $D_p - D_q$ will be consider. Recall in Section 2.4 where $p = D_p + \sqrt{N}$ and $q = \sqrt{N} - D_q$, $p + q = 2\sqrt{N} + D_p - D_q$, we have $D_p - D_q = p + q - 2\sqrt{N}$. As $h_t < D_p - D_q$, thus

$$h_t < D_p - D_q = p + q - 2\sqrt{N}. \tag{2}$$

We proceed with a lemma that shows $x = \lceil \sqrt{N} + \frac{h_t}{2} \rceil$ is always smaller than $\frac{p+q}{2}$.

Lemma 1. *Let $p = D_p + \sqrt{N}$ and $q = \sqrt{N} - D_q$. If $\frac{h_t}{k_t}$ be a fraction from the convergent list of continued fraction $\frac{1}{\sqrt{N}}$ with $h_t < D_p - D_q$, then $\sqrt{N} + \frac{h_t}{2} < \frac{p+q}{2}$.*

Proof. Suppose there exist $\frac{h_t}{k_t}$ from convergent list of continued fraction $\frac{1}{\sqrt{N}}$. If $h_t < D_p - D_q$, then substituting the Equation (2) into $\sqrt{N} + \frac{h_t}{2}$, thus it can be rewritten as follows.

$$\sqrt{N} + \frac{h_t}{2} = \frac{2\sqrt{N} + h_t}{2}$$
$$< \frac{2\sqrt{N} + D_p - D_q}{2}$$
$$= \frac{p+q}{2}$$

□

Observation 1. *Consider Lemma 1. As $\sqrt{N} + \frac{h_t}{2}$ is always smaller than $\frac{p+q}{2}$, therefore $\frac{h_t}{2}$ via EPF can be use as the additional extension λ. Furthermore, we can set $x = \sqrt{N} + \frac{h_t}{2}$ to serve as the (improved) initial value of x in the FFA1 algorithm.*

In this study, we discover two types of possible selected convergent on the list. The first is Type 1: $\frac{h_t}{k_t}$ from the convergent list of $\frac{1}{\sqrt{N}}$ where $k_t \leq N$. For Type 1, Wu et al. [1] mentioned that $\frac{h_t}{k_t}$ can be an indicator of convergent with index t to select the good candidate for initial value.

$$\left[\frac{h_1}{k_1}, \ldots, \frac{h_t}{k_t}, \frac{h_{t+1}}{k_{t+1}}, \ldots, \frac{h_i}{k_i} \right] \tag{3}$$

Next, is the Type 2: $\frac{h'_i}{k'_i}$ where it is the last convergent on the list, which will be elaborated further. Thus, h'_i will be selected for additional extension for potential initial value.

$$\left[\frac{h_1}{k_1}, \frac{h_2}{k_2}, \ldots, \frac{h'_i}{k'_i} \right] \tag{4}$$

The behaviour of Type 2 convergent selection is analysed via experiment on 50 distinct balanced prime moduli N. Figure 3 shows there are three possible positions on the potential initial value with additional extension that close to $\frac{p+q}{2}$ as follows.

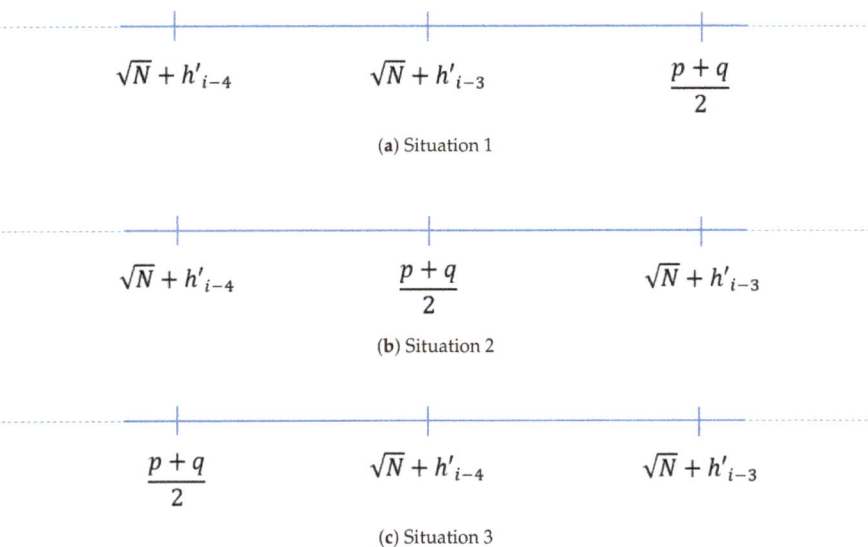

Figure 3. Three possible situations of Type 2 convergent selection.

Observation 2. *Figure 3 shows that Situation 1 happens when the value of $\frac{p+q}{2}$ is larger than the two initial values, Situation 2 shows that position $\frac{p+q}{2}$ is in the middle of potential initial values, and Situation 3 is where the value of $\frac{p+q}{2}$ is smaller than the potential initial values.*

The result via experimental analysis of 50 distinct moduli N indicates that 33 moduli are considered as Type 2 convergent selection and 42.42% of them have potential initial value with additional extension h'_{i-3} and h'_{i-4} larger than $\frac{p+q}{2}$ (i.e., Situation 3), while the rest might fall under Situation 1 or Situation 2. This experimental analysis suggests that the position of those potential initials values close to $\frac{p+q}{2}$ is undecidable. This study aims to provide solutions that covered unbalanced and balanced primes for all three situations. The in-depth analysis of Type 2 will be discussed in Section 4.3.

4. Results and Discussion

By the observation in Section 3, an improvement for FFA1 to increase the effectiveness of the algorithm to factor a modulus N is sought after. The aim is to focus on three important parts: (i) EPF's technique is used, (ii) establishing potential initial values for FFA1, and (iii) the alteration on the original FFA1 algorithm to suit the said potential initial value. Thus, we introduce mFFA1-EPF to improve the effectiveness of FFA1.

4.1. mFFA1-EPF: Unbalanced Prime

This section is dedicated for the modulus N with unbalanced primes factor. From Observation 1, we setting $x = \lceil \sqrt{N} + \frac{h_t}{2} \rceil$ as the improved initial value in FFA1 algorithm is which less than $\frac{p+q}{2}$. Let us start with the following question.

Question 1. *Is there any other potential initial values aside from $x = \sqrt{N} + \frac{h_t}{2}$ that can be selected to shorten the distance towards $\frac{p+q}{2}$?*

Answer. Interestingly, if we can find other candidates for initial values in which potentially reduces the distance d_{new}, then it can be useful to reduce the loop count to search for the value of $\frac{p+q}{2}$. Suppose the value of $\sqrt{N}+h_t$ is considered as the other candidate of potential values (i.e., the value of λ). Then, such value is supposed close to $\frac{p+q}{2}$. However, in general, it position is undecided whether $\sqrt{N}+h_t$ is larger (as shown by Figure 4) or is smaller than $\frac{p+q}{2}$ (i.e., similar to Lemma 1).

Figure 4. The position of $\sqrt{N}+h_t > \frac{p+q}{2}$.

Next, suppose there exist h_{t-1} such that $\frac{h_{t-1}}{k_{t-1}}$ is from the convergent list of $\frac{1}{\sqrt{N}}$ with $k_{t-1} < k_t \lesssim N$. By empirical evident, the value $\sqrt{N}+h_{t-1} < \frac{p+q}{2}$ can be considered as another candidate of additional extension for FFA1. Based on the empirical evident, it shows that the position of $\sqrt{N}+\frac{h_t}{2}$ and $\sqrt{N}+h_{t-1}$ are unpredictable as illustrated in Figure 5a,b.

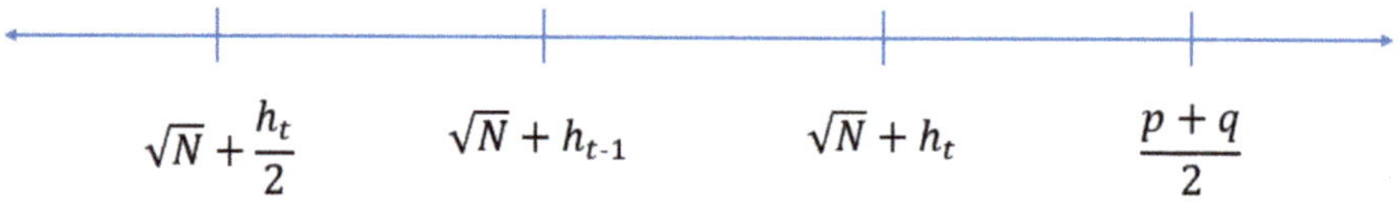

(a) The position with $\sqrt{N}+\frac{h_t}{2} < \sqrt{N}+h_{t-1}$

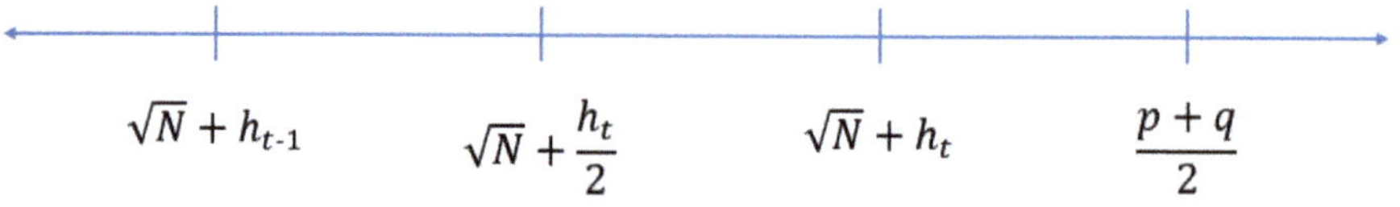

(b) The position with $\sqrt{N}+\frac{h_t}{2} > \sqrt{N}+h_{t-1}$

Figure 5. The possible position between $\sqrt{N}+\frac{h_t}{2}$, $\sqrt{N}+h_{t-1}$, $\sqrt{N}+h_t$, and $\frac{p+q}{2}$.

There are three candidates for the potential values for λ: $\frac{h_t}{2}$, h_{t-1}, and h_t. In answering Question 1, the algorithm will start to compute the first value, $a_1 = \max\left(\lceil\sqrt{N}+h_{t-1}\rceil, \lceil\sqrt{N}+\frac{h_t}{2}\rceil\right)$, while the second value is $a_2 = \lceil\sqrt{N}+h_t\rceil$. Once a_1 and a_2 are established, then two values—$y_1 = \sqrt{a_1^2 - N}$ and $y_2 = \sqrt{a_2^2 - N}$—are computed.

After establishing the values of a's and y's, three procedures run simultaneously:

- Procedure 1: The iteration with potential initial value a_1 and y_1. The value a_1 is increased by 1 until y_1 is integer.
- Procedure 2: The iteration with potential initial value a_2 and y_2. The value a_2 is increased by 1 until y_2 is integer.
- Procedure 3: The iteration with potential initial value a_2 and y_2. The value a_2 is decreased by 1 until y_2 is integer.

Remark 3. Note that the same value a_2 is applied in both Procedure 2 and Procedure 3. As the value a_2, might be larger than $\frac{p+q}{2}$, the main role of Procedure 3 is to prevent the mFFA1-EPF algorithm to keep running forever. All of the procedure is done by parallel computing, which means that the algorithm will completely be stopped whenever one of the procedures outputs y_1 or y_2 as an integer. Eventually, p and q will be obtained.

Unbalanced prime is demonstrated in Algorithm 1 as follows and flowchart on Figure A1 in Appendix B.1.

Algorithm 1: mFFA1-EPF: Unbalanced Prime

Input: Modulus N
Output: The prime p and q

1. Compute the continued fraction of $\frac{1}{\sqrt{N}}$.
2. Select $\frac{h_{t-1}}{k_{t-1}}$ and $\frac{h_t}{k_t}$ which is convergence to $\frac{1}{\sqrt{N}}$, where $k_{t-1} < k_t < N$
3. Compute $a_1 = \max\left(\lceil \sqrt{N} + h_{t-1} \rceil, \lceil \sqrt{N} + \frac{h_t}{2} \rceil\right)$ and $a_2 = \lceil \sqrt{N} + h_t \rceil$
4. Compute $y_1 = \sqrt{a_1^2 - N}$ and $y_2 = \sqrt{a_2^2 - N}$
5. **do in parallel**
6.   Procedure 1: **while** $y_1 \neq$ integer **do**
7.   Compute $a_1 \leftarrow a_1 + 1$
8.   Compute $y_1 = \sqrt{a_1^2 - N}$
9.   **end while**
10. $p = a_1 + y_1$ and $q = a_1 - y_1$
11.
12.   Procedure 2: **while** $y_2 \neq$ integer **do**
13.   Compute $a_2 \leftarrow a_2 + 1$
14.   Compute $y_2 = \sqrt{a_2^2 - N}$
15.   **end while**
16. $p = a_2 + y_2$ and $q = a_2 - y_2$
17.
18.   Procedure 3: **while** $y_2 \neq$ integer **do**
19.   Compute $a_2 \leftarrow a_2 - 1$
20.   Compute $y_2 = \sqrt{a_2^2 - N}$
21.   **end while**
22. $p = a_2 + y_2$ and $q = a_2 - y_2$
23. **return** (p, q)

Remark 4. *For the Type 2 case, Step 2 on Algorithm 1 is changed by selecting $\frac{h'_i}{k'_i}$ and $\frac{h'_{i-1}}{k'_{i-1}}$. Beside that, Step 3 will be modified with $a_1 = \max\left(\lceil \sqrt{N} + h'_{i-1} \rceil, \lceil \sqrt{N} + \frac{h'_i}{2} \rceil\right)$ and $a_2 = \lceil \sqrt{N} + h'_i \rceil$.*

Examples 1–4 are presented as illustrations of mFFA1-EPF for unbalanced primes. Example 1 demonstrates Type 1 of convergent-type selection, while Example 2 demonstrates Type 2 of convergent-type selection. Example 3 shows the importance of a_2 for this algorithm, and Example 4 shows the application of a previous example from Wu et al. [1].

Example 1. *Let $N = 707{,}896{,}463$. By the continued fraction method, the following list of fraction $\frac{1}{\sqrt{N}}$ is created*

$$\left[\ldots, \frac{34}{904{,}615}, \frac{139}{3{,}698{,}279}, \frac{33{,}811}{899{,}586{,}412}, \ldots\right]$$

We select $\frac{139}{3{,}698{,}279}$ as a candidate of $\frac{h_t}{k_t}$ and $\frac{34}{904{,}615}$ as a candidate of $\frac{h_{t-1}}{k_{t-1}}$ since $k_{t-1} < k_t \lesssim N$. Now, the potential initial values are computed as follows:

1. $a_1 = \max\left(\lceil \sqrt{N} + h_{t-1} \rceil, \lceil \sqrt{N} + \frac{h_t}{2} \rceil\right) = 26{,}676$
2. $a_2 = \lceil \sqrt{N} + h_t \rceil = 26{,}746$

With $a_1 = 26{,}676$ and $a_2 = 26{,}746$, Algorithm 1 is performed. The algorithm is stopped when Procedure 2 satisfy the searching on Algorithm 1 where y_2 become an integer ($y_2 = 66{,}126$). Finally, $p = a_2 + y_2 = 50{,}359$ and $q = a_2 - y_2 = 14{,}057$ are computed.

Example 2. Let $N = 7{,}665{,}365{,}527{,}725{,}431$. By continued fraction method, the following convergent list of $\frac{1}{\sqrt{N}}$ is created

$$\left[\ldots, \frac{1273}{111{,}453{,}789{,}228}, \frac{1554}{136{,}055{,}921{,}807}, \frac{2827}{247{,}509{,}711{,}035}\right]$$

$\frac{2827}{247{,}509{,}711{,}035}$ is selected as $\frac{h'_i}{k'_i}$ (the last convergent on the list) and $\frac{1554}{136{,}055{,}921{,}807}$ as $\frac{h'_{i-1}}{k'_{i-1}}$ as $k'_i < N$. The potential initial values are computed as follows:

1. $a_1 = \max\left(\lceil\sqrt{N}+h'_{i-1}\rceil, \lceil\sqrt{N}+\frac{h'_i}{2}\rceil\right) = 87{,}553{,}628$
2. $a_2 = \lceil\sqrt{N}+h'_i\rceil = 87{,}554{,}901$

With $a_1 = 87{,}553{,}628$ and $a_2 = 87{,}554{,}901$, Algorithm 1 is performed. The algorithm is stopped when Procedure 2 satisfies the searching on Algorithm 1 where y_2 become an integer ($y_2 = 96{,}393{,}384$). Last, $p = a_2 + y_2 = 136{,}721{,}029$ and $q = a_2 - y_2 = 56{,}065{,}739$ are computed.

Example 3. Suppose $N = 2{,}927{,}489{,}533$. By continued fraction method, the following convergent list of $\frac{1}{\sqrt{N}}$ is created

$$\left[\ldots, \frac{5329}{288{,}332{,}366}, \frac{7097}{383{,}992{,}269}, \frac{12426}{672{,}324{,}635}\right]$$

$\frac{12{,}426}{672{,}324{,}635}$ is selected as $\frac{h'_i}{k'_i}$ and $\frac{7097}{383{,}992{,}269}$ as $\frac{h'_{i-1}}{k'_{i-1}}$ as $k'_i < N$. Now, the potential initial values are computed as follows:

1. $a_1 = \max\left(\lceil\sqrt{N}+h'_{i-1}\rceil, \lceil\sqrt{N}+\frac{h'_i}{2}\rceil\right) = 61{,}204$
2. $a_2 = \lceil\sqrt{N}+h'_i\rceil = 66{,}533$

With $a_1 = 29{,}682$ and $a_2 = 36{,}369$, Algorithm 1 is performed. The algorithm is stopped when Procedure 3 satisfies the searching on Algorithm 1 where y_2 become integer ($y_2 = 65{,}903$). Last, $p = a_2 + y_2 = 49{,}307$ and $q = a_2 - y_2 = 10{,}723$ are computed.

Example 4. Suppose $N = 1{,}783{,}647{,}329$ which adapted from the numerical example of Wu et al. [1]. By continued fraction method, the following convergent list of $\frac{1}{\sqrt{N}}$ is created

$$\left[\ldots, \frac{2758}{11{,}6479{,}301}, \frac{10{,}205}{430{,}990{,}307}, \frac{12{,}963}{547{,}469{,}608}, \ldots\right]$$

$\frac{12{,}963}{547{,}469{,}608}$ is selected as a candidate of $\frac{h_t}{k_t}$ and $\frac{10{,}205}{430{,}990{,}307}$ as a candidate of $\frac{h_{t-1}}{k_{t-1}}$ since $k_{t-1} < k_t < N$. Two potential initial values are computed as follows:

1. $a_1 = \max\left(\lceil\sqrt{N}+h_{t-1}\rceil, \lceil\sqrt{N}+\frac{h_t}{2}\rceil\right) = 52{,}439$
2. $a_2 = \lceil\sqrt{N}+h_t\rceil = 55{,}197$

Algorithm 1 is performed and stop when Procedure 1 satisfies that y_1 is an integer ($y_1 = 31{,}664$). Last, $p = a_1 + y_1 = 84{,}449$ and $q = a_1 - y_1 = 21{,}121$ are computed.

4.2. Discussion of Algorithm 1 (mFFA1-EPF: Unbalanced Primes)

This section presents a comparative analysis between the mFFA1-EPF and the previous technique, based on loop count and computational time. Note that all experimental results were using a computer running on 2.1 GHz on Intel® Core i3 with 4 GB of RAM.

According to Table 2, the loop count on Procedure 2 of Examples 1 and 2 is the shortest one. It shows that the mFFA1-EPF has the smallest loop count compared to the other methods. For Example 3, Procedure 3 has the shortest path (630), at the same time it indicates $\sqrt{N} + h_t$ is larger than $\frac{p+q}{2}$. Thus far, the results give a good visualization representing the factorization of modulus N by our method experimentally. Example 4 is the same example as given in Wu et al. [1]. Procedure 1 obtained the smallest loop count (346) compared to other methods. This N/A (Not Applicable) means that all the procedure in Algorithm 1 is stopped when one of the y's from any procedures got the integer first.

Table 2. The comparison on loop count between FFA1, FFA2, and FFA2-EPF with our propose method toward Examples 1–4.

$N = pq$	FFA1	FFA2	FFA2-EPF	mFFA1-EPF		
				Procedure 1	Procedure 2	Procedure 3
Example 1	128,439	386,939	342,627	N/A	120,265	N/A
Example 2	8,841,310	98,337,910	97,340,072	N/A	8,838,483	N/A
Example 3	11,796	98,844	33,091	N/A	N/A	630
Example 4 [1]	11455	42,215	10,551	346	N/A	N/A

The mFFA1-EPF is performed by parallel computing, which means Procedures 1–3 were run simultaneously. We recorded the computational time for the three different procedures. In Table 3, mFFA1-EPF is faster than FFA1 by 4 numerical examples. This seems to be a slight improvement for FFA1 as there is an involvement of additional extension. mFFA1-EPF is good in term of loop count, running without failure and computational time (compared to FFA1).

Table 3. The comparison on computational time in second (s) between FFA1, FFA2, and FFA2-EPF with mFFA1-EPF toward Examples 1–4.

$N = pq$	FFA1	FFA2	FFA2-EPF	mFFA1-EPF		
				Procedure 1	Procedure 2	Procedure 3
Example 1	2.86	0.64	0.50	N/A	2.01	N/A
Example 2	338.12	215.89	199.27	N/A	301.98	N/A
Example 3	0.63	0.17	0.09	N/A	N/A	0.33
Example 4 [1]	0.55	0.12	0.10	0.23	N/A	N/A

To make mFFA1-EPF more convincing, there are numerical examples from Somsuk [6] provided in Table 4.

Table 4. The comparison loop count between FFA1, FFA2, FFA2-EPF, and FFA-Euler with our proposed method by Somsuk's Example (Tables 2 and 3 [6]).

Modulus N	FFA1	FFA2	FFA2-EPF	FFA-Euler	mFFA1-EPF		
					Procedure 1	Procedure 2	Procedure 3
1,047,329,636,821,139,813 = 1,971,074,143 · 531,349,691	227,820,673	1,895,365,798	1,893,402,196	227,820,673	N/A	227,819,732	N/A
788,582,867,650,121,563 = 1,066,200,463 · 739,619,701	14,888,197	356,357,156	307,600,540	14,888,197	N/A	14,236,836	N/A

This comparison highlights the improvement made by mFFA1-EPF compared to the method FFA-Euler, provided by Somsuk [6]. By two examples from in [6], the exhaustive search is improved with shortest loop counts, and the potential initial values are shorter than the FFA-Euler loop count. This shows that our method can be compatible with all unbalanced prime.

4.3. mFFA1-EPF: Balanced Prime

Previously, in the case of a modulus with unbalanced primes, three candidates are determined as the λ. Only two potential initial values that possibly shorten the d_{new} were selected. In this section, we will explore the case of a modulus with balanced primes. The aim is to dictate the proper candidates from the convergent list (i.e., mFFA1-EPF) for the potential initial values via a similar approach.

Recall that $\frac{h_t}{k_t}$ with index t, where $k_t \lesssim N$, deemed as the indicator for selecting a good convergent to additional extension of initial values for FFA2-EPF [1]. When the EPF technique applied for the balanced prime case on FFA1, by empirical evidence, it shows that such indicator leads to the initial value $(\sqrt{N} + h_t)$ relatively far away from the target value (exceeded by $\frac{p+q}{2}$). Therefore, we conjecture that the EPF method seems not to be an effective method to factor the modulus N with a balanced prime. Furthermore, the result in Somsuk [5] agreed that EPF is only suitable for unbalanced prime. It failed to address the convergent with index t as a suitable index to improve the initial value.

Therefore, in this section, we provide the strategies to address such drawback of factoring the modulus N with a balanced prime by imposing modification in the mFFA1-EPF algorithm. The strategies involve convergent selection and modification of potential initial values. Therefore, to enhance the effectiveness via mFFA1-EPF, the additional extension h_t until h_{t-5} is observe empirically to determine the smallest value of d_{new}. The result of the observation is presented on Figure 6, and the discussion follows.

Suppose $\frac{h_t}{k_t}$ with index t where $k_t \lesssim N$ is chosen via EPF. Note that for the modulus N with balanced primes case, the value $\sqrt{N} + h_t \gg \frac{p+q}{2}$. Therefore, the additional extension from h_t to h_{t-5} is analysed. Interestingly, the additional extension $\sqrt{N} + h_{t-j}$ for $j = 0, 1, 2, 3, 4, 5$ can be a potential initial values as it moves closer to the value of $\frac{p+q}{2}$. Figure 6a–f shows comparison of potential initial values between the additional extension of $\sqrt{N} + h_{t-j}$ for $j = 0, 1, 2, 3, 4, 5$ and $\frac{p+q}{2}$, respectively. The potential initial values decrease, because the value of additional extension is become smaller from h_t to h_{t-5} (i.e., $h_t > h_{t-1} > h_{t-2} > h_{t-3} > h_{t-4} > h_{t-5}$).

Question 2. *What are the suitable initial values that need to be implemented on mFFA1-EPF with balanced primes?*

Answer. Based on Figure 6, the line graph between the initial value (represented by the blue dots) starts closer to the target value $\frac{p+q}{2}$ (represented by the red dots) as the value of initial values changes. A hindrance to the development process for this approach is that we can not determine the smallest value of d_{new} via additional extension h_t to h_{t-5}. In other words, the "closeness" of the potential initial values with additional extension unable to be decided simply from the results of Figure 6. This is because the additional extension a random value from the generation of convergent list of $\frac{1}{\sqrt{N}}$. It requests further analysis. A statistical analysis of 50 distinct moduli N with balanced primes is conducted to determine the closeness of potential initial value through index t to $t - 5$, as follows.

In this work, a measurement called Mahalanobis Distance (MD) is implemented. MD is the distance between two points in multivariate space. According to Çakmakçı et al. [15], MD measures the distance between a multidimensional point of probability distribution and distribution of distance. The smaller the value of MD, the closer the mean of candidate of potential initial values to the mean of the target value.

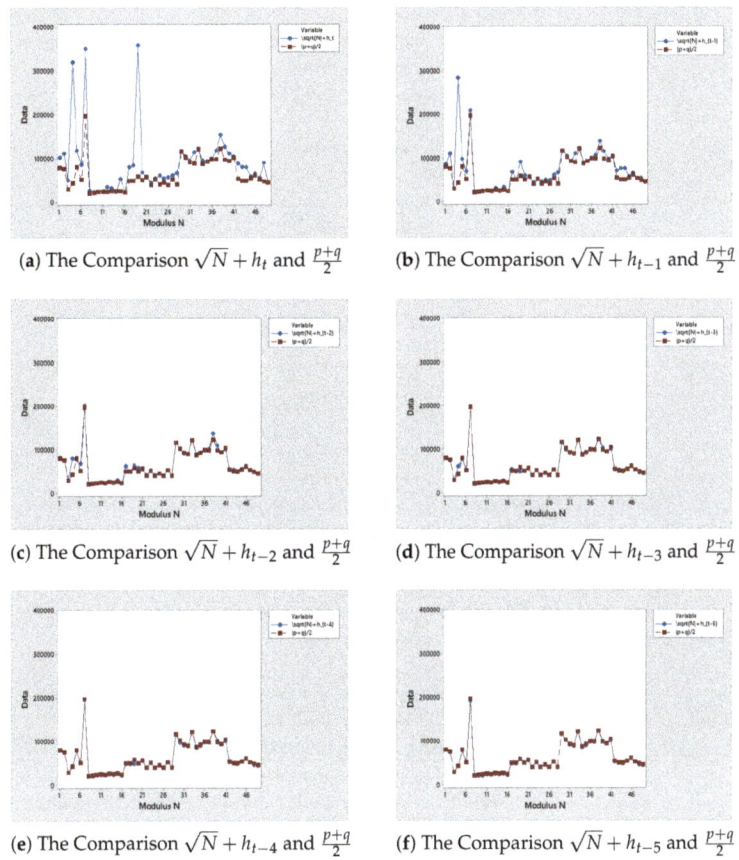

Figure 6. The comparison value of 50 data (distinct modulus N) between h_t, h_{t-1}, h_{t-2}, h_{t-3}, h_{t-4} and $\frac{p+q}{2}$.

In the one-dimensional case on the mFFA1-EPF for balanced prime, MD is used to calculate the normalized distance between the mean of each h_t to h_{t-5} and the mean of the target value $\frac{p+q}{2}$. The measurement formula is

$$\text{MD} = \frac{|\mu_{IV} - \mu_{AV}|}{\sigma_{IV+AV}} \tag{5}$$

where μ_{IV} is a mean of each data potential initial value of $\sqrt{N} + h_t$ to $\sqrt{N} + h_{t-5}$ while μ_{AV} is mean of actual value $\frac{p+q}{2}$. The value σ_{IV+AV} is calculated from combination data from potential initial value and the actual value $\frac{p+q}{2}$. The following formula is represented for MD between $\sqrt{N} + h_t$ and $\frac{p+q}{2}$,

$$\text{MD}_{(\sqrt{N}+h_t)} = \frac{\left|\mu_{IV(\sqrt{N}+h_t)} - \mu_{AV}\right|}{\sigma_{IV+AV}}$$

We calculate MD for $\sqrt{N} + h_t$ to $\sqrt{N} + h_{t-5}$ by same data of 50 moduli N and represent the MD value on Table 5.

Table 5 shows the comparison of the MD index of "closer distance" between several potential initial values from h_t to h_{t-5} and $\frac{p+q}{2}$. Table 5 reported that the MD index value of $\sqrt{N}+h_{t-3}$ and $\sqrt{N}+h_{t-4}$ are the smallest MD values among other potential initial values, that is, 0.0114 and 0.0116, respectively. It means that $\sqrt{N}+h_{t-3}$ and $\sqrt{N}+h_{t-4}$ are the most suitable candidates for potential initial values, because they have the smallest d_{new} on average with respect to MD measurement.

Table 5. The comparison on the closeness of value between the candidate of potential initial value with $\frac{p+q}{2}$ by MD.

Mahalanobis Distance (MD)	Value
$MD_{(\sqrt{N}+h_t)}$	0.4965
$MD_{(\sqrt{N}+h_{t-1})}$	0.2932
$MD_{(\sqrt{N}+h_{t-2})}$	0.0756
$MD_{(\sqrt{N}+h_{t-3})}$	0.0114
$MD_{(\sqrt{N}+h_{t-4})}$	0.0116
$MD_{(\sqrt{N}+h_{t-5})}$	0.0257

Observation 3. *The candidates $\sqrt{N}+h_{t-3}$ and $\sqrt{N}+h_{t-4}$ have the smallest value of MD. Therefore, it is highly suggested to select convergents with index $t-3$ and $t-4$ to improve the initial values.*

Based on Observation 3, two potential initial values are set as follows:
1. $b_1 = \sqrt{N}+h_{t-3}$
2. $b_2 = \sqrt{N}+h_{t-4}$

Remark that the FFA1 algorithm requires an initial value less than the target value and will keep increasing by one (i.e., +1) until it reaches $\frac{p+q}{2}$. Therefore, in mFFA1-EPF, we use the variation technique, which means the value of b_1 and b_2 need to be increased and decreased by 1 simultaneously. Next, the following values are established:

1. $y_1 = \sqrt{b_1^2 - N}$
2. $y_2 = \sqrt{b_2^2 - N}$

Four procedures are introduced using the above values, with the variation technique as follows:

- Procedure 1: The iteration with potential initial values b_1 and y_1. The value of b_1 is increased by 1 until it is the same as the y_1 becomes an integer.
- Procedure 2: The iteration with potential initial values b_2 and y_2. The value of b_2 is increased by 1 until it is the same as y_2 becomes an integer.
- Procedure 3: The iteration with potential initial values b_1 and y_1. The value of b_1 is **decreased** by 1 until it is the same as the y_1 becomes an integer.
- Procedure 4: The iteration with potential initial values b_2 and y_2. The value of b_2 is **decreased** by 1 until it is the same as the y_2 becomes an integer.

Note that these four procedures were run simultaneously by parallel computing which will stop when one of the y's become the first integer. Algorithm 2 shows how the workflow runs.

Algorithm 2: mFFA1-EPF: Balanced Prime

Input: Modulus N
Output: The prime p and q

1. Compute the continued fraction of $\frac{1}{\sqrt{N}}$.
2. Select $\frac{h_{t-3}}{k_{t-3}}$ and $\frac{h_{t-4}}{k_{t-4}}$ which is convergence to $\frac{1}{\sqrt{N}}$, where $k_{t-4} < k_{t-3} < N$
3. Compute $b_1 = \lceil \sqrt{N} + h_{t-3} \rceil$ and $b_2 = \lceil \sqrt{N} + h_{t-4} \rceil$
4. Compute $y_1 = \sqrt{b_1^2 - N}$ and $y_2 = \sqrt{b_2^2 - N}$
5. **do in parallel**
6.   Procedure 1: **while** $y_1 \neq$ integer **do**
7.   Compute $b_1 \leftarrow b_1 + 1$
8.   Compute $y_1 = \sqrt{b_1^2 - N}$
9.   **end while**
10.   Compute $p = b_1 + y_1$ and $q = b_1 - y_1$
11.
12.   Procedure 2: **while** $y_2 \neq$ integer **do**
13.   Compute $b_2 \leftarrow b_2 + 1$
14.   Compute $y_2 = \sqrt{b_2^2 - N}$
15.   **end while**
16.   Compute $p = b_2 + y_2$ and $q = b_2 - y_2$
17.
18.   Procedure 3: **while** $y_1 \neq$ integer **do**
19.   Compute $b_1 \leftarrow b_1 - 1$
20.   Compute $y_1 = \sqrt{b_1^2 - N}$
21.   **end while**
22.   Compute $p = b_1 + y_1$ and $q = b_1 - y_1$
23.
24.   Procedure 4: **while** $y_2 \neq$ integer **do**
25.   Compute $b_2 \leftarrow b_2 - 1$
26.   Compute $y_2 = \sqrt{b_2^2 - N}$
27.   **end while**
28.   Compute $p = b_2 + y_2$ and $q = b_2 - y_2$
29. **return** (p, q)

4.4. Discussion on Algorithm 2 (mFFA1-EPF: Balanced Primes)

Algorithm 2 is also illustrated as a flowchart in Figure A2 in Appendix B.2. The experimental result is represented using the mFFA1-EPF via balanced prime on Example 5 while applying mFFA1-EPF is represented on Somsuk's numerical example [6] in Example 6.

Example 5. *(Procedure 1 satisfies on Example 5). Let $N = 616{,}696{,}115{,}591$. By continued fraction method, the following convergent list $\frac{1}{\sqrt{N}}$ is created*

$$\left[\ldots, \frac{61}{47{,}903{,}301}, \frac{123}{96{,}591{,}902}, \frac{184}{144{,}495{,}203}, \frac{491}{385{,}582{,}308}, \ldots \right]$$

$\frac{491}{385{,}582{,}308}$ *is selected as a candidate of* $\frac{h_{t-3}}{k_{t-3}}$ *and* $\frac{184}{144{,}495{,}203}$ *as a candidate of* $\frac{h_{t-4}}{k_{t-4}}$ *since* $k_{t-4} < k_{t-3} < k_t \lesssim N$. *Now, there are two candidates of potential initial value and 2 y's are computed as follows:*

1. $b_1 = \lceil \sqrt{N} + h_{t-3} \rceil = 785{,}792$
2. $b_2 = \lceil \sqrt{N} + h_{t-4} \rceil = 785{,}485$

Algorithm 2 is performed because y_1 and y_2 not integers. The values b_1 and b_2 in Procedure 1 and 2 are increased by 1 while initial values in Procedure 3 and 4 are decreased by 1. The algorithm stop where y_1 from Procedure 1 is integer ($y_1 = 801{,}204$). Finally, compute $p = b_1 + y_1 = 960{,}049$ and $q = b_1 - y_1 = 642{,}359$.

Example 6. [6] *(Procedure 1 satisfies on Example 6) Say $N = 340{,}213$. By continued fraction method, a list of fraction $\frac{1}{\sqrt{N}}$ is created*

$$\left[\ldots, \frac{1}{583}, \frac{3}{1750}, \frac{4}{2333}, \frac{7}{4083}, \ldots\right]$$

$\frac{3}{1750}$ is selected as a candidate of $\frac{h_{t-3}}{k_{t-3}}$ and $\frac{1}{583}$ as a candidate of $\frac{h_{t-4}}{k_{t-4}}$ since $k_{t-4} < k_{t-3} < k_t < N$. Now, there are two candidates of potential initial value and 2 y's are computed as follows:

1. $b_1 = \lceil \sqrt{N} + h_{t-3} \rceil = 587$
2. $b_2 = \lceil \sqrt{N} + h_{t-4} \rceil = 585$

Algorithm 2 is performed because y_1 and y_2 not integer. The values b_1 and b_2 in Procedure 1 and 2 are increased by 1 while initial values in Procedure 3 and 4 are decreased by 1. The algorithm stop where y_1 from Procedure 1 is integer ($y_1 = 587$). We compute $p = b_1 + y_1 = 653$ and $q = b_1 - y_1 = 521$.

Table 6 shows the comparison on count loop and computational time in seconds (s), between several FFAs with our proposed method toward Example 5. The loop count on Procedure 1 (15412) is the least number of loop count compared to previous methods. Besides, FFA2-EPF can not undergo the process and the loop count is not available since the initial value exceeded the value of $p + q$ and $p - q$. When it goes on computational time, the algorithm is not shown the fastest one but it still improves from FFA1.

Table 6. The comparison on loop count and computational time in second (s) between several FFAs with our propose method toward Example 5.

Example 5	FFA1	FFA2	FFA2-EPF	mFFA1-EPF			
				Procedure 1	Procedure 2	Procedure 3	Procedure 4
Loop count	15,903	174,748	N/A	15,412	N/A	N/A	N/A
Computational time, s	0.71	0.2	N/A	0.48	N/A	N/A	N/A

For Table 7, mFFA1-EPF is applied toward [6] to compare the loop count and computational time. It shows the shortest loop count even the initial value is exactly the value of $\frac{p+q}{2}$ ($\frac{p+q}{2} = \sqrt{N} + h_{t-3} = 587$). Using Algorithm 2, the loop count reduced significantly, which result in the exhaustive search to run without fail.

Table 7. The comparison on loop count and computational time in second (s) between several FFAs with our propose method toward Example 6 [6].

Example 6	FFA1	FFA2	FFA2-EPF	FFA-Euler	mFFA1-EPF			
					Procedure 1	Procedure 2	Procedure 3	Procedure 4
Loop count	3	6	N/A	3	0	N/A	N/A	N/A
Computational time, s	2.12×10^{-2}	1.45×10^{-2}	N/A	7.89×10^{-3}	4.92×10^{-2}	N/A	N/A	N/A

Remark 5. *Consider Type 2 of the continued fraction convergent selection of the modulus N with balanced primes. For Type 2, we use Algorithm 2 with a changes in Step 2 where h'_{i-3} and h'_{i-4} and $k'_i < N$.*

Now, we replicate a numerical example from the Algorithm 2 with respect to Remark 5 on Example 7.

Example 7. (Procedure 4 satisfies on Example 7). Suppose $N = 9,355,908,869$. By continued fraction method, a convergent list of $\frac{1}{\sqrt{N}}$ is created

$$\left[\ldots, \frac{1611}{155,825,501}, \frac{2009}{194,322,428}, \frac{3620}{350,147,929}, \frac{9249}{894,618,286}, \frac{22,118}{2,139,384,501}\right]$$

As $\frac{h'_i}{k'_i}$ is the last convergent on the list, $\frac{2009}{194,322,428}$ is selected as a candidate of $\frac{h'_{i-3}}{k'_{i-3}}$ and $\frac{1611}{155,825,501}$ as a candidate of $\frac{h'_{i-4}}{k'_{i-4}}$ as $k'_{i-4} < k'_{i-3} < k'_i < N$. We compute two candidates of initial value of x, y_1 and y_2 as follows:

1. $b_1 = \lceil \sqrt{N} + h'_{i-3} \rceil = 98{,}735$
2. $b_2 = \lceil \sqrt{N} + h'_{i-4} \rceil = 98{,}337$

Since y_1 and y_2 are not integer, the values b_1 and b_2 in Procedures 1 and 2 are increased by 1 while initial values in Procedures 3 and 4 are decreased by 1. The algorithm stop where y_2 in Procedure 4 is integer ($y_2 = 97{,}245$). We compute $p = b_2 + y_2 = 107{,}279$ and $q = b_2 - y_2 = 87{,}211$.

Table 8 shows the comparison on loop count and computational time in second between several FFAs with our propose method toward Example 7. Procedure 4 shows the smallest loop count with 1092 compared to FFA1 (1590) and FFA2 (10,553). The loop count of FFA2-EPF is unavailable as the initial values are exceeded the value of $p + q$ and $p - q$. On computational time, our algorithm is slightly better than FFA1. Procedure 4 plays it crucial part to achieve the value $\frac{p+q}{2}$, and, at the same time, Procedure 4 obtains the indicators of whether the value is larger or smaller than $\frac{p+q}{2}$. Therefore, Algorithm 2 with Remark 5 helps to search for the value of $\frac{p+q}{2}$ without failure.

Table 8. The comparison on loop count and computational time in second (s), between several FFAs with our propose method toward Example 7.

Example 7	FFA1	FFA2	FFA2-EPF	mFFA1-EPF			
				Procedure 1	Procedure 2	Procedure 3	Procedure 4
Loop count	1590	10553	N/A	N/A	N/A	N/A	1092
Computational time, s	8.90×10^{-2}	1.90×10^{-2}	N/A	N/A	N/A	N/A	4.07×10^{-2}

Recall that there is $d_{new} = \frac{p+q}{2} - (\sqrt{N} + \lambda)$. For mFFA1-EPF, the λ varies according to type of modulus N; $\lambda = h_t$ and $\lambda = h_{t-1}$ for a modulus with unbalanced primes while $\lambda = h_{t-3}$ and $\lambda = h_{t-4}$ for a modulus with balanced primes. Multiple λ can lead to the shortest path toward $\frac{p+q}{2}$. For comparison on mFFA1-EPF with FFA1, both methods use the same process of calculating the square roots to reach the target value $\frac{p+q}{2}$. However, mFFA1-EPF uses the additional extension on its potential initial values $d_{new} < d_0$ where d_0 is the loop count of FFA1. Based on the empirical results in this work, the loop count and computational time of mFFA1-EPF are improved compared to FFA1. Consequently, it reduces the cost of running the exhaustive search.

The uniqueness of FFA2 is that it uses multiplication operation as the main process, it has less cost in computational time compared to the mFFA1-EPF which uses square root operation. Alas, FFA2 requires a greater number of iterations to achieve $p + q$ and $p - q$. In this regard, the mFFA1-EPF uses less cost in terms of computational memory and less space to run the iteration compared to FFA2.

The objective of establishing additional extension on FFA2-EPF is the same as mFFA1-EPF, to shorten the path toward $p + q$ and $p - q$. mFFA1-EPF has a shorter loop count than

FFA2-EPF because the main operation comes from FFA2, which uses a huge number of iteration to achieve its target values: $p+q$ and $p-q$. Thus, mFFA1-EPF requires less cost in terms of space compared to FFA2-EPF.

5. Conclusions and Future Works

In this study, we discovered two types of convergent list selections. The first is Type 1: $\frac{h_t}{k_t}$ is selected from the convergent list of $\frac{1}{\sqrt{N}}$ where $k_t \lesssim N$. For Type 1, Wu et al. [1] mentioned that $\frac{h_t}{k_t}$ can be an indicator of convergent with index t to select a good candidate for initial value. Next, Type 2: $\frac{h'_i}{k'_i}$ where k'_i is the last convergent on the list (as illustrated by Figure 4) where h'_i will be selected for additional extension for potential initial value. This paper proposed two improved factoring algorithms called mFFA1-EPF for unbalanced primes and mFFA1-EPF for balanced primes. The general idea for designing the algorithms is due to a modification made to EPF and then implemented to (improved) FFA1. The resulting study shows a significant improvement that reduces the loop count of FFA1 via (improved) EPF compared to previous methods (FFA1, FFA2, FFA2-EPF, and FFA-Euler).

An interesting limitation to our work is that the computational time of mFFA1-EPF is still far beyond efficient to factor a modulus with 1024-bit size of balanced primes, with the current technology. We foresee that the mFFA1-EPF might be useful once a large quantum computer with stable qubits is available. The mFFA1-EPF is a type of searching algorithm, thus it might take advantage of making fine adjustments or manipulating the mathematical nature within Grover's searching quantum algorithm [16]. The mFFA1-EPF can be used in machine architecture with low power such as the Internet of Things-based devices, which requires to factor small composites integer [1,9]. Furthermore, we expect mFFA1-EPF to be an assistive tool to increase the effort on the machine learning and artificial intelligence approaches, such as in [17], have been introduced in the literature to deal with similar problems as ours.

Author Contributions: Conceptualization, M.A.A.; Formal Analysis, R.R.M.T.; Funding Acquisition, M.A.A.; Investigation, R.R.M.T. and Z.M.; Methodology, R.R.M.T.; Project Administration, M.A.A. and M.R.K.A.; Software, Z.M.; Supervision, M.A.A. and M.R.K.A.; Validation, M.A.A., M.R.K.A. and Z.M.; Writing—Original Draft, R.R.M.T.; Writing—Review & Editing, R.R.M.T. and M.A.A. All authors have read and agreed to the published version of the manuscript.

Funding: The present research was supported by Universiti Putra Malaysia under Putra Grant—IPM with project number GP-IPM/2017/9519200.

Institutional Review Board Statement: Not applicable.

Informed Consent Statement: Not applicable.

Data Availability Statement: Not applicable.

Acknowledgments: Not applicable.

Conflicts of Interest: The authors declare no conflict of interest.

Appendix A. The Proving on Estimated Prime Factor (EPF)

We discuss the original study by Wu et al. [1,9]. There is a distance between p and q with $\sqrt{N}$ that written as

$$D_p = p - \sqrt{N} \tag{A1}$$

$$D_q = \sqrt{N} - q \tag{A2}$$

Derive (A1) and (A2) to become

$$p = D_p + \sqrt{N} \tag{A3}$$
$$q = \sqrt{N} - D_q \tag{A4}$$

Denote that $N = pq$. We substitute (A3) to p and (A4) to q and yield

$$\begin{aligned} N = pq &= (\sqrt{N} + D_p)(\sqrt{N} - D_q) \\ &= N + \sqrt{N}(D_p - D_q) - D_p D_q \end{aligned} \tag{A5}$$

N is eliminated on the both side which generate Equation (A6) and lead to Equation (A7).

$$\begin{aligned} N &= N + \sqrt{N}(D_p - D_q) - D_p D_q \\ N - N &= \sqrt{N}(D_p - D_q) - D_p D_q \\ D_p D_q &= \sqrt{N}(D_p - D_q) \end{aligned} \tag{A6}$$
$$\frac{1}{\sqrt{N}} = \frac{D_p - D_q}{D_p D_q} \tag{A7}$$

We do not have any informations about the value of $D_p - D_q$ and $D_p D_q$. However, from Equation (A7), $\frac{1}{\sqrt{N}}$ can be useful to get $D_p - D_q$ and $D_p D_q$ as N is publically known. Now, a convergent list $\frac{1}{\sqrt{N}}$ is produced by continued fraction.

Suppose there is a convergent list $\frac{1}{\sqrt{N}}$, assign as $\frac{h_i}{k_i}$ with $h_i, k_i \in \mathbb{Z}$ and i be a number of the convergent produced. From the continued fraction, we know that $\frac{h_i}{k_i} \to \frac{1}{\sqrt{N}}$ as $i \to \infty$. As the size h_i and k_i increase as i increase, there exist t such that

$$h_t < D_p - D_q < h_{t+1} \tag{A8}$$

We use h_t and k_t to correspond the estimation of D_p and D_q that is

$$h_t \approx D_p - D_q$$
$$k_t \approx D_p D_q$$

The convergent with index t be the selection fraction for improving the FFA as it give an advantage on shorten the exhasutive search on $p + q$ and $p - q$.

Appendix B. Flowchart mFFA1-EPF

Appendix B.1. mFFA1-EPF on Unbalanced Prime

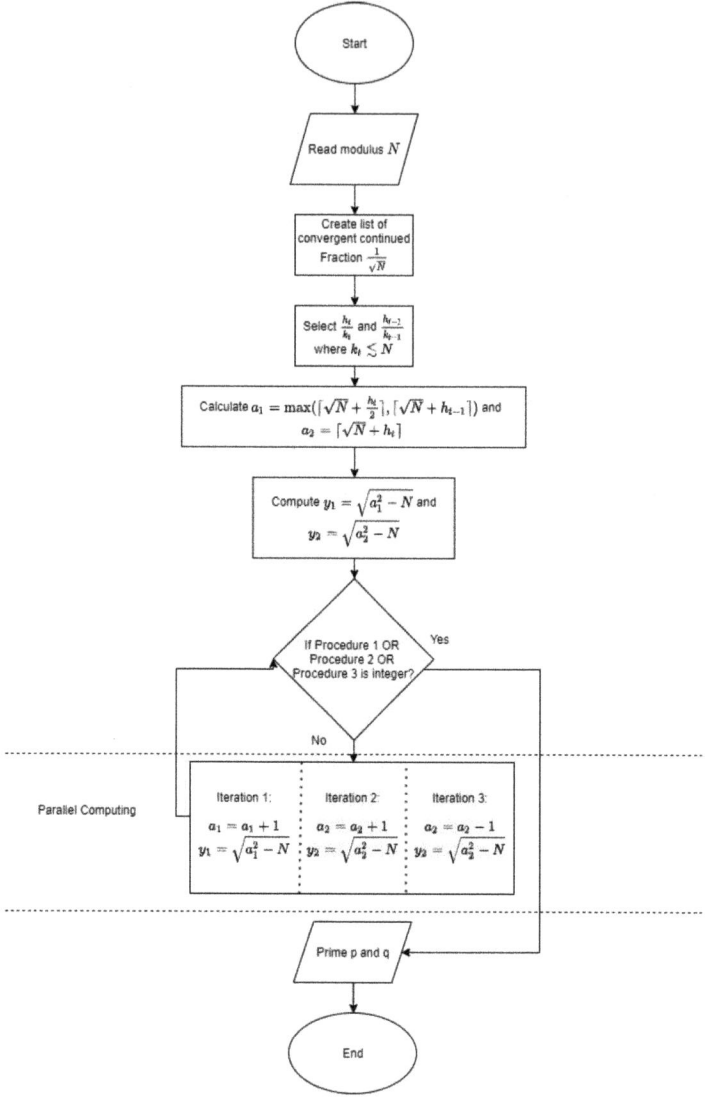

Figure A1. The flowchart of mFFA1-EPF on unbalanced prime.

Appendix B.2. mFFA1-EPF on Balanced Prime

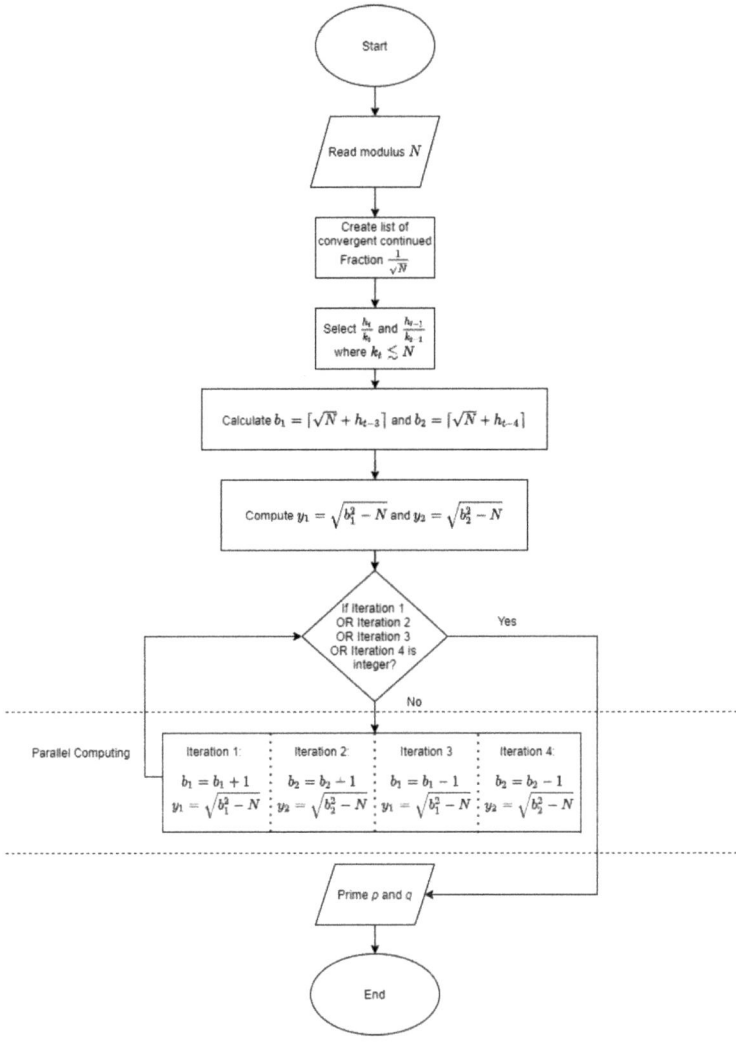

Figure A2. The flowchart of mFFA1-EPF on balanced prime.

References

1. Wu, M.E.; Tso, R.; Sun, H.M. On the improvement of Fermat Factorization using a Continued Fraction technique. *Future Gener. Comput. Syst.* **2014**, *30*, 162–168. [CrossRef]
2. Ghafar, A.H.A; Ariffin, M.R.K.; Asbullah, M.A. A New LSB Attack on Special-Structured RSA Primes. *Symmetry* **2020**, *12*, 838. [CrossRef]
3. Ambedkar, B.R.; Gupta, A.; Gautam, P.; Bedi, S.S. An efficient method to factorize the RSA public key encryption. In Proceedings of the 2011 International Conference on Communication Systems and Network Technologies, Katra, India, 3–5 June 2011; pp. 108–111.
4. Somsuk, K.; Tientanopajai, K. An Improvement of Fermat's Factorization by Considering the Last m Digits of Modulus to Decrease Computation Time. *IJ Netw. Secur.* **2017**, *19*, 99–111.
5. Somsuk, K. The improvement of initial value closer to the target for Fermat's factorization algorithm. *J. Discret. Math. Sci. Cryptogr.* **2018**, *21*, 1573–1580. [CrossRef]

6. Somsuk, K. The new integer factorization algorithm based on fermat's factorization algorithm and euler's theorem. *Int. J. Electr. Comput. Eng.* **2020**, *10*, 1469–1476. [CrossRef]
7. Somsuk, K.; Kasemvilas, S. MFFV2 and MNQSV2: Improved factorization Algorithms. In Proceedings of the 2013 International Conference on Information Science and Applications (ICISA), Pattaya, Thailand, 24–26 June 2013; pp. 1–3.
8. Somsuk, K.; Chiawchanwattana, T.; Sanemueang, C. Estimating the new Initial Value of Trial Division Algorithm for Balanced Modulus to Decrease Computation Loops. In Proceedings of the 2019 16th International Joint Conference on Computer Science and Software Engineering (JCSSE), Chonburi, Thailand, 10–12 July 2019; pp. 143–147.
9. Wu, M.E.; Chen, C.M.; Lin, Y.H.; Sun, H.M. On the improvement of Wiener attack on RSA with small Private Exponent. *Sci. World J.* **2014**, *2014*, 650537. [CrossRef] [PubMed]
10. Asbullah, M.A.; Ariffin, M.R.K. Another Proof Of Wiener's Short Secret Exponent. *Malays. J. Sci.* **2019**, *38* (Suppl. 1), 67–73. [CrossRef]
11. Ruzai, W.N.A.; Ariffin, M.R.K.; Asbullah, M.A.; Mahad, Z.; Nawawi, A. On the Improvement Attack Upon Some Variants of RSA Cryptosystem via the Continued Fractions Method. *IEEE Access* **2020**, *8*, 80997–81006. [CrossRef]
12. De Weger, B. Cryptanalysis of RSA with small prime difference. *Appl. Algebra Eng. Commun. Comput.* **2002**, *13*, 17–28. [CrossRef]
13. Bressoud, D.M. *Factorization and Primality Testing*; Springer: New York, NY, USA, 1989; pp. 58–61.
14. Chung, H.; Kim, M.; Al Badawi, A.; Aung, K.M.M.; Veeravalli, B. Homomorphic Comparison for Point Numbers with User-Controllable Precision and Its Applications. *Symmetry* **2020**, *12*, 788. [CrossRef]
15. Çakmakçı, S.D.; Kemmerich, T.; Ahmed, T.; Baykal, N. Online DDoS attack detection using Mahalanobis distance and Kernel-based learning algorithm. *J. Netw. Comput. Appl.* **2020**, *168*, 102756. [CrossRef]
16. Chuang, I.L.; Gershenfeld, N.; Kubinec, M. Experimental implementation of fast quantum searching. *Phys. Rev. Lett.* **1998**, *80*, 3408. [CrossRef]
17. Huang, X.L.; Ma, X.; Hu, F. Machine Learning and Intelligent Communications. *Mob. Netw. Appl.* **2018**, *23*, 68–70. [CrossRef]

Article

Improving Multivariate Microaggregation through Hamiltonian Paths and Optimal Univariate Microaggregation

Armando Maya-López [1], Fran Casino [2] and Agusti Solanas [1,*]

1 Department of Computer Engineering and Mathematics, Universitat Rovira i Virgili, Av. Països Catalans 26, 43007 Tarragona, Catalonia, Spain; armando.maya@estudiants.urv.cat
2 Department of Informatics, University of Piraeus, Karaoli kai dimitriou 80, 18534 Pireas, Greece; francasino@unipi.gr
* Correspondence: agusti.solanas@urv.cat

Citation: Maya-López, A.; Casino, F.; Solanas, A. Improving Multivariate Microaggregation through Hamiltonian Paths and Optimal Univariate Microaggregation. *Symmetry* 2021, 13, 916. https://doi.org/10.3390/sym13060916

Academic Editor: Egon Schulte

Received: 1 April 2021
Accepted: 14 May 2021
Published: 21 May 2021

Publisher's Note: MDPI stays neutral with regard to jurisdictional claims in published maps and institutional affiliations.

Copyright: © 2021 by the authors. Licensee MDPI, Basel, Switzerland. This article is an open access article distributed under the terms and conditions of the Creative Commons Attribution (CC BY) license (https://creativecommons.org/licenses/by/4.0/).

Abstract: The collection of personal data is exponentially growing and, as a result, individual privacy is endangered accordingly. With the aim to lessen privacy risks whilst maintaining high degrees of data utility, a variety of techniques have been proposed, being microaggregation a very popular one. Microaggregation is a family of perturbation methods, in which its principle is to aggregate personal data records (i.e., microdata) in groups so as to preserve privacy through k-anonymity. The multivariate microaggregation problem is known to be NP-Hard; however, its univariate version could be optimally solved in polynomial time using the Hansen-Mukherjee (HM) algorithm. In this article, we propose a heuristic solution to the multivariate microaggregation problem inspired by the Traveling Salesman Problem (TSP) and the optimal univariate microaggregation solution. Given a multivariate dataset, first, we apply a TSP-tour construction heuristic to generate a Hamiltonian path through all dataset records. Next, we use the order provided by this Hamiltonian path (i.e., a given permutation of the records) as input to the Hansen-Mukherjee algorithm, virtually transforming it into a multivariate microaggregation solver we call Multivariate Hansen-Mukherjee (MHM). Our intuition is that good solutions to the TSP would yield Hamiltonian paths allowing the Hansen-Mukherjee algorithm to find good solutions to the multivariate microaggregation problem. We have tested our method with well-known benchmark datasets. Moreover, with the aim to show the usefulness of our approach to protecting location privacy, we have tested our solution with real-life trajectories datasets, too. We have compared the results of our algorithm with those of the best performing solutions, and we show that our proposal reduces the information loss resulting from the microaggregation. Overall, results suggest that transforming the multivariate microaggregation problem into its univariate counterpart by ordering microdata records with a proper Hamiltonian path and applying an optimal univariate solution leads to a reduction of the perturbation error whilst keeping the same privacy guarantees.

Keywords: microaggregation; statistical disclosure control; graph theory; traveling salesman problem; data privacy; location privacy

1. Introduction

Knowledge retrieval and data processing are catalysts for innovation. The continuous advances in information and communication technologies (ICT) and the efficient processing of data allow the extraction of new knowledge by discovering non-obvious patterns and correlations in the data. Nevertheless, such knowledge extraction procedures may threaten individuals' privacy if the proper measures are not implemented to protect it [1–3]. For instance, an attacker may use publicly available datasets to obtain insights about individuals and extract knowledge by exploiting correlations that were not obvious from examining a single dataset [4]. Therefore, before disclosing any data, privacy protection procedures (e.g., anonymization, pseudonymization, aggregation, generalization) must be applied. A wide variety of privacy models and protection mechanisms have been proposed in the literature so as to guarantee anonymity (at different levels depending on the utilized model) when disclosing data [5]. Since most privacy protection methods are based on modifying/perturbing/deleting original data, their main drawback is that they negatively affect the utility of the data. Hence, there is a need for finding a proper trade-off between data utility and privacy.

One of the most well-known disciplines studying methods to protect individuals' private information is Statistical Disclosure Control (SDC [6]), which seeks to anonymize microdata sets (i.e., datasets consisting of multiple records corresponding to individual respondents) in a way that it is not possible to re-identify the respondent corresponding to any particular record in the published microdata set. Microaggregation [7], which perturbs microdata sets by aggregating the attributes' values of groups of k records so as to reduce re-identification risk by achieving k−anonymity, stands out among the most widely used families of SDC methods. It is usually applied by statistical agencies to limit the disclosure of sensitive microdata, and it has been used to protect data in a variety of fields, namely healthcare [8], smart cities [9], or collaborative filtering applications [10], to name a few.

Although the univariate microaggregation problem can be optimally solved in polynomial time, optimal multivariate microaggregation is an NP-hard problem [11]. Thus, finding a solution for the multivariate problem requires heuristic approaches that aim to minimize the amount of data distortion (often measured in terms of information loss), whilst guaranteeing a desired privacy level (typically determined by a parameter k that defines the cardinality of the aggregated groups).

1.1. Contribution and Research Questions

In this article, we propose a novel solution for the multivariate microaggregation problem, inspired by the heuristic solutions of the Traveling Salesman Problem (TSP) and the use of the optimal univariate microaggregation algorithm of Hansen and Mukherjee (HM) [12]. Given an ordered numerical vector, the HM algorithm creates the optimal k-partition (i.e., the optimal univariate microaggregation solution). Hence, our intuition is that, if we feed the HM algorithm with a good ordering of the records in a multivariate dataset, it would output a good k-partition of the multivariate dataset (although not necessarily optimal).

Ordering the records of a univariate dataset is trivial. However, ordering those records in a multivariate dataset, in which every record has p attributes, is not obvious since it is not apparent how to determine the precedence of an element over another. Thus, the primary question is:

Q1: *How to create this ordering, when the records are in $\mathbb{R}^p$.*

We suggest that a possible order for the records in $\mathbb{R}^p$ is determined by the Hamiltonian path resulting from solving the Traveling Salesman Problem, in which the goal is to find the path that travels through all elements of a set only once, whilst minimizing the total length of the path. Optimally solving the TSP is known to be NP-Hard, but very good heuristic solutions are available. Hence, our intuition is that good heuristic solutions of the TSP (i.e., those with shorter path lengths) would provide a Hamiltonian path, that could be used as an ordered vector for the HM optimal univariate microaggregation algorithm, resulting in a good multivariate microaggregation solution.

The quality of a TSP solution is measured in terms of "path length", the shorter the length the better the solution. However, the quality of the microaggregation is measured in terms of information loss. Given a cardinality parameter k (which sets the minimum size of the aggregation clusters), the lower the information loss, the better the microaggregation. Hence, the next questions that we aim to answer are:

Q2: *Are the length of the Hamiltonian path and the information loss of the microaggregation related?, or Do shorter Hamiltonian paths lead to microaggregation solutions with lower information loss?*

and

Q3: *Is the length of the Hamiltonian path the only factor affecting information loss or does the particular construction of the path (regardless of the length) affect the information loss?*

Overall, the key question is:

Q4: *Does this approach provide better solutions (in terms of information loss) than the best performing microaggregation methods in the literature?*

In order to answer these questions, we have tested seven TSP solvers, combined with the HM algorithm (virtually applied in a multivariate manner, or Multivariate HM (MHM)). Particularly, we have tested the "Concorde" heuristic, which, to the best of our knowledge, is the first time it is used for microaggregation. In addition, we have tested well-known classic microaggregation methods (i.e., MDAV and V-MDAV), and an advanced refinement of the former (i.e., MDAV-LK-MHM).

With the aim to test all the aforementioned approaches on a variety of datasets, we have used three classical benchmarks (i.e., Census, Tarragona, and EIA) and three novel datasets containing trajectory data retrieved from public sources, which lead to our last research question:

Q5: *Do TSP-based microaggregation methods perform better than current solutions on trajectories datasets?*

1.2. Plan of the Article

The rest of the article aims to answer the research questions above, and it is organized as follows: Section 2 provides the reader with some fundamental knowledge on Statistical Disclosure Control and microaggregation. In addition, it introduces the basics of the Traveling Salesman Problem and an overview of the existing heuristics to solve it. Next, Section 3 analyzes related work and highlights the novelty of our proposal compared with the state of the art. Section 4 describes our proposal, which is later thoroughly tested and compared with well-known classical and state-of-the-art microaggregation methods in Section 5. Section 6 discusses the research questions and the main benefits of our proposal. The article concludes in Section 7 with some final remarks and comments on future research lines.

2. Background

2.1. Statistical Disclosure Control and Microaggregation

Statistical disclosure control (SDC) has the goal of preserving the statistical properties of datasets, whilst minimizing the privacy risks related to the disclosure of confidential information from individual respondents. Microaggregation is a family of SDC methods for microdata, which use data perturbation as a protection strategy.

Given an original data file D and a privacy parameter k, microaggregation can be defined as follows: Let us assume a microdata set D with p continuous numerical attributes and n records. Clusters (also referred to as groups or subsets in this context) of D are formed with n_i records in the i-th cluster ($n_i \geq k$ and $n = \sum_{i=1}^{g} n_i$), where g is the number of resulting clusters, and k a cardinality constraint. Optimal microaggregation is defined as the one yielding a k-partition maximizing the within-clusters homogeneity. Optimal microaggregation requires heuristic approaches since it is an NP-hard problem [11] for multivariate data. Microaggregation heuristics can be classified into two main families:

- Fixed-size microaggregation: These heuristics cluster the elements of D into k-partitions where all clusters have size k, except perhaps one group which has a size between k and $2k - 1$, when the total number of records is not divisible by k.
- Variable-size microaggregation: These heuristics cluster the elements of D into k-partitions where all clusters have sizes in $(k, 2k - 1)$. Note that it is easy to show that any cluster with size larger than $(2k - 1)$ could be divided in several smaller clusters

of size between k and $2k-1$ in which its overall within-cluster homogeneity is better than that of the single larger cluster.

Therefore, a microaggregation process consists in constructing a k-partition of the dataset, this is a set of disjoint clusters (in which the cardinality is between k and $2k-1$) and replacing each original data record by the centroid (i.e., the average vector) of the cluster to which it belongs, hence creating a k-anonymous dataset D'. With the aim to reduce the information loss caused by the aggregation, the clusters are created so that the records in each cluster are similar.

2.2. Data Utility and Information Loss

The sum of square error (SSE) is commonly used for measuring the homogeneity in each group. In terms of sums of squares, maximizing within-groups homogeneity is equivalent to finding a k-partition minimizing the within-groups sum of square error (SSE) [13] defined as:

$$SSE = \sum_{i=1}^{g} \sum_{j=1}^{n_i} (x_{i,j} - \bar{x}_i)(x_{i,j} - \bar{x}_i)', \tag{1}$$

where $x_{i,j}$ is the j-th record in group i, and $\bar{x}_i$ is the average record of group i. The total sum of squares (SST), an upper bound on the partitioning information loss, can be computed as follows:

$$SST = \sum_{i=1}^{n} (x_i - \bar{x})(x_i - \bar{x})', \tag{2}$$

where x_i is the i-th record in D, and $\bar{x}_i$ is the average record of D. Note that all the above equations use vector notation, so $x_i \in \mathbb{R}^p$.

The microaggregation problem consists in finding a k-partition with minimum SSE, this is, the set of disjoint subsets of D so that $D = \bigcup_{m=1}^{g} s_m$, where s_m is the m-th subset, and g is the number of subsets, with minimum SSE. However, a normalized measure of information loss (expressed in percentage) is also used:

$$I_{loss} = \frac{SSE}{SST} \times 100. \tag{3}$$

In terms of information loss, the worst case scenario for microaggregation would happen when all records in D are replaced in D' by the average of the dataset (i.e., $SSE = SST \rightarrow I_{loss} = 100$), and the best case scenario implies that $D = D'$ (i.e., $k = 1$, no aggregation), which leads to $SSE = I_{loss} = 0$. Obviously, the latter case is optimal in terms of information loss, but it offers no privacy protection, at all. Hence, values for the protection parameter k are greater than one, typically: $k = 3, 4, 5,$ or 6, and are chosen by privacy experts in statistical agencies so as to adapt to the needs of each particular dataset.

2.3. Basics on the Traveling Salesman Problem

The Traveling Salesman Problem (TSP) [14] consists of finding a particular *Hamiltonian cycle*. The problem can be stated as follows: a salesman leaves from one city and wants to visit (exactly once) each other city in a given group and, finally, return to the starting city. The salesman wonders in what order he should visit these cities so as to travel the shortest possible total distance.

In terms of graph theory, the TSP can be modeled by a graph $G = (V, E)$, where cities are the nodes in set $V = \{v_1, v_2, ..., v_n\}$ and each edge $e_{ij} \in E$ has an associated weight w_{ij} representing the distance between nodes i and j. The goal is to find a *Hamiltonian cycle*, i.e., a cycle which visits each node in the graph exactly once, with the least total weight. An alternative approach to the *Hamiltonian cycle* to solve the TSP is finding the *Shortest Hamiltonian path* through a graph (i.e., a path which visits each node in the graph exactly once). As an example, Figure 1 shows a short Hamiltonian path for the *Eurodist* dataset, which contains the distance (in km) between 21 cities in Europe.

Figure 1. A Hamiltonian path for the Eurodist dataset.

Finding an optimal solution to the TSP is known to be NP-Hard. Hence, several heuristics to find good but sub-optimal solutions have been developed. TSP heuristics typically fall into two groups: those involving minimum spanning trees for tour construction and those with edge exchanges to improve existing tours. There are numerous heuristics to solve the TSP [15,16]. In this article, we have selected a representative sample of heuristics, including well-known approaches and top performers from the state-of-the-art:

- **Nearest Neighbor algorithm**: The algorithm starts with a tour containing a randomly chosen node and appends the next nearest node iteratively.
- **Repetitive Nearest Neighbor**: The algorithm is an extension of the Nearest Neighbor algorithm. In this case, the tour is computed n times, each one considering a different starting node and then selecting the best tour as the outcome.
- Insertion Algorithms: All insertion algorithms start with a tour that originated from a random node. In each step, given two nodes already inserted in the tour, the heuristic selects a new node that minimizes the increase in the tour's length when inserted between such two nodes. Depending on the way such the next node is selected, one can find different variants of the algorithm. For instance, **Nearest Insertion**, **Farthest Insertion**, **Cheapest Insertion**, and **Arbitrary Insertion**.
- **Concorde**: This method is currently one of the best implementations for solving the symmetric TSP. It is based on the *Branch-and-Cut* method to search for optimal solutions.

3. Related Work on Microaggregation

There is a wide variety of heuristics to solve the multivariate microaggregation problem in the literature. One of the most well-known methods is the Maximum Distance to Average Vector (MDAV), proposed by Domingo-Ferrer et al. [17]. This method iteratively creates clusters of k members considering the furthest records from the dataset centroid. A variant of MDAV was proposed by Laszlo et al., namely the Centroid-Based Fixed Size method (CBFS) [18], which also has optimized versions based on kd-tree neighborhood search, such as KD-CBFS and KD-CBFSapp [19]. The Two Fixed Reference Points (TFRP) method was proposed by Chang et al. [20]. It uses the two most extreme points of the dataset at each iteration as references to create clusters. Differential Privacy-based microaggregation was explored by Yang et al. [21], which created a variant of the MDAV algorithm that uses the correlations between attributes to select the minimum required noise to achieve the desired privacy level. In addition, V-MDAV, a variable group-size heuristic based on the MDAV method was introduced by Solanas et al. in Reference [13] with the aim to relax the cardinality constraints of fixed-size microaggregation and allow clusters to better adapt to the data and reduce the SSE.

Laszlo and Mukherjee [18] approached the microaggregation problem through minimum spanning trees, aimed at creating graph structures that can be pruned according to each node's associated weights to create the groups. Lin et al. proposed a Density-Based Algorithm (DBA) [22], which first forms groups of records in density descending order, and then fine-tunes these groups in reverse order. The successive Group Selection based on sequential Minimization of SSE (GSMS) method [23], proposed by Panagiotakis et al., optimizes the information loss by discarding the candidate cluster that minimizes the current SSE of the remaining records. Some methods are built upon the HM algorithm. For example, Mortazavi et al. proposed the IMHM method [24]. Domingo-Ferrer et al. [17] proposed a grouping heuristic that combines several methods, such as Nearest Point Next (NPN-MHM), MDAV-MHM, and CBFS-MHM.

Other approaches have focused on the efficiency of the microaggregation procedure, for example, the Fast Data-oriented Microaggregation (FDM) method proposed by Mortazavi et al. [25] efficiently anonymizes large multivariate numerical datasets for multiple successive values of k. The interested readers can find more detailed information about microaggregation in Reference [5,26].

The most similar work related to ours is the one presented in Reference [27] by Heaton and Mukherjee. The authors use TSP tour optimization heuristics (e.g., 2-opt, 3-opt) to refine a path created with the information of a multivariate microaggregation method (e.g., MDAV, MD, CBFS). Notice that, in our proposed method (described in the next section), we use tour construction TSP heuristics instead of optimization heuristics; thus, we eliminate the need for using a multivariate microaggregation method as a pre-processing step, and we decrease the computational time without hindering data utility.

4. Our Method

Our proposal is built upon two main building blocks: a TSP tour construction heuristic (H), and the optimal univariate microaggregation algorithm of Hansen and Mukherjee (HM). As we have already explained in Section 2, the HM algorithm is applied to univariate numerical data, because it requires the input elements to be in order. However, we virtually use it with multivariate data; thus, when we do that, we refer to it as Multivariate Hansen-Mukherjee (MHM), although, in practice, the algorithm is univariate. Since our proposal is based on a Heuristic (H) to obtain a Hamiltonian Path and the MHM algorithm, we have come to call it HMHM-microaggregation or $(HM)^2$-Micro, for short.

Given a multivariate microdata set (D) with p columns and r rows, we model it as a complete graph $G(N, E)$, where we assume that each row is represented by a node $n_i \in N$ (or a city, if we think in terms of the TSP), and each edge $e_{ij} \in E$ represents the Euclidean distance between n_i and n_j (or the distance between cities in TSP terms). Hence, we have a set of nodes $N = \{n_1, n_2, \ldots n_r\}$ each representing rows of the microdata set in a multivariate space $\mathbb{R}^p$.

In a nutshell, we use H over G to create a Hamiltonian path (H_{path}) that travels across all nodes. H_{path} is a permutation ($\Pi^N = \{\pi_1^N, \pi_2^N, \ldots \pi_r^N\}$) of the nodes in N, and *de facto* it determines an order for the nodes (i.e., it provides a sense of precedence between nodes). Hence, although D is multivariate, its rows represented as nodes in N can be sorted in a univariant permutation H_{path} that we use as input to the MHM algorithm. As a result, the MHM algorithm returns the optimal univariate k-partition of H_{path}, this is, the set of disjoint subsets $S = \{s_1, s_2, \ldots s_t\}$ defining the clusters of N. Hence, since each node n_i represents a row in D, which is indeed multivariate, we have obtained a multivariate microaggregation of the rows in D and provided a solution for the multivariate microaggregation. Notice that, although MHM returns the optimal k-partition of H_{path}, it does not imply that the resulting microaggregation of D is optimal A schematic of our solution is depicted in Figure 2.

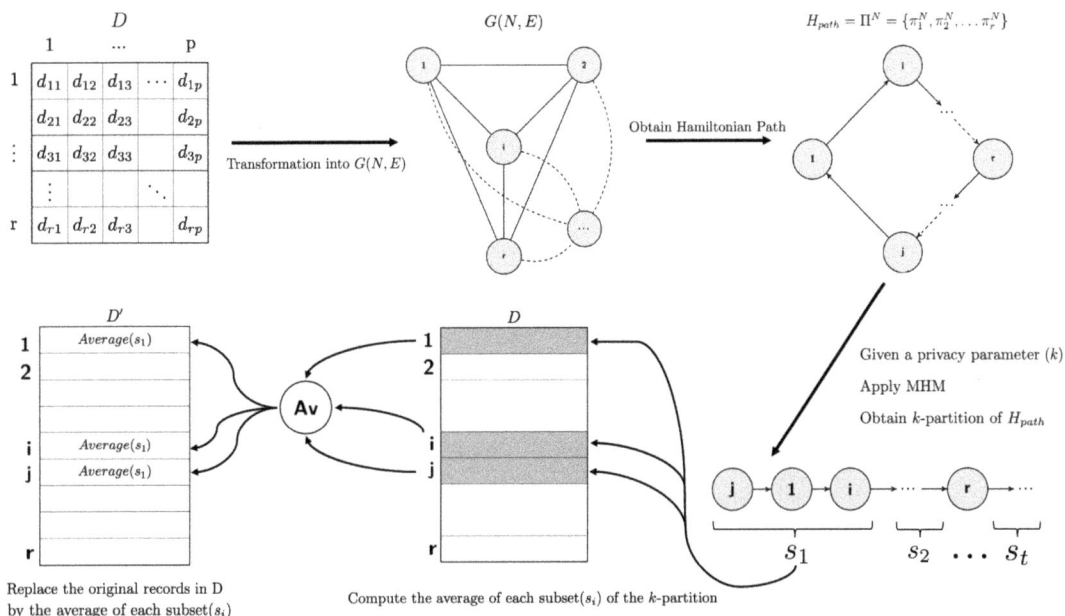

Figure 2. Given a microdata dataset, we use a tour construction heuristic to generate a Hamiltonian path, which will be used as the input of the MHM method to generate the groups.

Although the foundation of our proposal described above is pretty straightforward, it has the beauty of putting together complex mathematical building blocks from the multivariate and univariate worlds in a simple yet practical manner. In addition, our solution is very flexible, since it allows the use of any heuristic H to create the Hamiltonian path H_{path}, and it allows for comprehensive studies, such as the one we report in the next section.

Note that most TSP heuristics output a Hamiltonian cycle. However, since we need a Hamiltonian path, we use the well-known solution of adding a dummy node in the graph (i.e., a theoretical node in which its distance to all other nodes is zero), and we cut the cycle by eliminating this node, so as to obtain a Hamiltonian path.

For the sake of completeness, we summarize our proposal step-by-step in Algorithm 1, and we next comment on it. Our solution can be seen as a meta-heuristic to solve the multivariate microaggregation problem, since it can accommodate any Heuristic (H) able to create a Hamiltonian cycle from a complete graph (G), and it could deal with any privacy parameter (k). Thus, our algorithm receives as input a numerical multivariate microdata set D with p columns (attributes) and r rows, that have to be microaggregated, a Heuristic H, and a privacy parameter k (see Algorithm 1: line 1). In order to avoid bias towards higher magnitude variables, the original dataset D (understood as a matrix) is standardized by subtracting to each element the average of its column and dividing it by the standard deviation of the column. The result is a standardized dataset D_{std} in which each column has zero mean and unitary standard deviation (see Algorithm 1: line 2). Next, the distance matrix M_{dist} is computed. Each element $m_{ij} \in M_{dist}$ contains the Euclidean distance between row i and row j in D_{std}; hence, M_{dist} is a square matrix ($r \times r$) (see Algorithm 1: line 3). In order to be able to cut the Hamiltonian Cycle and obtain a Hamiltonian path, we add a dummy node to the dataset by adding a zero column and a zero row to M_{dist} and generate M_{dist}^{dum}, which is a square matrix ($r+1 \times r+1$) (see Algorithm 1: line 4). M_{dist}^{dum} is, in fact, a weighted adjacency matrix that defines a graph $G(N, E)$ with nodes $N = \{n_1, \ldots, n_{r+1}\}$ and edges $E = \{e_{11}, \ldots e_{i,j} \ldots e_{r+1,r+1}\} = \{M_{dist\ 1,1}^{dum}, \ldots M_{dist\ r+1,r+1}^{dum}\}$. With this matrix as an input, we could compute a Hamiltonian Cycle H_{cycle} on G by applying a TSP heuristic H (see Algorithm 1: line 5). Notice that this Heuristic H could be anyone that gets as input a weighted graph and returns a Hamiltonian cycle. Some examples are: Concorde, Nearest Neighbor, Repetitive Nearest Neighbor, and Insertion Algorithms. After obtaining H_{cycle}, we cut it by removing the dummy node (see Algorithm 1: line 6), and we obtain a Hamiltonian path H_{path} that defines a permutation ($\Pi^N = \{\pi_1^N, \pi_2^N, \ldots \pi_r^N\}$) of the nodes in N, as well as determines an order for the nodes that can be inputted to the MHM algorithm to obtain its optimal k-partition (S) (see Algorithm 1: line 7). S is a set of disjoint subsets $S = \{s_1, s_2, \ldots s_t\}$ defining the clusters of nodes in N. Hence, with S and D, we could create a microaggregated dataset D' by replacing each row in D by the average vector of the k-partition subset to which it belongs (see Algorithm 1: line 8).

After applying the algorithm, we have transformed the original dataset D into a dataset D' that has been microaggregated so as to guarantee the privacy criteria established by k.

Algorithm 1 $(HM)^2$-Micro

1: **function** $(HM)^2$-MICRO(Microdata set D, TSP-Heuristic H, Privacy Parameter k)

2:     D_{std} = StandardizeDataset(D)

3:     M_{dist} = ComputeDistanceMatrix(D_{std})

4:     M_{dist}^{dum} = InsertDummyNode(M_{dist})

5:     H_{cycle} = CreateHamiltonianCycle(M_{dist}^{dum}, H)

6:     H_{path} = CutDummyNode(H_{cycle})

7:     S = MHM(H_{path}, D_{std}, k)

8:     D' = BuildMicroaggregatedDataSet(D, S);

9: **return** D'

10: **end function**

5. Experiments

With the aim to practically validate the usefulness of our multivariate microaggregation proposal, we have thoroughly tested it on six datasets (described in Section 5.1) that serve as benchmarks. In addition, we are interested in knowing (if and) to what extend our method outperforms the best performing microaggregation methods in the literature. Hence, we have compared our proposal with these methods (described in Section 5.2), and the results of all these tests are summarized in Section 5.3. Overall, considering four different values for the privacy parameter $k \in \{3, 4, 5, 6\}$, ten microaggregation algorithms, 50 repetitions per case, and six datasets, we have run over 12.000 microaggregation tests, which allow us to provide a statistically solid set of results.

5.1. Datasets

We used six datasets as benchmarks for our experiments. We can classify those datasets into two main groups: The first group comprises three well-known SDC microdata sets that have been used for years as benchmarks in the literature, namely "Census", "EIA", and "Tarragona". The second group comprises three mobility datasets containing real GPS traces from three Spanish cities, namely "Barcelona", "Madrid", and "Tarraco". Notice that we use the term "Tarraco", the old Roman name for the city of Tarragona, in order to avoid confusion with the classic benchmark dataset "Tarragona". The features of each dataset are next summarized:

The Census dataset was obtained using the public *Data Extraction System of the U.S. Census Bureau*. It contains 1080 records with 13 numerical attributes. The Tarragona dataset was obtained from the Tarragona Chamber of Commerce. It contains information on 834 companies in the Tarragona area with 13 variables per record. The EIA dataset was obtained from the U.S. Energy Information Authority, and it consists of 4092 records with 15 attributes. More details on the aforementioned datasets can be obtained in Reference [28].

The Barcelona, Madrid, and Tarraco datasets consist of OpenStreetMap [29] GPS traces collected from those cities: Barcelona contains the GPS traces of the city of Barcelona within the area determined by the parallelogram formed by latitude (41.3726866, 41.4078446) and longitude (2.1268845, 2.1903992). The dataset has 969 records with 30 GPS locations each. Madrid contains the GPS traces of the city of Madrid within the area determined by the parallelogram formed by latitude (40.387613, 40.483515) and longitude (−3.7398145, −3.653985). The dataset has 959 records with 30 GPS locations each. Tarraco contains the GPS traces of the city of Tarragona within the area determined by the parallelogram formed by latitude (41.0967083, 41.141174) and longitude (1.226008, 1.2946691). The dataset has 932 records with 30 GPS locations each.

In all trajectories datasets, each record consists of 30 locations represented as (latitude and longitude). Hence, each record has 60 numerical values. These locations were extracted from each corresponding parallelogram according to the amount of recorded tracks and their length.

All datasets are available in our website: https://www.smarttechresearch.com/publications/symmetry2021-Maya-Casino-Solanas/ (accessed on 1 May 2021).

Table 1. Comparing methods and features. For Concorde, M is a bound on the time to explore subproblems, b is a branching factor, and d is a search depth.

Method		Cardinality	Computational Cost	Reference
MDAV		fixed	$O(n^2/2k)$	[17]
V-MDAV		variable	$O(n^2)$	[13]
MDAV-LK-MHM		variable	$O(n^2/2k)$	[27]
$(HM)^2$-Micro	**TSP Heuristic + MHM**			
	Nearest Neighbor	variable	$O(n^2)$	[15]
	Repetitive Nearest-Neighbor	variable	$O(n^2 \log n)$	[15]
	Nearest Insertion	variable	$O(n^2)$	[30]
	Farthest Insertion	variable	$O(n^2)$	[30]
	Cheapest Insertion	variable	$O(n^2)$	[30]
	Arbitrary Insertion	variable	$O(n^2)$	[30]
	Concorde	variable	$O(Mb^d)$	[31]

5.2. Compared Methods

We have selected a representative set of well-known and state-of-the-art methods to assess the value of our approach. We have selected two classic microaggregation methods (i.e., **MDAV** and **V-MDAV**), as baselines. In the case of V-MDAV, the method was run for several values of $\gamma \in \{0,2\}$, and the best result is reported. Although some other newer methods might have achieved better results, they are still landmarks that deserve to be included in any microaggregation comparison.

For newer and more sophisticated methods, we have considered the work of Heaton and Mukherjee [27], in which they study a variety of microaggregation heuristics, including methods, such as CBFS and MD. Thus, instead of comparing our proposal with all those methods, we have taken the method that Heaton and Mukherjee reported as the best performer, namely the **MDAV-LK-MHM** method. This method, which is based on MDAV, first creates a microaggregation using MDAV, next improves the result of MDAV by applying the LK heuristic, and it finally applies MHM to obtain the resulting microaggregation (cf. Reference [27] for further details on the algorithm).

Regarding our proposal (i.e., $(HM)^2$-**Micro**), as we already discussed, it can be understood as a meta-heuristic able to embody any heuristic H that returns a Hamiltonian Cycle. Hence, with the aim to determine the best heuristic, we have analyzed seven alternatives, namely **Nearest Neighbor**, **Repetitive Nearest Neighbor**, **Nearest Insertion**, **Farther Insertion**, **Cheapest Insertion**, **Arbitrary Insertion**, and (our suggestion) **Concorde**. Table 1 summarizes some features of all selected methods, including the reference to the original article where the method was described. For our method, each reference points to the article describing the TSP heuristic.

The implementation of all these methods have used the R package *sdcMicro* [28], the TSP heuristics implemented in Reference [32], and the LK heuristics implemented in Reference [33]. LK has been configured so that the algorithm runs once at each iteration parameter RUN=1 until a local optimum is reached. This same criteria was followed for the other TSP heuristics. In this regard, the heuristics we used consider a random starting node at each run. Hence, each experiment has been repeated 50 times to guarantee statistically sound outcomes regardless of this random starting point.

5.3. Results Overview

By using the datasets and methods described above, we have analyzed the Information Loss (expressed in percentage), as a measure of data utility (cf. Section 2 for details). It is assumed that, given a privacy parameter k that guarantees that the microaggregated dataset is k-anonymous, the lower the Information Loss the better the result and performance of the microaggregation method. The results are reported in Tables 2–7 with the best (lowest) information loss highlighted in green.

Overall, it can be observed that our method, $(HM)^2$-Micro, with the Concorde heuristic is the best performer in 79% of the experiments, and it is the second best in the remaining 21% (for which the MDAV-LK-MHM outperforms it by a narrow margin of less than 2%). Interestingly enough, although $(HM)^2$-Micro, with both Nearest Insertion and Farthest-Insertion, is not the best performer in any experiment, it outperforms MDAV-LK-HMH 50% of the times. The rest of the methods obtain less consistent results and highly depend on the dataset.

When we analyze the results more closely for each particular dataset, we observe that, in the case of the "Census" dataset (cf. Table 2), our method with Concorde outperforms all methods for all values of k. In addition, despite the random nature of TSP-heuristics, the values of σ are very stable, denoting the robustness of all methods, yet slightly higher on average in the case of the methods with higher Information Loss. It is worth emphasizing though, that, in all runs, our method with Concorde and the MDAV-LK-MHM method obtained better results than MDAV and V-MDAV (i.e., the max values obtained in all runs are lower than the outcomes obtained by MDAV and V-MDAV).

Table 2. Information Loss obtained on the Census dataset.

Method	Census															
	$k=3$				$k=4$				$k=5$				$k=6$			
	Average	σ	min	max	Average	σ	min	max	Average	σ	min	max	Average	σ	min	max
MDAV	5.6922	NA	NA	NA	7.4947	NA	NA	NA	9.0884	NA	NA	NA	10.3847	NA	NA	NA
V-MDAV	5.6619	NA	NA	NA	7.4947	NA	NA	NA	9.0070	NA	NA	NA	10.2666	NA	NA	NA
MDAV-LK-MHM	5.1085	0.0398	5.0256	5.1877	6.9131	0.0526	6.7774	7.0227	8.5199	0.0842	8.3100	8.7030	9.9752	0.1284	9.7675	10.2527
Nearest Insertion-MHM	5.6561	0.1369	5.3596	6.0695	7.4818	0.1579	7.1946	7.9318	8.9617	0.2539	8.5190	9.4727	10.3005	0.2927	9.7624	11.2086
Farthest Insertion-MHM	5.5638	0.0956	5.3300	5.8995	7.3485	0.0990	7.1723	7.5853	8.8234	0.1322	8.5784	9.1748	10.1250	0.1932	9.6970	10.7363
Cheapest Insertion-MHM	5.7044	0.0719	5.5669	5.8766	7.4625	0.1155	7.2674	7.8052	9.0340	0.1236	8.7212	9.3847	10.3787	0.1305	10.1706	10.9089
Arbitrary Insertion-MHM	5.5883	0.0976	5.4235	5.8763	7.3723	0.1438	7.1272	7.8250	8.8696	0.1788	8.5072	9.2867	10.2011	0.2475	9.7081	10.7794
Nearest Neighbor-MHM	6.9718	0.3508	6.1978	7.7291	9.2433	0.3702	8.6744	10.2246	11.3287	0.3854	10.5230	12.3958	13.1357	0.4053	12.4711	13.9421
Repetitive NN-MHM	6.2888	0.2192	5.8811	6.6841	8.6779	0.2799	7.9941	9.3345	10.7518	0.2472	10.3421	11.4554	12.5882	0.3143	11.9360	13.2915
Concorde-MHM	5.0563	0.0377	4.9917	5.1169	6.8846	0.0555	6.7895	7.0217	8.4576	0.0903	8.2372	8.6614	9.8440	0.1232	9.5542	10.2517

For the "EIA" dataset (cf. Table 3), MDAV-LK-MHM is the best performer for all values of k except $k = 5$, for which our proposal with Concorde performs better. In this case, the results obtained by these two methods are very close. Similarly to the results in "Census", the max values obtained by these two methods outperform MDAV and V-MDAV. In the case of "Tarragona" (cf. Table 4), our method with Concorde outperforms all other methods. Surprisingly, both MDAV and V-MDAV obtain better results than MDAV-LK-MHM, which performs poorly in this dataset.

Table 3. Information Loss obtained on the EIA dataset.

	EIA															
	$k=3$				$k=4$				$k=5$				$k=6$			
Method	Average	σ	min	max	Average	σ	min	max	Average	σ	min	max	Average	σ	min	max
MDAV	0.4829	NA	NA	NA	0.6713	NA	NA	NA	1.6667	NA	NA	NA	1.3078	NA	NA	NA
V-MDAV	0.4829	NA	NA	NA	0.6713	NA	NA	NA	1.2771	NA	NA	NA	1.2320	NA	NA	NA
MDAV-LK-MHM	**0.3741**	0.0075	0.3659	0.4097	**0.5251**	0.0116	0.5117	0.5693	0.7890	0.0336	0.7502	0.8932	**1.0430**	0.0289	1.0033	1.1113
Nearest Insertion-MHM	0.4061	0.0114	0.3831	0.4238	0.5781	0.0241	0.5441	0.6179	0.8621	0.0456	0.8032	0.9760	1.1254	0.0837	0.9976	1.3334
Farthest Insertion-MHM	0.4070	0.0119	0.3872	0.4207	0.5878	0.0251	0.5524	0.6277	0.8764	0.0522	0.8190	0.9747	1.1776	0.0359	1.1245	1.2484
Cheapest Insertion-MHM	0.5254	0.0358	0.4692	0.5651	0.7321	0.0641	0.6322	0.8477	1.0868	0.0689	0.9910	1.2264	1.4061	0.1147	1.2605	1.6329
Arbitrary Insertion-MHM	0.4281	0.0300	0.3921	0.4944	0.6092	0.0376	0.5566	0.6699	0.9048	0.0840	0.8194	1.0621	1.1928	0.1077	1.0652	1.3476
Nearest Neighbor-MHM	0.9028	0.1455	0.5089	1.1023	1.1510	0.1675	0.7056	1.3776	1.4015	0.1788	0.9451	1.6767	1.6792	0.1107	1.4635	1.9139
Repetitive NN-MHM	0.5110	0.0532	0.4725	0.6599	0.7192	0.0557	0.6646	0.8619	1.0072	0.0701	0.9274	1.1126	1.3101	0.1521	1.1561	1.4825
Concorde-MHM	0.3889	0.0203	0.3673	0.4210	0.5288	0.0170	0.5087	0.5576	**0.7802**	0.0267	0.7581	0.8501	1.0476	0.0282	1.0009	1.0904

Table 4. Information Loss obtained on the Tarragona dataset.

	Tarragona															
	$k=3$				$k=4$				$k=5$				$k=6$			
Method	Average	σ	min	max	Average	σ	min	max	Average	σ	min	max	Average	σ	min	max
MDAV	16.9326	NA	NA	NA	19.5460	NA	NA	NA	22.4619	NA	NA	NA	26.3252	NA	NA	NA
V-MDAV	16.6603	NA	NA	NA	19.5460	NA	NA	NA	22.4619	NA	NA	NA	26.3252	NA	NA	NA
MDAV-LK-MHM	18.7969	1.8738	15.0595	23.0830	22.8523	1.7576	19.1195	26.2806	26.2432	1.5066	23.0421	28.9522	28.5244	1.7742	25.1703	30.9656
Nearest Insertion-MHM	15.9687	0.8360	15.1107	20.1835	19.3677	1.3141	17.8032	24.5286	23.7323	1.4376	21.8365	28.9753	26.9018	1.5674	24.6538	33.0785
Farthest Insertion-MHM	15.7634	0.2062	15.4743	16.6623	19.0323	0.5521	18.1062	20.2105	22.8316	0.7636	21.3313	24.1988	25.7627	0.4496	24.9004	26.9613
Cheapest Insertion-MHM	16.3142	1.4861	15.2169	22.0271	19.7784	1.6060	18.3103	25.8916	23.9017	1.7155	22.3121	30.0828	27.5572	1.6611	25.2394	32.7082
Arbitrary Insertion-MHM	16.0918	0.7527	15.1310	18.9668	19.5461	1.3436	18.2072	25.8572	23.7685	1.3985	21.7333	29.1863	27.0419	1.6872	25.0093	33.2382
Nearest Neighbor-MHM	22.3019	0.8866	19.9620	23.5496	27.1002	1.2234	24.2527	29.5117	30.4478	1.5455	27.7026	33.3513	34.5445	1.2088	31.3302	37.5350
Repetitive NN-MHM	17.6981	1.2157	15.7435	20.9981	22.1232	1.9138	20.0839	28.7399	27.9089	1.7946	25.1434	32.5729	30.4085	1.9216	28.0648	35.2458
Concorde-MHM	**14.7677**	0.0858	14.6294	14.9633	**17.9957**	0.1241	17.7528	18.2211	**21.9895**	0.2164	21.6712	22.3479	**25.3459**	0.2061	24.8045	25.6564

So, it can be concluded that the overall winner for the classical benchmarks (i.e., Census, EIA, and Tarragona) is our method, $(HM)^2$-Micro, with the Concorde heuristic, that is only marginally outperformed by MDAV-LK-MHM in the EIA dataset.

Regarding the other three datasets containing GPS traces (i.e., Barcelona, Madrid and Tarraco), our method, $(HM)^2$-Micro, with the Concorde heuristic, is the best performer in 83% of the cases and comes second best in the remaining 17%. For the Barcelona dataset (cf. Table 5), MDAV-LK-MHM and $(HM)^2$-Micro, with the Concorde heuristic, perform very well and similarly. The methods with the worst Information Loss are MDAV and V-MDAV. Our method, $(HM)^2$-Micro, with the Insertion heuristics, have a remarkable performance, obtaining values similar to those of MDAV-LK-MHM and Concorde. Nevertheless, it is worth noting that the max (worst) values obtained by MDAV-LK-MHM and Concorde are still better than the averages obtained by the other methods. In the case of the Madrid dataset (cf. Table 6), our method, $(HM)^2$-Micro, with the Concorde heuristic, achieves the minimum (best) value of Information Loss for all values of k. We can also observe that our method with Insertion heuristics offers higher performance than MDAV-LK-MHM. Finally, the results for the Tarraco dataset (cf. Table 7) show that the minimum (best) Information Loss value is obtained by our method with the Concorde heuristic in all cases. In this case, MDAV-LK-MHM performs poorly, and, for $k = 3$ and $k = 4$, MDAV and V-MDAV are better.

Table 5. Information Loss obtained on the Barcelona dataset.

	Barcelona															
	$k=3$				$k=4$				$k=5$				$k=6$			
Method	Average	σ	min	max	Average	σ	min	max	Average	σ	min	max	Average	σ	min	max
MDAV	2.5667	NA	NA	NA	3.5023	NA	NA	NA	4.2849	NA	NA	NA	5.1873	NA	NA	NA
V-MDAV	2.5667	NA	NA	NA	3.3193	NA	NA	NA	4.2849	NA	NA	NA	5.1873	NA	NA	NA
MDAV-LK-MHM	1.6251	0.0362	1.5637	1.7425	2.1913	0.0339	2.1170	2.2738	2.6798	0.0607	2.5156	2.8067	3.2120	0.0664	3.0731	3.3825
Nearest Insertion-MHM	1.8022	0.0656	1.6857	1.9438	2.3526	0.0842	2.1754	2.5050	2.8405	0.1008	2.6417	3.0411	3.3316	0.1083	3.1093	3.5103
Farthest Insertion-MHM	1.7838	0.0525	1.6967	1.8980	2.3575	0.0698	2.1919	2.4681	2.8386	0.0751	2.6654	2.9670	3.3189	0.1131	3.1112	3.6445
Cheapest Insertion-MHM	1.8156	0.0565	1.6887	1.9293	2.3880	0.0912	2.2354	2.5473	2.8887	0.0792	2.7807	3.0405	3.4118	0.1238	3.1938	3.6247
Arbitrary Insertion-MHM	1.8061	0.0635	1.6823	1.9469	2.3593	0.0749	2.1808	2.5414	2.8231	0.0911	2.6338	3.0251	3.3331	0.1085	3.1031	3.5584
Nearest Neighbor-MHM	2.2019	0.1202	1.9165	2.4476	2.9274	0.1778	2.5276	3.3377	3.4733	0.2168	3.0611	3.9399	4.1053	0.2590	3.5159	4.6420
Repetitive NN-MHM	2.0091	0.0563	1.8899	2.2547	2.7474	0.0611	2.6108	3.0130	3.2318	0.1001	3.1176	3.5701	3.8877	0.1220	3.7106	4.1982
Concorde-MHM	1.6829	0.0375	1.6210	1.7848	2.2132	0.0534	2.1138	2.3426	2.6786	0.0627	2.4974	2.8268	3.1075	0.0718	2.9588	3.2348

Table 6. Information Loss obtained on the Madrid dataset.

	Madrid															
	$k=3$				$k=4$				$k=5$				$k=6$			
Method	Average	σ	min	max	Average	σ	min	max	Average	σ	min	max	Average	σ	min	max
MDAV	3.1876	NA	NA	NA	4.3353	NA	NA	NA	5.2883	NA	NA	NA	5.8235	NA	NA	NA
V-MDAV	3.1876	NA	NA	NA	4.3353	NA	NA	NA	5.2883	NA	NA	NA	5.8235	NA	NA	NA
MDAV-LK-MHM	2.9872	0.1285	2.7200	3.1946	4.0536	0.1398	3.6804	4.3314	4.8541	0.1664	4.4680	5.1856	5.5703	0.2163	5.0931	6.0088
Nearest Insertion-MHM	2.7511	0.0814	2.5782	2.9116	3.7039	0.1122	3.4304	3.9623	4.4522	0.1535	4.1533	4.8463	5.1544	0.1549	4.8661	5.5510
Farthest Insertion-MHM	2.6683	0.0558	2.5319	2.8280	3.6187	0.0742	3.4605	3.7755	4.3338	0.1131	4.1260	4.5668	5.0598	0.1172	4.8391	5.3372
Cheapest Insertion-MHM	2.7833	0.0749	2.6517	2.9789	3.7531	0.0804	3.5253	3.9830	4.4752	0.1140	4.3163	4.7356	5.2496	0.1345	5.0147	5.5609
Arbitrary Insertion-MHM	2.7476	0.0757	2.6009	2.9160	3.7156	0.0986	3.5213	3.9828	4.4149	0.1420	4.0583	4.7078	5.1070	0.1437	4.7687	5.3754
Nearest Neighbor-MHM	3.4257	0.1714	3.0816	3.9040	4.7553	0.2116	4.2823	5.3736	5.7671	0.2194	5.1807	6.3191	6.7615	0.2507	6.1871	7.4355
Repetitive NN-MHM	3.1236	0.1345	2.8799	3.5430	4.4141	0.1482	4.1254	5.0012	5.3911	0.2127	5.0894	6.1676	6.4865	0.2223	6.1764	7.3492
Concorde-MHM	2.4845	0.0336	2.4053	2.5728	3.4302	0.0466	3.3249	3.5664	4.1124	0.0774	3.9816	4.3228	4.8066	0.1065	4.6538	5.0534

Table 7. Information Loss obtained on the Tarraco dataset.

	Tarraco															
	$k=3$				$k=4$				$k=5$				$k=6$			
Method	Average	σ	min	max	Average	σ	min	max	Average	σ	min	max	Average	σ	min	max
MDAV	0.9988	NA	NA	NA	1.4180	NA	NA	NA	1.7683	NA	NA	NA	2.0260	NA	NA	NA
V-MDAV	0.9988	NA	NA	NA	1.3093	NA	NA	NA	1.7182	NA	NA	NA	2.0051	NA	NA	NA
MDAV-LK-MHM	1.1365	0.0154	1.0979	1.1465	1.4216	0.0203	1.4115	1.4723	1.7201	0.0401	1.6995	1.8257	2.0238	0.0404	2.0061	2.1247
Nearest Insertion-MHM	0.9113	0.0345	0.8490	1.0100	1.2634	0.0745	1.1052	1.4306	1.5988	0.1160	1.4220	1.8839	1.9105	0.1517	1.7018	2.2870
Farthest Insertion-MHM	0.9190	0.0368	0.8582	1.0268	1.2217	0.0490	1.1123	1.3755	1.5040	0.0581	1.3965	1.7118	1.8346	0.0612	1.7533	2.1299
Cheapest Insertion-MHM	0.9500	0.0406	0.8975	1.0962	1.2951	0.0557	1.2270	1.4637	1.6200	0.0870	1.5225	1.8677	1.9704	0.1094	1.8584	2.2471
Arbitrary Insertion-MHM	0.9258	0.0455	0.8589	1.0269	1.2530	0.0753	1.1419	1.4538	1.5695	0.0971	1.4454	1.8312	1.9051	0.1265	1.7475	2.3396
Nearest Neighbor-MHM	1.5080	0.1937	1.1624	2.0189	2.1341	0.2232	1.5881	2.6725	2.6499	0.2671	2.0802	3.2271	3.3041	0.4123	2.6557	4.3884
Repetitive NN-MHM	1.2177	0.1286	1.0276	1.5906	1.7806	0.1599	1.4244	2.1131	2.2545	0.1882	1.9146	2.7394	2.7384	0.2209	2.3073	3.4314
Concorde-MHM	0.8482	0.0179	0.8167	0.9005	1.1031	0.0324	1.0739	1.2348	1.3805	0.0556	1.3275	1.6813	1.7280	0.0652	1.6610	2.1308

We have already discussed that all studied methods (with the exception of MDAV and V-MDAV) have a non-deterministic component emerging from the random selection of the initial node. This random selection affects the performance of the final microaggregation obtained. With the aim to analyze the effect of this non-deterministic behavior, we have studied the standard deviation of all methods for all values of k and for all datasets. In addition, we have visually inspected the variability of the results by using box plot diagrams.

Since the results are quite similar and consistent across all datasets, for the sake of clarity, we only reproduce here the box plots for the "Census" dataset (see Figure 3), and we leave the others in Appendix A for the interested reader.

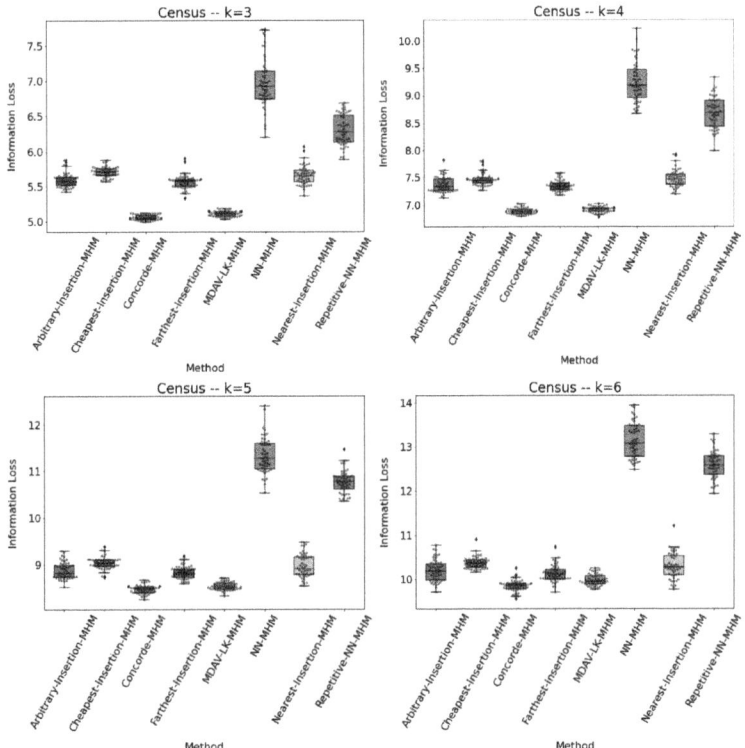

Figure 3. Information Loss variability for each value of k over the Census dataset.

In Figure 3, we can observe that the Information Loss values increase with k, but all methods have the same behavior regardless of the value of k. In addition, it is clear that the most stable methods are $(HM)^2$-Micro, with Concorde, and MDAV-LK-MHM.

Overall, we observe some expected differences depending on the datasets. However, the behavior of the best performing methods is stable. Particularly, the datasets with GPS traces (i.e., Barcelona, Madrid, and Tarraco) show more stable results. In summary, the best method was our $(HM)^2$-Micro with Concorde, exhibiting the most stable results across all datasets.

6. Discussion

Over the previous sections, we have presented our microaggregation method, $(HM)^2$-Micro, its rationale, and its performance against other classic and state-of-the-art methods on a variety of datasets. In the previous section, we have reported the main results, and we will discuss them next by progressively answering the research questions that we posed in the Introduction of the article.

Q1: How to create a suitable ordering for a univariate microaggregation algorithm, when the records are in $\mathbb{R}^p$.

A main takeaway of this article is that, by using a combination of TSP tour construction heuristics (e.g., Concorde) and an optimal univariate microaggregation algorithm, we are properly ordering multivariate datasets in a univariate fashion that leads to excellent multivariate microaggregation solutions. Other approaches to order $\mathbb{R}^p$ points might consider projecting them over the principal component. However, the information loss associated with this approach makes it unsuitable. In addition, other more promising approaches, like the one used in MDAV-LK-MHM, first create a k-partition and set an order based on maximum distance criteria. Although this approach might work well in some cases, we have clearly seen that Hamiltonian paths created by TSP-heuristics, like Concorde, outperform this approach. Hence, based on the experiments of Section 5, we can conclude that TSP-heuristics, like Concorde, provide an order for elements in $\mathbb{R}^p$ that is suitable for an optimal univariate microaggregation algorithm to output a consistent multivariate microaggregation solution with low Information Loss (i.e., high data utility). Moreover, from all analyzed heuristics, it is clear that the best performer is Concorde, followed by insertion heuristics.

Q2: Are the length of the Hamiltonian path and the information loss of the microaggregation related?, or Do shorter Hamiltonian paths lead to microaggregation solutions with lower information loss?

When we started this research, our intuition was that good heuristic solutions of the TSP (i.e., those with shorter path lengths) would provide a Hamiltonian path, that could be used as an ordered vector for the HM optimal univariate microaggregation algorithm, resulting in a good multivariate microaggregation solution. From this intuition, we assumed that shorter Hamiltonian paths would lead to lower Information Loss in microaggregated datasets.

In order to validate (or disproof) this intuition, we have analyzed the Pearson correlation between the Hamiltonian path length obtained by all studied heuristics (i.e., Nearest Neighbor, Repetitive Nearest Neighbor, Nearest Insertion, Farther Insertion, Cheapest Insertion, Arbitrary Insertion, and Concorde) and the SSE of the resulting microaggregation. We have done so for all studied datasets and k values. The results are summarized in Table 8, and all plots along with a trend line are available in Appendix B.

Table 8. Summary of the Pearson correlation between Path Length and SSE.

Dataset	$k = 3$	$k = 4$	$k = 5$	$k = 6$
Census	0.48	0.39	0.32	0.28
EIA	0.62	0.67	0.74	0.76
Tarragona	0.70	0.72	0.82	0.71
Barcelona	0.83	0.81	0.81	0.80
Madrid	0.84	0.81	0.80	0.78
Tarraco	0.80	0.82	0.82	0.80

From the correlation analysis, it can be concluded that there is a positive correlation between the Hamiltonian path length and the SSE. This is, the shorter the path length the lower the SSE. This statement holds for all k and for all datasets (although Census exhibits a lower correlation). Hence, although this result is not a causality proof, it can be safely said that good solutions of the TSP problem lead to good solutions of the multivariate microaggregation problem. In fact, the best heuristic (i.e., Concorde) always results in the lowest (best) SSE.

Interested readers can find all plots in Appendix B. However, for the sake of clarity, let us illustrate this result by discussing the case of the Madrid dataset with $k = 6$, depicted in Figure 4. In the figure, the positive correlation is apparent. In addition, it is clear that heuristics tend to form clusters. In a nutshell, the best heuristic is Concorde, followed by the insertion family of methods (i.e., Nearest Insertion, Furthest Insertion, Cheapest Insertion, and Arbitrary Insertion), followed by Repetitive Nearest Neighbor and Nearest Neighbor.

Although Figure 4 clearly illustrates the positive correlation between the path length and the SSE, it also shows that heuristics tend to cluster and might indicate that not only the path but the heuristic (per se) plays a role in the reduction of the SSE. This indication leads us to our next research question.

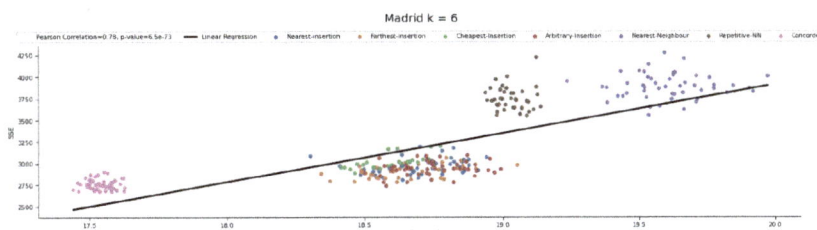

Figure 4. Relation between SSE and Path Length for Madrid and $k = 6$.

Q3: Is the length of the Hamiltonian path the only factor affecting information loss or does the particular construction of the path (regardless of the length) affect the information loss?

In the previous question, we have found clear positive correlation between the path length and the SSE. However, we have also observed apparent clusters suggesting that the very heuristics could be responsible for the minimization of the SSE. In other words, although the path length and SSE are positively correlated when all methods are analyzed together, would this correlation hold when heuristics are analyzed one at a time? In order to answer this question, we have analyzed the results of each heuristic individually, and we have observed that there is still positive correlation between path length and SSE, but it is very weak or almost non-existent (i.e., very close to 0), as Figure 5 illustrates.

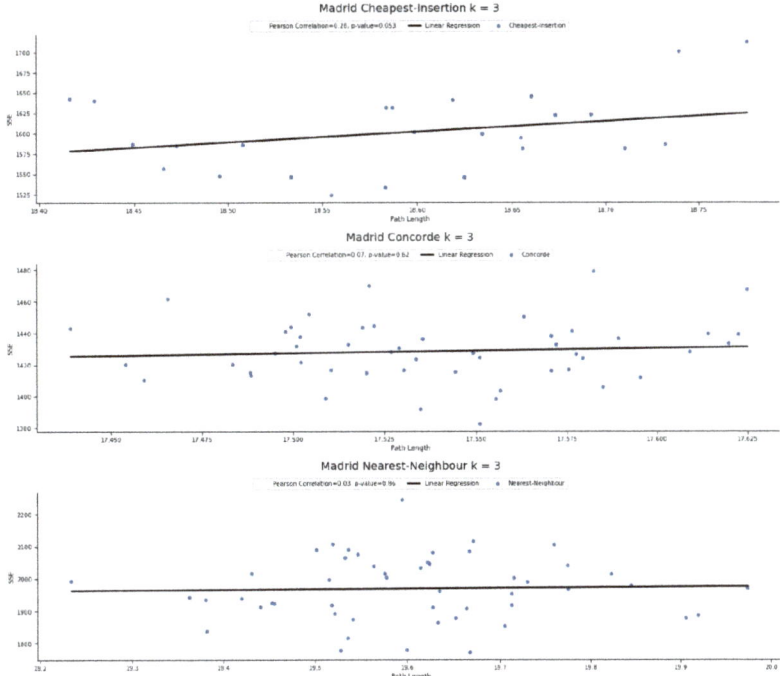

Figure 5. Correlation between path length and SSE for each individual method (from top to bottom: Cheapest Insertion, Concorde, and Nearest Neighbor) for $k = 3$ over the Madrid dataset.

The results shown in Figure 5 are only illustrative, and a deeper analysis that is out of the scope of this paper would be necessary. However, our initial results indicate that, although there is positive correlation between path length and SSE globally, this correlation weakens significantly when analyzed on each heuristic individually. This result suggests that it is not only the length of the path but the way in which this path is constructed what affects the SSE. This would explain why similar methods (e.g., those based on insertion) behave similarly in terms of SSE although their paths' length varies.

Q4: Does $(HM)^2$-Micro provide better solutions (in terms of information loss) than the best performing microaggregation methods in the literature?

This question has been already answered in Section 5.3. However, for the sake of completeness, we summarize it here: The results obtained after executing more than 12,000 tests suggest that our solution $(HM)^2$-Micro obtains better results than classic microaggregation methods, such as MDAV and V-MDAV. Moreover, when $(HM)^2$-Micro uses the Concorde heuristic to determine the Hamiltonian path, it outperforms the best state-of-the-art methods consistently. In our experiments, $(HM)^2$-Micro with Concorde was the best performer 79% of the times and was the second best in the remaining 21%.

Q5: Do TSP-based microaggregation methods perform better than current solutions on trajectories datasets?

$(HM)^2$-Micro with Concorde is the best overall performer. Moreover, if we focus on those datasets with trajectory data (i.e., Barcelona, Madrid, and Tarraco), the results are even better. It is the best performer in 83% of the tests and the second best in the remaining 17%. This good behavior of the method could result from the very foundations of the TSP; however, there is still plenty of research to do in this line to reach more solid conclusions. Location privacy is a very complex topic that encompasses many nuances beyond *k*-anonymity models (such as the one followed in this article). However, this result is an invigorating first step towards the analysis of novel microaggregation methods applied to trajectory analysis and protection.

7. Conclusions

Microaggregation has been studied for decades now, and, although finding the optimal microaggregation is NP-Hard and a polynomial-time microaggregation algorithm has not been found, steady improvements over microaggregation heuristics have been made. Hence, after such a long research and polishing process, finding new solutions that improve the best methods is increasingly difficult. In this article, we have presented $(HM)^2$-Micro, a meta-heuristic that leverages the advances in TSP solvers and combines them with the optimal univariate microaggregation to create a flexible and robust multivariate microaggregation solution.

We have studied our method and thoroughly compared it to classic and state-of-the-art microaggregation algorithms over a variety of classic benchmarks and trajectories datasets. Overall, we have executed more than 12,000 tests, and we have shown that our solution embodying the Concorde heuristic outperforms the others. Hence, we have shown that our TSP-inspired method could be used to guarantee *k*-anonymity of trajectories datasets whilst reducing the Information Loss, thus increasing data utility. Furthermore, our proposal is very stable, i.e., it does not change significantly its performance regardless of the random behavior associated with initial nodes selection.

In addition to proposing $(HM)^2$-Micro, we have found clear correlations between the length of Hamiltonian Paths and the SSE introduced by microaggregation processes, and we have shown the importance of the Hamiltonian Cycle construction algorithms over the overall performance of microaggregation.

Despite these relevant results, there is still much to do in the study of microaggregation and data protection. Future work will focus on scaling up $(HM)^2$-Micro to high-dimensional and very-large datasets. Considering the growing importance of Big Data and Cloud Computing, adapting our solution to distributed computation environments is paramount. Moreover, adjusting TSP heuristics to leverage lightweight microaggregation-based approaches is an interesting research path to follow. In addition, although the values of the privacy parameter *k* are typically low (i.e., 3, 4, 5, 6), we plan to study the effect of larger values of k on our solution. Last but not least, since microaggregation is essentially a data-oriented procedure, we will study how our solution adapts to data structures from specific domains, such as healthcare, transportation, energy, and the like.

All in all, with $(HM)^2$-Micro, we have set the ground for the study of multivariate microaggregation meta-heuristics from a new perspective, that might continue in the years to come.

Author Contributions: Conceptualization, A.M.-L. and A.S.; methodology, A.S.; software, A.M.-L.; validation, A.M.-L., F.C. and A.S.; formal analysis, A.S.; investigation, A.M.-L., F.C. and A.S.; resources, A.S.; data curation, A.M.-L.; writing—original draft preparation, A.M.-L. and A.S.; writing—review and editing, F.C. and A.S.; visualization, F.C.; supervision, F.C. and A.S.; project administration, A.S.; funding acquisition, A.S. All authors have read and agreed to the published version of the manuscript.

Funding: This work was supported by the European Commission under the Horizon 2020 Programme (H2020), as part of the project *LOCARD* (https://locard.eu (accessed on 1 May 2021)) (Grant Agreement no. 832735), and by the Spanish Ministry of Science & Technology with project IoTrain RTI2018-095499-B-C32. The content of this article does not reflect the official opinion of the European Union. Responsibility for the information and views expressed therein lies entirely with the authors.

Institutional Review Board Statement: Not applicable.

Informed Consent Statement: Not applicable.

Data Availability Statement: Dataset and high resolution data can be found in: https://www.smarttechresearch.com/publications/symmetry2021-Maya-Casino-Solanas/ (accessed on 1 May 2021).

Conflicts of Interest: The authors declare no conflict of interest.

Appendix A. Information Loss Variability Box Plots

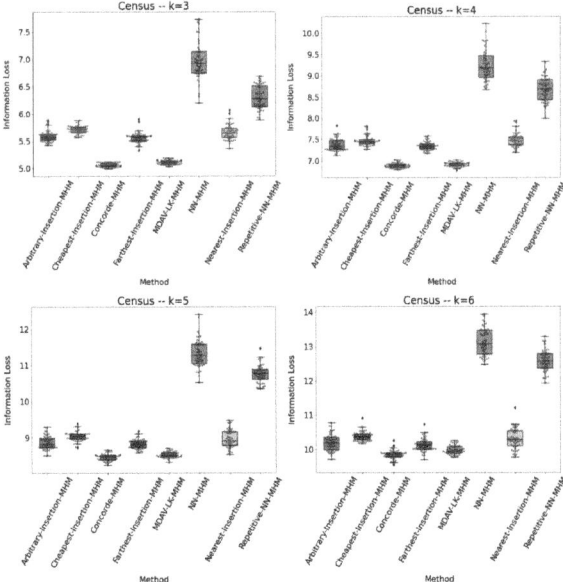

Figure A1. Information Loss variability for each value of k over the Census dataset.

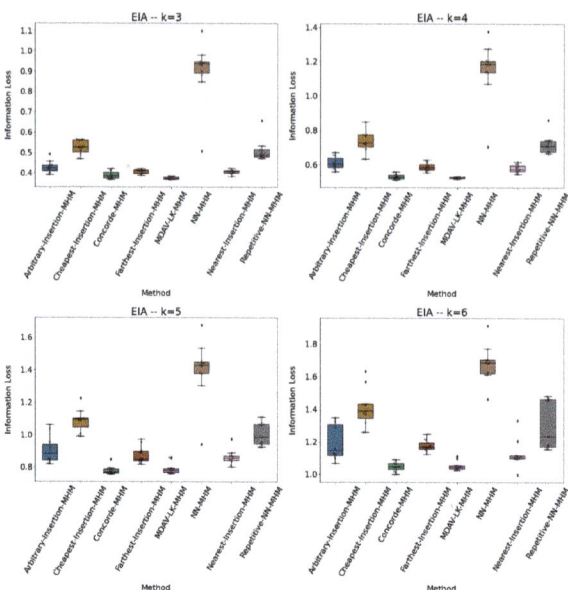

Figure A2. Information Loss variability for each value of *k* over the EIA dataset.

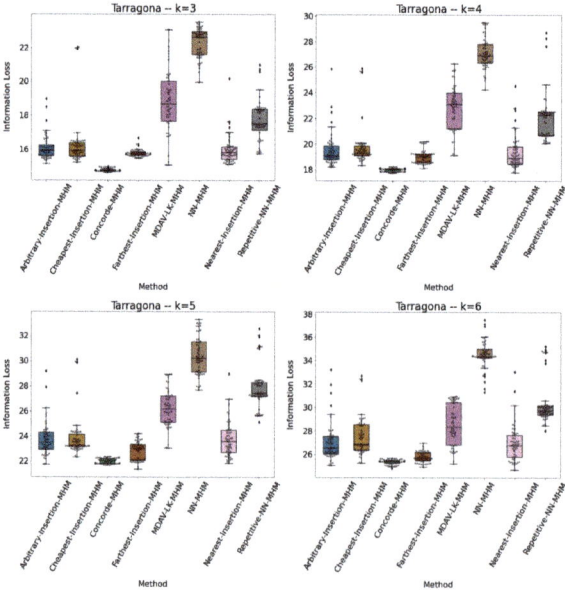

Figure A3. Information Loss variability for each value of *k* over the Tarragona dataset.

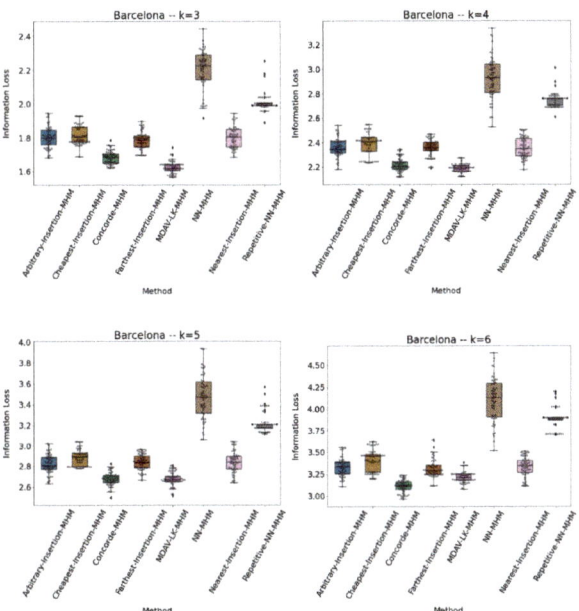

Figure A4. Information Loss variability for each value of k over the Barcelona dataset.

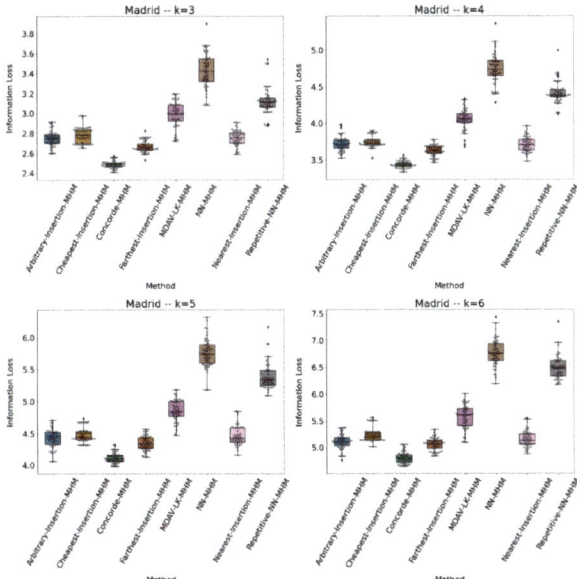

Figure A5. Information Loss variability for each value of k over the Madrid dataset.

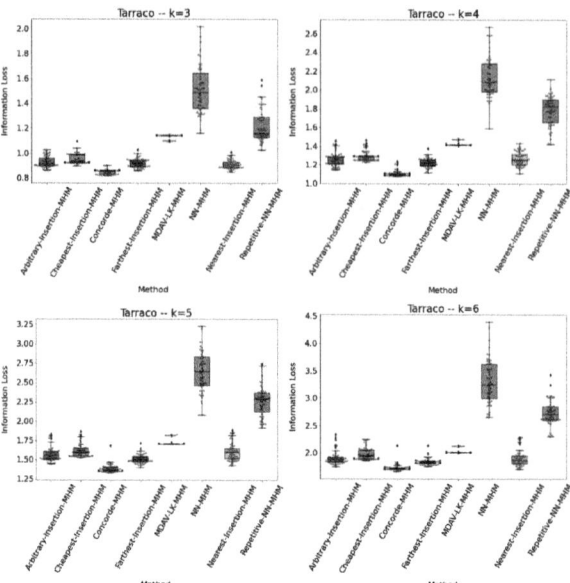

Figure A6. Information Loss variability for each value of *k* over the Tarraco dataset.

Appendix B. Correlation Analysis between "Path Length" and "SSE"

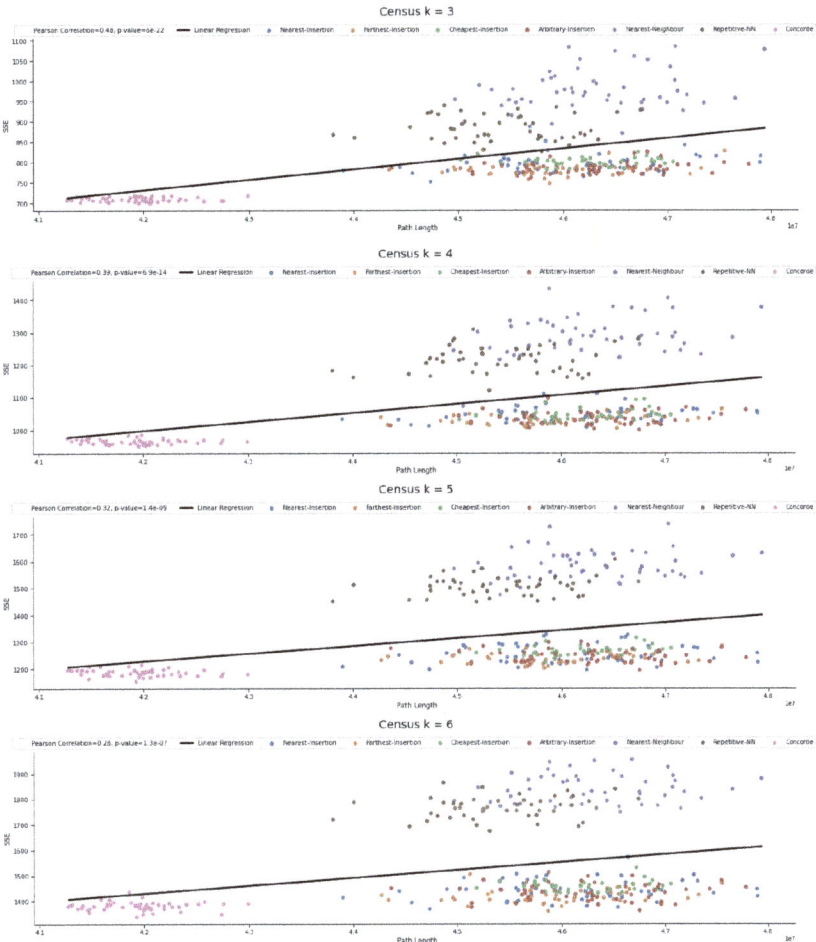

Figure A7. Relation between SSE and Path Length for Census and $k \in \{3, 4, 5, 6\}$.

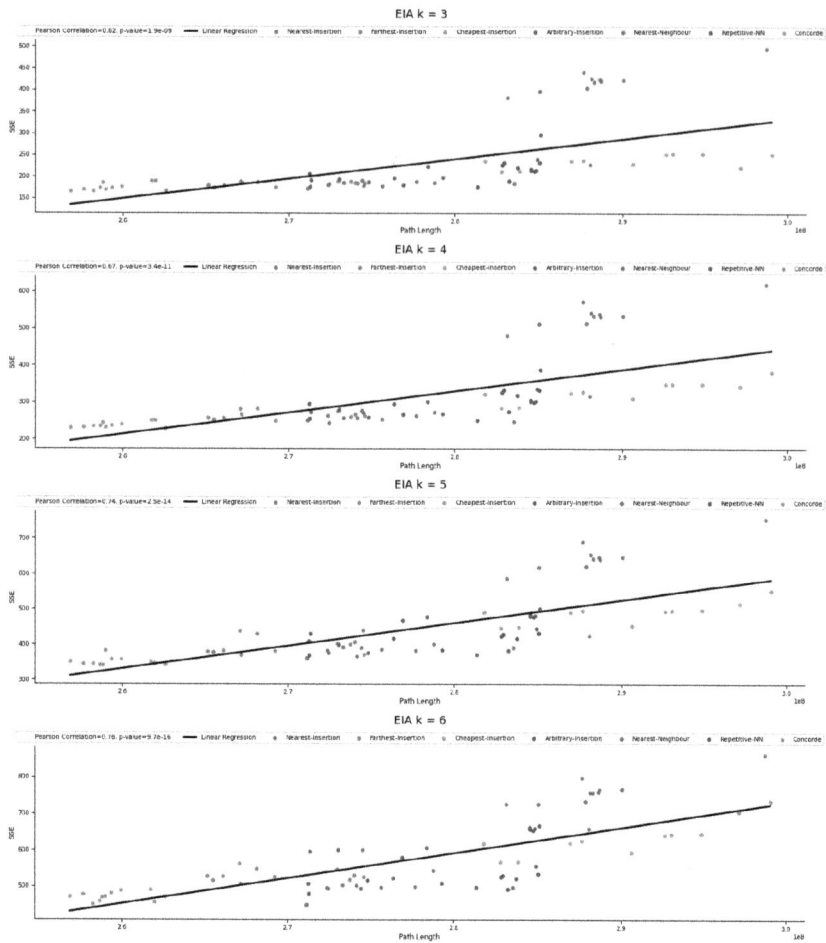

Figure A8. Relation between SSE and Path Length for EIA and $k \in \{3, 4, 5, 6\}$.

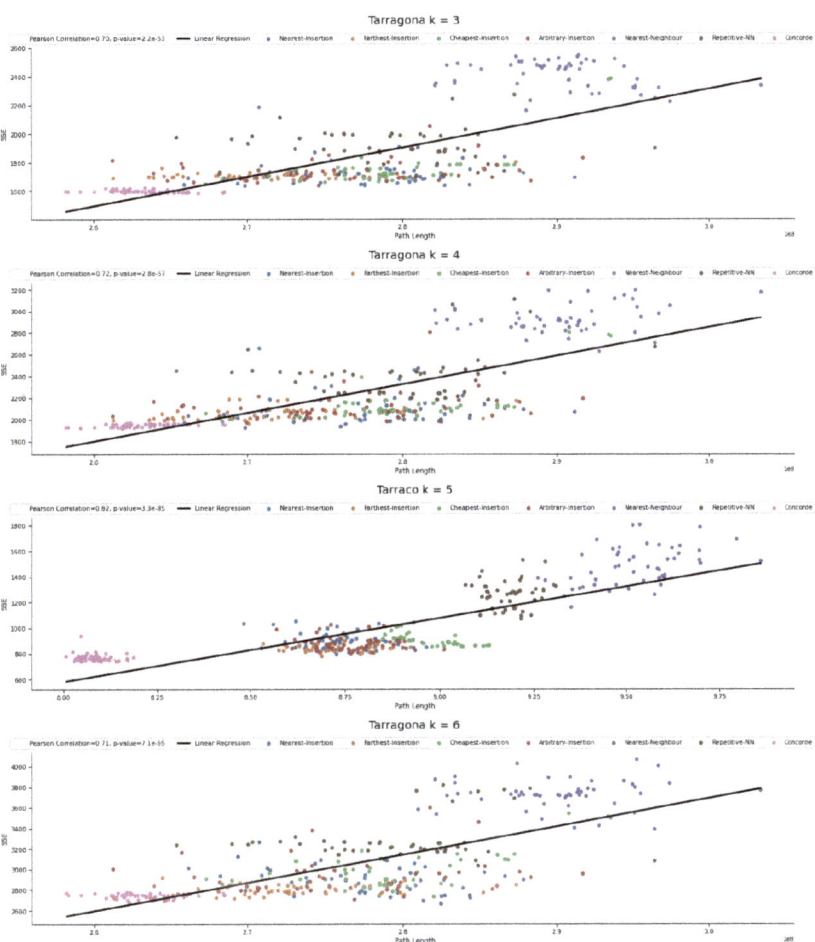

Figure A9. Relation between SSE and Path Length for Tarragona and $k \in \{3, 4, 5, 6\}$.

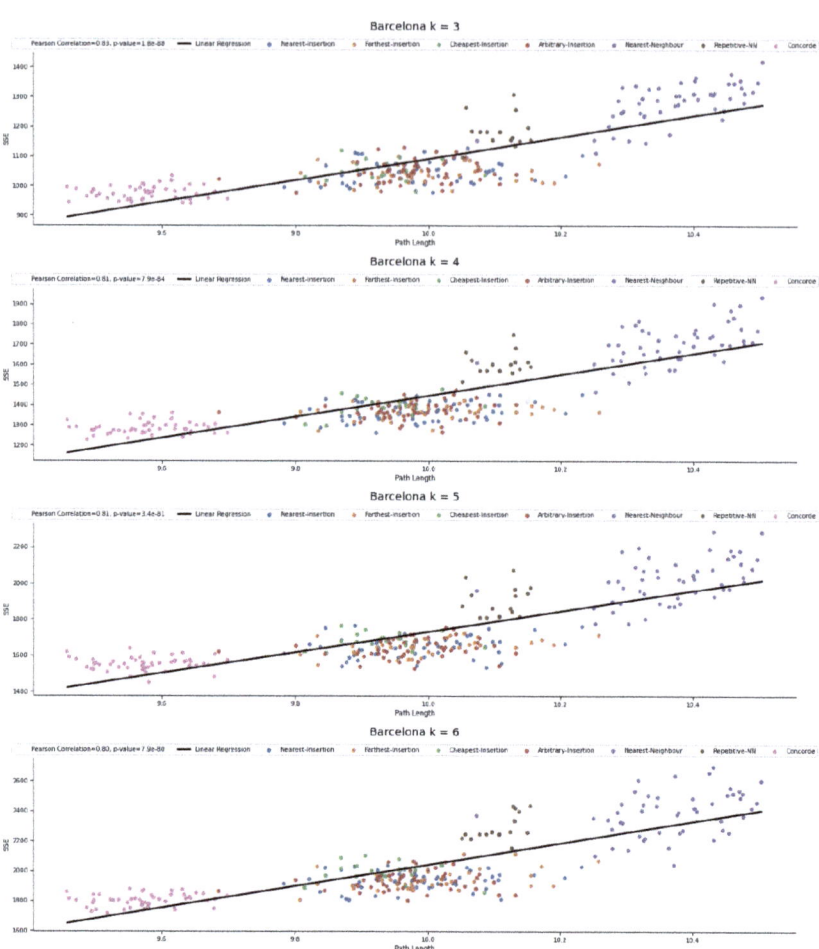

Figure A10. Relation between SSE and Path Length for Barcelona and $k \in \{3, 4, 5, 6\}$.

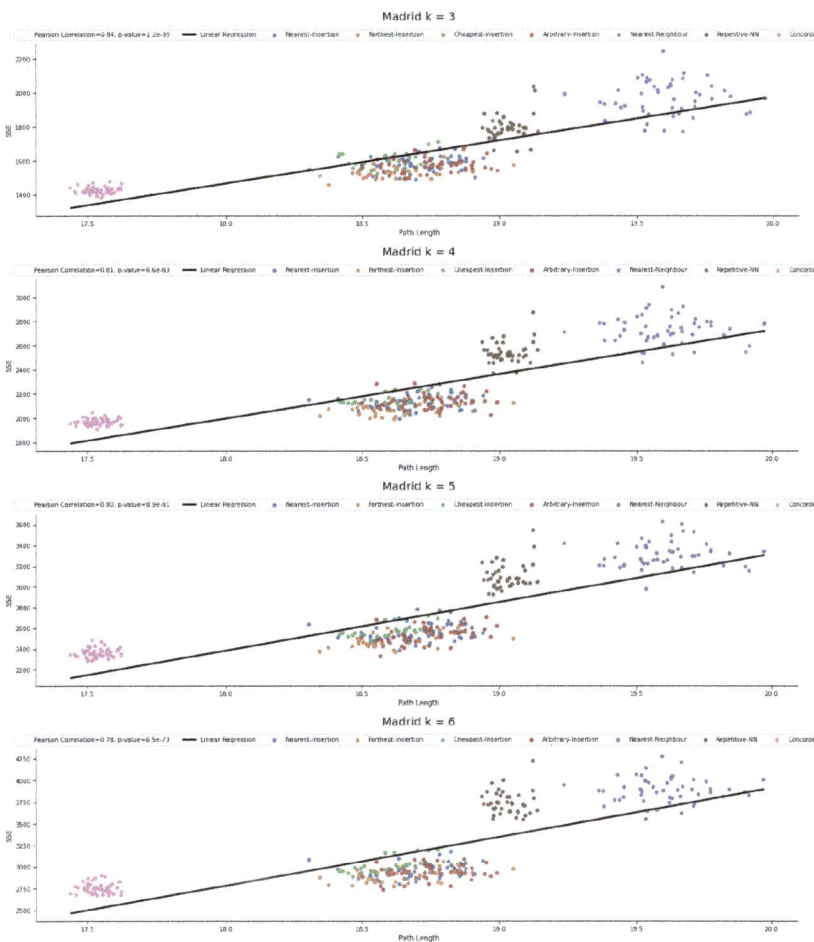

Figure A11. Relation between SSE and Path Length for Madrid and $k \in \{3, 4, 5, 6\}$.

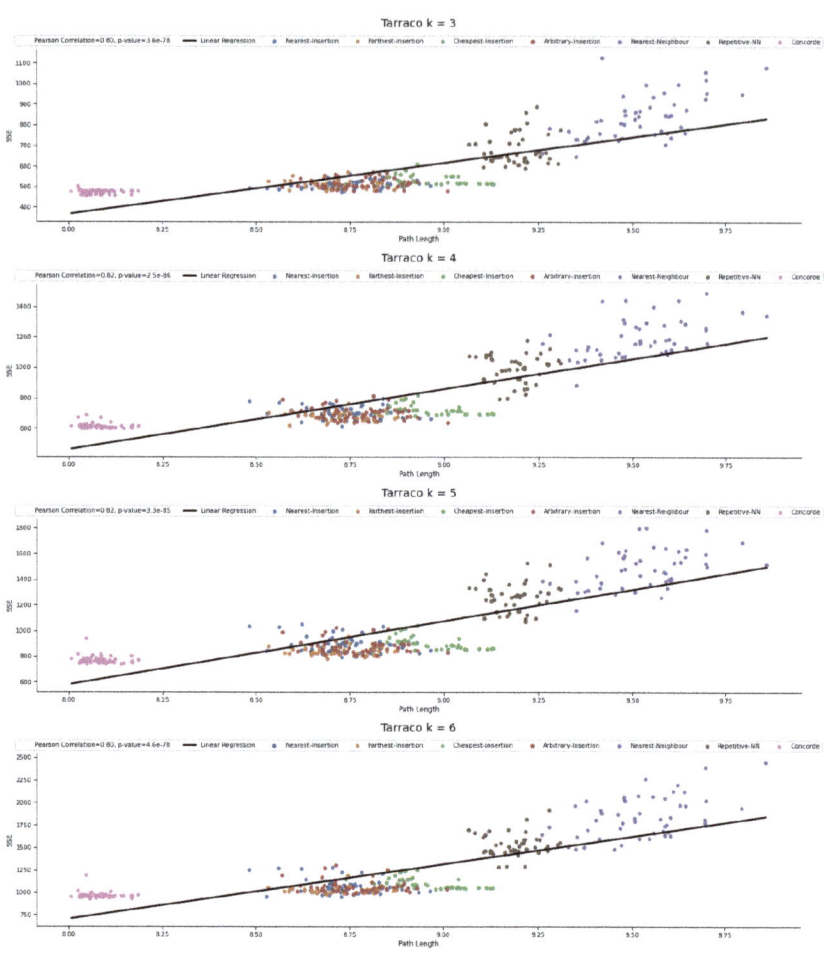

Figure A12. Relation between SSE and Path Length for Tarraco and $k \in \{3, 4, 5, 6\}$.

References

1. Ye, H.; Cheng, X.; Yuan, M.; Xu, L.; Gao, J.; Cheng, C. A survey of security and privacy in big data. In *Proccedings of the 2016 16th International Symposium on Communications and Information Technologies (ISCIT)*, Qingdao, China, 26–28 September 2016; pp. 268–272.
2. Vatsalan, D.; Sehili, Z.; Christen, P.; Rahm, E., Privacy-Preserving Record Linkage for Big Data: Current Approaches and Research Challenges. In *Handbook of Big Data Technologies*; Springer International Publishing: Cham, Switzerlands, 2017; pp. 851–895.
3. Mehmood, A.; Natgunanathan, I.; Xiang, Y.; Hua, G.; Guo, S. Protection of big data privacy. *IEEE Access* **2016**, *4*, 1821–1834. [CrossRef]
4. Barbaro, M.; Zeller, T.; Hansell, S. A face is exposed for AOL searcher no. 4417749. *New York Times*, 9 August 2006.
5. Zigomitros, A.; Casino, F.; Solanas, A.; Patsakis, C. A Survey on Privacy Properties for Data Publishing of Relational Data. *IEEE Access* **2020**, *8*, 51071–51099. [CrossRef]
6. Domingo-Ferrer, J.; Mateo-Sanz, J.M. Practical data-oriented microaggregation for statistical disclosure control. *Knowl. Data Eng. IEEE Trans.* **2002**, *14*, 189–201. [CrossRef]
7. Torra, V. Microaggregation for Categorical Variables: A Median Based Approach. In *Privacy in Statistical Databases*; Springer: Berlin/Heidelberg, Germany, 2004; pp. 162–174.
8. Solanas, A.; Martinez-Balleste, A.; Mateo-Sanz, J.M. Distributed architecture with double-phase microaggregation for the private sharing of biomedical data in mobile health. *IEEE Trans. Inf. Forensics Secur.* **2013**, *8*, 901–910. [CrossRef]
9. Martinez-Balleste, A.; Perez-Martines, P.A.; Solanas, A. The pursuit of citizens' privacy: A privacy-aware smart city is possible. *IEEE Commun. Mag.* **2013**, *51*, 136–141. [CrossRef]

10. Casino, F.; Domingo-Ferrer, J.; Patsakis, C.; Puig, D.; Solanas, A. A k-anonymous approach to privacy preserving collaborative filtering. *J. Comput. Syst. Sci.* **2015**, *81*, 1000–1011. [CrossRef]
11. Domingo-Ferrer, J.; Sebé, F.; Solanas, A. A polynomial-time approximation to optimal multivariate microaggregation. *Comput. Math. Appl.* **2008**, *55*, 714–732. [CrossRef]
12. Hansen, S.L.; Mukherjee, S. A polynomial algorithm for optimal univariate microaggregation. *IEEE Trans. Knowl. Data Eng.* **2003**, *15*, 1043–1044. [CrossRef]
13. Solanas, A.; Martinez, A. VMDAV: A Multivariate Microaggregation With Variable Group Size. In Proceedings of the 17th COMPSTAT Symposium of the IASC, Rome, Italy, 28 August–1 September 2006; pp. 917–925.
14. Shmoys, D.B.; Lenstra, J.; Kan, A.R.; Lawler, E.L. *The Traveling Salesman Problem*; John Wiley & Sons, Incorporated: Hoboken, NJ, USA, 1985; Volume 12.
15. Rosenkrantz, D.J.; Stearns, R.E.; Lewis, P.M., II. An analysis of several heuristics for the traveling salesman problem. *SIAM J. Comput.* **1977**, *6*, 563–581. [CrossRef]
16. Applegate, D.L.; Bixby, R.E.; Chvatal, V.; Cook, W.J. *The Traveling Salesman Problem: A Computational Study*; Princeton University Press: Princeton, NJ, USA, 2006.
17. Domingo-Ferrer, J.; Martinez-Ballesté, A.; Mateo-Sanz, J.; Sebé, F. Efficient multivariate data-oriented microaggregation. *VLDB J.* **2006**, *15*, 355–369. [CrossRef]
18. Laszlo, M.; Mukherjee, S. Minimum spanning tree partitioning algorithm for microaggregation. *IEEE Trans. Knowl. Data Eng.* **2005**, *17*, 902–911. [CrossRef]
19. Solé, M.; Muntés-Mulero, V.; Nin, J. Efficient microaggregation techniques for large numerical data volumes. *Int. J. Inf. Secur.* **2012**, *11*, 253–267. [CrossRef]
20. Chang, C.; Li, Y.; Huang,W. TFRP: An efficient microaggregation algorithm for statistical disclosure control. *J. Syst. Softw.* **2007**, *80*, 1866–1878. [CrossRef]
21. Yang, G.; Ye, X.; Fang, X.; Wu, R.; Wang, L. Associated attribute-aware differentially private data publishing via microaggregation. *IEEE Access* **2020**, *8*, 79158–79168. [CrossRef]
22. Lin, J.L.; Wen, T.H.; Hsieh, J.C.; Chang, P.C. Density-based microaggregation for statistical disclosure control. *Expert Syst. Appl.* **2010**, *37*, 3256–3263. [CrossRef]
23. Panagiotakis, C.; Tziritas, G. Successive group selection for microaggregation. *IEEE Trans. Knowl. Data Eng.* **2011**, *25*, 1191–1195. [CrossRef]
24. Mortazavi, R.; Jalili, S.; Gohargazi, H. Multivariate microaggregation by iterative optimization. *Appl. Intell.* **2013**, *39*, 529–544. [CrossRef]
25. Mortazavi, R.; Jalili, S. Fast data-oriented microaggregation algorithm for large numerical datasets. *Knowl. Based Syst.* **2014**, *67*, 195–205. [CrossRef]
26. Fayyoumi, E.; Oommen, B.J. A survey on statistical disclosure control and micro-aggregation techniques for secure statistical databases. *Softw. Pract. Exp.* **2010**, *40*, 1161–1188. [CrossRef]
27. Heaton, B.; Mukherjee, S. Record Ordering Heuristics for Disclosure Control through Microaggregation. In Proceedings of the International Conference on Advances in Communication and Information Technology, Amsterdam, The Netherlands, 1–2 December 2011.
28. Templ, M. Statistical disclosure control for microdata using the R-package sdcMicro. *Trans. Data Priv.* **2008**, *1*, 67–85.
29. OpenStreetMap Contributors. 2017. Available online: https://planet.osm.org (accessed on 1 May 2021).
30. Nilsson, C. *Heuristics for the Traveling Salesman Problem*; Technical Report; Linköping University: Linköping, Sweden, 2003. Available online: http://www.ida.liu.se/~TDDB19/reports_2003/htsp.pdf (accessed on 1 May 2021).
31. Morrison, D.R.; Jacobson, S.H.; Sauppe, J.J.; Sewell, E.C. Branch-and-bound algorithms: A survey of recent advances in searching, branching, and pruning. *Discret. Optim.* **2016**, *19*, 79–102. [CrossRef]
32. Hahsler, M.; Hornik, K. TSP-Infrastructure for the Traveling Salesperson Problem. *J. Stat. Softw.* **2007**, *23*, 1–21. [CrossRef]
33. Helsgaun, K. An effective implementation of the Lin–Kernighan traveling salesman heuristic. *Eur. J. Oper. Res.* **2000**, *126*, 106–130. [CrossRef]

Article

On the Outer-Independent Roman Domination in Graphs

Abel Cabrera Martínez [1,*], **Suitberto Cabrera García** [2], **Andrés Carrión García** [3] and **Angela María Grisales del Rio** [3]

1. Departament d'Enginyeria Informàtica i Matemàtiques, Universitat Rovira i Virgili, Av. Països Catalans 26, 43007 Tarragona, Spain
2. Departamento de Estadística e Investigación Operativa Aplicadas y Calidad, Universitat Politécnica de Valencia, Camino de Vera s/n, 46022 Valencia, Spain; suicabga@eio.upv.es
3. Centro de Gestión de la Calidad y del Cambio, Universitat Politécnica de Valencia, Camino de Vera s/n, 46022 Valencia, Spain; acarrion@eio.upv.es (A.C.G.); angride1@upv.es (A.M.G.d.R.)
* Correspondence: abel.cabrera@urv.cat

Received: 21 October 2020; Accepted: 6 November 2020; Published: 9 November 2020

Abstract: Let G be a graph with no isolated vertex and $f : V(G) \to \{0, 1, 2\}$ a function. Let $V_i = \{v \in V(G) : f(v) = i\}$ for every $i \in \{0, 1, 2\}$. The function f is an outer-independent Roman dominating function on G if V_0 is an independent set and every vertex in V_0 is adjacent to at least one vertex in V_2. The minimum weight $\omega(f) = \sum_{v \in V(G)} f(v)$ among all outer-independent Roman dominating functions f on G is the outer-independent Roman domination number of G. This paper is devoted to the study of the outer-independent Roman domination number of a graph, and it is a contribution to the special issue "Theoretical Computer Science and Discrete Mathematics" of *Symmetry*. In particular, we obtain new tight bounds for this parameter, and some of them improve some well-known results. We also provide closed formulas for the outer-independent Roman domination number of rooted product graphs.

Keywords: outer-independent Roman domination; Roman domination; vertex cover; rooted product graph

1. Introduction

Throughout this paper, we consider $G = (V(G), E(G))$ as a simple graph with no isolated vertex. Given a vertex v of G, $N(v)$ and $N[v]$ represent the open neighbourhood and the closed neighbourhood of v, respectively. We also denote by $\deg(v) = |N(v)|$ the degree of vertex v. For a set $D \subseteq V(G)$, its open neighbourhood and closed neighbourhood are $N(D) = \cup_{v \in D} N(v)$ and $N[D] = N(D) \cup D$, respectively. Moreover, the subgraph of G induced by $D \subseteq V(G)$ will be denoted by $G[D]$.

Domination theory is an interesting topic in the theory of graphs, as well as one of the most active topic of research in this area. A set $D \subseteq V(G)$ is a dominating set of G if $N[D] = V(G)$. The domination number of G, denoted by $\gamma(G)$, is the minimum cardinality amongst all dominating sets of G. Numerous results on this issue obtained in the previous century are shown in [1,2]. We define a $\gamma(G)$-set as a dominating set of cardinality $\gamma(G)$. The same terminology will be assumed for optimal parameters associated with other sets or functions defined in the paper.

Moreover, in the last two decades, the interest in the domination theory in graphs has increased. In that sense, a very high number of variants of domination parameters have been studied, many of which are combinations of two or more parameters. Next, we expose some of them.

- A set $S \subseteq V(G)$ is an independent set of G if the subgraph induced by S is edgeless. The maximum cardinality among all independent sets of G is the independence number of G, and is denoted by $\beta(G)$. In some kind of "opposed" side of an independent set, we find a vertex cover, which is a set $D \subseteq V(G)$ such that $V(G) \setminus D$ is an independent set of G. The vertex cover number of G, denoted by $\alpha(G)$, is the minimum cardinality among all vertex covers of G. It is well-known that for any graph G of order n, $\alpha(G) + \beta(G) = n$ (see [3]).
- A set $S \subseteq V(G)$ is an independent dominating set of G if S is an independent and dominating set at the same time. The independent domination number of G is the minimum cardinality among all independent dominating sets of G and is denoted by $i(G)$. Independent domination in graphs was formally introduced in [4,5]. However, a fairly complete survey on this topic was recently published in [6].
- A function $f : V(G) \to \{0,1,2\}$ is called a Roman dominating function on G, if every $v \in V(G)$ for which $f(v) = 0$ is adjacent to at least one vertex $u \in V(G)$ for which $f(u) = 2$. The Roman domination number of G, denoted by $\gamma_R(G)$, is the minimum weight $w(f) = \sum_{v \in V(G)} f(v)$ among all Roman dominating functions f on G. This parameter was introduced in [7]. Let $V_i = \{v \in V(G) : f(v) = i\}$ for $i \in \{0,1,2\}$. We will identify a Roman dominating function f with the subsets V_0, V_1, V_2 of $V(G)$ associated with it, and so we will use the unified notation $f(V_0, V_1, V_2)$ for the function and these associated subsets.
- A Roman dominating function $f(V_0, V_1, V_2)$ is called an outer-independent Roman dominating function, abbreviated OIRDF, if V_0 is an independent set of G. Notice that then $V_1 \cup V_2$ is a vertex cover of G. The outer-independent Roman domination number of G is the minimum weight among all outer-independent Roman dominating functions on G, and is denoted by $\gamma_{oiR}(G)$. This parameter was introduced in [8] and also studied in [9–11].

All the previous parameters are, in one way or another, related to each other. Next, we show the most natural relationships that exist between them, which are easily deductible by definition.

Remark 1. *For any graph G of order n with no isolated vertex,*

(i) $\gamma(G) \leq i(G) \leq \beta(G) = n - \alpha(G)$.
(ii) $\gamma(G) \leq \gamma_R(G) \leq \gamma_{oiR}(G)$.

For the graphs shown in Figure 1 we have the following.

- $\gamma(G_1) = 2 < i(G_1) < 4 = \alpha(G_1) = \gamma_R(G_1) < \beta(G_1) < \gamma_{oiR}(G_1) = 6$.
- $\gamma(G_2) = i(G_2) = \alpha(G_2) = 2 < \gamma_R(G_2) = \gamma_{oiR}(G_2) = 3 < \beta(G_2) = 5$.

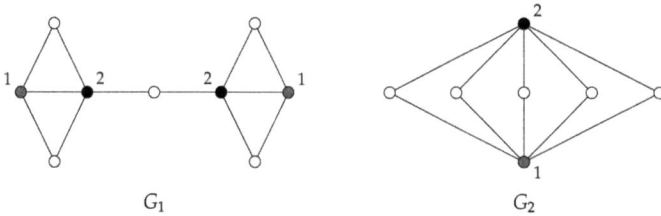

Figure 1. The labels of (gray and black) coloured vertices describe the positive weights of a $\gamma_{oiR}(G_i)$-function, for $i \in \{1,2\}$.

In this paper, we continue the study of the outer-independent Roman domination number of graphs. For instance, in Section 2 we give some new relationships between this parameter and the others mentioned above. Several of these results improve other bounds previously given. Finally, in Section 3 we provide closed formulas for this parameter in rooted product graphs. In particular, we show that there are four possible expressions for the outer-independent Roman domination number of a rooted product graph, and we characterize the graphs reaching these expressions.

2. Bounds and Relationships with Other Parameters

Abdollahzadeh Ahangar et al. [8] in 2017, established the following result.

Theorem 1 ([8]). *For any graph G with no isolated vertex,*

$$\alpha(G) + 1 \leq \gamma_{oiR}(G) \leq 2\alpha(G).$$

Observe that any graph G with no isolated vertex, order n and maximum degree Δ, satisfies that $1 \leq \left\lceil \frac{n-\alpha(G)}{\Delta} \right\rceil$. It is also well-know that $\gamma(G) \leq \alpha(G)$, which implies $\alpha(G) + \gamma(G) \leq 2\alpha(G)$. With the above inequalities in mind, we state the following theorem, which improves the bounds given in Theorem 1.

Theorem 2. *For any graph G with no isolated vertex, order n and maximum degree Δ,*

$$\alpha(G) + \left\lceil \frac{n - \alpha(G)}{\Delta} \right\rceil \leq \gamma_{oiR}(G) \leq \alpha(G) + \gamma(G).$$

Proof. We first prove the upper bound. Let D be a $\gamma(G)$-set and S an $\alpha(G)$-set. Let $g(W_0, W_1, W_2)$ be a function defined by $W_0 = V(G) \setminus (D \cup S)$, $W_1 = (D \cup S) \setminus (D \cap S)$ and $W_2 = D \cap S$. We claim that g is an OIRDF on G. Without loss of generality, we may assume that $W_0 \neq \emptyset$. Notice that $W_0 = V(G) \setminus (D \cup S)$ is an independent set of G as S is a vertex cover. Now, we prove that every vertex in W_0 has a neighbour in W_2. Let $x \in W_0 = V(G) \setminus (D \cup S)$. Since S is a vertex cover and D is a dominating set, we deduce that $N(x) \subseteq S$ and $N(x) \cap D \neq \emptyset$, respectively. Hence $N(x) \cap D \cap S \neq \emptyset$, or equivalently, $N(x) \cap W_2 \neq \emptyset$. Thus, g is an OIRDF on G, as required. Therefore, $\gamma_{oiR}(G) \leq \omega(g) = |(D \cup S) \setminus (D \cap S)| + 2|D \cap S| = \alpha(G) + \gamma(G)$.

We now proceed to prove the lower bound. Let $f(V_0, V_1, V_2)$ be a $\gamma_{oiR}(G)$-function. By definition, we have that V_0 is an independent set, and so, $V_1 \cup V_2$ is a vertex cover. Moreover, we note that every vertex in V_2 has at most Δ neighbours in V_0. Hence, $|V_0| \leq \Delta |V_2|$. By inequality above, and the fact that $n - \alpha(G) = \beta(G) \geq |V_0|$, we have

$$
\begin{aligned}
\Delta \gamma_{oiR}(G) &= \Delta(|V_1| + 2|V_2|) \\
&= \Delta(|V_1| + |V_2|) + \Delta|V_2| \\
&\geq \Delta(n - |V_0|) + |V_0| \\
&= n\Delta - (\Delta - 1)|V_0| \\
&\geq n\Delta - (\Delta - 1)(n - \alpha(G)) \\
&= \Delta \alpha(G) + (n - \alpha(G)).
\end{aligned}
$$

Therefore, $\gamma_{oiR}(G) \geq \alpha(G) + \left\lceil \frac{n - \alpha(G)}{\Delta} \right\rceil$, which completes the proof. □

The bounds above are tight. To see this, let us consider the vertex cover Roman graphs G. These graphs were defined in [8] and satisfy the equality $\gamma_{oiR}(G) = 2\alpha(G)$. Since $\gamma(G) \leq \alpha(G)$, we deduce that for every vertex cover Roman graph G it follows that $\gamma_{oiR}(G) = \alpha(G) + \gamma(G)$. Note also that both bounds are achieved for the graph G_1 given in Figure 1, i.e., $\alpha(G_1) + \left\lceil \frac{|V(G_1)| - \alpha(G_1)}{\Delta(G_1)} \right\rceil = \gamma_{oiR}(G_1) = \alpha(G_1) + \gamma(G_1)$.

The following result is an immediate consequence of Theorem 2.

Corollary 1. *If G is a graph such that $\gamma(G) = 1$, then*

$$\gamma_{oiR}(G) = \alpha(G) + 1.$$

However, the graphs G with $\gamma(G) = 1$ are not the only ones that satisfy the equality $\gamma_{oiR}(G) = \alpha(G) + 1$. For instance, the path P_4 satisfies that $\gamma(P_4) = 2$ and $\gamma_{oiR}(P_4) = 3 = \alpha(P_4) + 1$. In such a sense, we next give a theoretical characterization of the graphs that satisfy this equality above.

Theorem 3. *If G is a graph with no isolated vertex, then the following statements are equivalent.*

(i) $\gamma_{oiR}(G) = \alpha(G) + 1$.
(ii) *There exist an $\alpha(G)$-set S and a vertex $v \in S$ such that $V(G) \setminus S \subseteq N(v)$.*

Proof. We first suppose that (i) holds, i.e., $\gamma_{oiR}(G) = \alpha(G) + 1$. Let $f(V_0, V_1, V_2)$ be a $\gamma_{oiR}(G)$-function such that $|V_2|$ is maximum. Hence, $V_2 \neq \emptyset$. Let $v \in V_2$. Since $V_1 \cup V_2$ is a vertex cover of G, it follows that $\alpha(G) + 1 \le (|V_1| + |V_2|) + |V_2| = \gamma_{oiR}(G) = \alpha(G) + 1$. Hence, we have equalities in the previous inequality chain, which implies that $S = V_1 \cup V_2$ is an $\alpha(G)$-set and $V_2 = \{v\}$. So, $V(G) \setminus S = V_0 \subseteq N(V_2) = N(v)$. Therefore, (ii) follows.

On the other hand, suppose that (ii) holds, i.e., suppose there exist an $\alpha(G)$-set S and $v \in S$ such that $V(G) \setminus S \subseteq N(v)$. Observe that the function $g(W_0, W_1, W_2)$, defined by $W_2 = \{v\}$, $W_1 = S \setminus \{v\}$ and $W_0 = V(G) \setminus S$, is an OIRDF on G. Therefore, and using the lower bound given in the Theorem 1, we obtain that $\alpha(G) + 1 \le \gamma_{oiR}(G) \le w(g) = |S| + 1 = \alpha(G) + 1$. Hence, $\gamma_{oiR}(G) = \alpha(G) + 1$, which completes the proof. □

A tree T is an acyclic connected graph. A leaf vertex of T is a vertex of degree one. The set of leaves is denoted by $L(T)$. We say that a vertex $v \in V(T)$ is a support vertex (strong support vertex) if $|N(v) \cap L(T)| \ge 1$ ($|N(v) \cap L(T)| \ge 2$). The set of support vertices and strong support vertices are denoted by $S(T)$ and $S_s(T)$, respectively.

With this notation in mind, we next characterize the trees T with $\gamma_{oiR}(T) = \alpha(T) + 1$. Before we do this, we shall need to state the following useful lemma, in which $diam(T)$ represents the diameter of T.

Lemma 1. *If T is a tree such that $\gamma_{oiR}(T) = \alpha(T) + 1$, then the following statements hold.*

(i) $diam(T) \le 4$.
(ii) $V(T) = L(T) \cup S(T)$.

Proof. We first proceed to prove (i). By Theorem 3 there exist an $\alpha(T)$-set S and $v \in S$ such that $V(T) \setminus S \subseteq N(v)$. Now, we suppose that $k = diam(T) \ge 5$. Let $P = v_0 v_1 \cdots v_{k-1} v_k$ be a diametrical path of T. Hence, $\emptyset \neq \{v_0, v_1, v_{k-1}, v_k\} \cap (V(T) \setminus S) \not\subseteq N(v)$, which is a contradiction. Therefore, $diam(T) \le 4$, as desired.

Finally, we proceed to prove (ii). By (i) we have that $diam(T) \le 4$. If $V(T) \setminus (L(T) \cup S(T)) \neq \emptyset$, then for every $\alpha(T)$-set S and $v \in S$ it follows that $V(T) \setminus S \not\subseteq N(v)$, which is a contradiction with Theorem 3. Hence, $V(T) = L(T) \cup S(T)$, which completes the proof. □

Let $\mathcal{T}$ be the family of trees $T_{r,s}$ of order $r + s + 1$ with $r \ge 1$ and $r - 1 \ge s \ge 0$, obtained from a star $K_{1,r}$ by subdividing s edges exactly once. In Figure 2 we show the tree $T_{5,3}$.

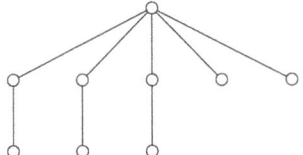

Figure 2. The tree $T_{5,3}$.

Theorem 4. *Let T be a nontrivial tree. Then $\gamma_{oiR}(T) = \alpha(T) + 1$ if and only if $T \in \mathcal{T}$.*

Proof. If $T \in \mathcal{T}$, then it is easy to check that $\gamma_{oiR}(T) = \alpha(T) + 1$. Now, we prove the converse. Let T be a nontrivial tree such that $\gamma_{oiR}(T) = \alpha(T) + 1$. By Lemma 1-(i) we have that $diam(T) \leq 4$. If $diam(T) \leq 2$, then $T \cong T_{r,0} \in \mathcal{T}$. If $diam(T) = 3$, then $T \cong T_{r,1} \in \mathcal{T}$. We now suppose that $diam(T) = 4$. By Lemma 1-(ii) we have that $V(T) = L(T) \cup S(T)$. We claim that for any diametrical path $P = v_0 v_1 v_2 v_3 v_4$ of T, it follows that $v_1, v_3 \in S(T) \setminus S_s(T)$. First, we observe that $v_1, v_3 \in S(T)$. Without loss of generality, suppose that $v_1 \in S_s(T)$. Hence, v_1 belongs to every $\alpha(T)$-set. By Theorem 3 there exist an $\alpha(T)$-set S and $v \in S$ such that $V(T) \setminus S \subseteq N(v)$. Since $v_0 \in V(T) \setminus S$, then $v = v_1$. Notice also that $\emptyset \neq \{v_3, v_4\} \cap (V(T) \setminus S) \not\subseteq N(v_1)$, which is a contradiction. Therefore, $v_1, v_3 \in S(T) \setminus S_s(T)$, as desired. From above, we deduce that $T \cong T_{r,s} \in \mathcal{T}$, where $r \geq 3$ and $r - 1 \geq s \geq 2$. Therefore, the proof is complete. □

The following result is another consequence of Theorem 2.

Theorem 5. *Let G be a graph with no isolated vertex. For any $\gamma_R(G)$-function $f(V_0, V_1, V_2)$,*

$$\gamma_{oiR}(G) \leq \gamma_R(G) + \alpha(G) - |V_2|.$$

Proof. Let $f(V_0, V_1, V_2)$ be a $\gamma_R(G)$-function. Since $V_1 \cup V_2$ is a dominating set of G, it follows that $\gamma(G) \leq |V_1| + |V_2| = \gamma_R(G) - |V_2|$. Therefore, Theorem 2 leads to $\gamma_{oiR}(G) \leq \alpha(G) + \gamma(G) \leq \gamma_R(G) + \alpha(G) - |V_2|$, which completes the proof. □

The bound above is tight. For instance, in the corona graph $G \odot N_r$ with $r \geq 3$, the unique $\gamma_R(G \odot N_r)$-function $f(V_0, V_1, V_2)$, defined by $V_2 = V(G)$ and $V_1 = \emptyset$, is also a $\gamma_{oiR}(G \odot N_r)$-function, and so, $\gamma_R(G \odot N_r) = \gamma_{oiR}(G \odot N_r) = \gamma_R(G \odot N_r) + \alpha(G \odot N_r) - |V_2| = 2|V(G)|$. The following result, which is a consequence of Remark 1 and Theorem 5, generalizes the previous example.

Proposition 1. *If there exists a $\gamma_R(G)$-function $f(V_0, V_1, V_2)$ such that $|V_2| = \alpha(G)$, then*

$$\gamma_{oiR}(G) = \gamma_R(G).$$

We now relate the outer-independent Roman domination number with other domination parameters of graphs. Before, we shall state the following proposition.

Proposition 2. *For any graph G with no isolated vertex, there exists a $\gamma_{oiR}(G)$-function $f(V_0, V_1, V_2)$ such that V_0 is an independent dominating set of G.*

Proof. Let $f(V_0, V_1, V_2)$ be a $\gamma_{oiR}(G)$-function such that $|V_2|$ is maximum. By definition we have that V_0 is an independent set. We next prove that V_0 is a dominating set of G. It is clear that $V_2 \subseteq N(V_0)$. Let $v \in V_1$. If $N(v) \subseteq V_1 \cup V_2$, then the function $f'(V_0', V_1', V_2')$, defined by $f'(v) = 0$, $f'(u) = f(u) + 1$ for some vertex $u \in N(v) \cap V_1$ and $f'(x) = f(x)$ whenever $x \in V(G) \setminus \{v, u\}$, is a $\gamma_{oiR}(G)$-function and

$|V_2'| > |V_2|$, which is a contradiction. Hence, $N(v) \cap V_0 \neq \emptyset$, which implies that V_0 is an independent dominating set of G, as desired. □

Theorem 6. *For any graph G with no isolated vertex, order n, minimum degree δ and maximum degree Δ,*

$$\left\lceil \frac{i(G)\delta}{\Delta} \right\rceil + 1 \leq \gamma_{oiR}(G) \leq n - i(G) + \gamma(G).$$

Proof. The upper bound follows by Theorem 2 and the fact that $\alpha(G) = n - \beta(G) \leq n - i(G)$. Now, we proceed to prove the lower bound. Let $f(V_0, V_1, V_2)$ be a $\gamma_{oiR}(G)$-function which satisfies Proposition 2. Since every vertex in $V_1 \cup V_2$ has at most Δ neighbours in V_0 and V_0 is an independent dominating set, it follows that $\delta|V_0| \leq \Delta(|V_1| + |V_2|)$ and $|V_0| \geq i(G)$. Hence,

$$\begin{aligned}\gamma_{oiR}(G) &= (|V_1| + |V_2|) + |V_2| \\ &\geq \frac{|V_0|\delta}{\Delta} + |V_2| \\ &\geq \frac{i(G)\delta}{\Delta} + 1.\end{aligned}$$

Therefore, the proof is complete. □

The bounds above are tight. For example, the lower bound is achieved for the complete bipartite graphs $K_{r,r}$, where $\gamma_{oiR}(K_{r,r}) = r + 1 = \left\lceil \frac{r^2}{r} \right\rceil + 1 = \left\lceil \frac{i(K_{r,r})\delta(K_{r,r})}{\Delta(K_{r,r})} \right\rceil + 1$. In addition, the upper bound is achieved for the case of complete graphs, and in connection with this fact, we pose the following question.

Open question: Is it the case that $\gamma_{oiR}(G) = n - i(G) + \gamma(G)$ if and only if G is a complete graph?

Next, we give new bounds for the outer-independent Roman domination number of triangle-free graphs. Recall that in these graphs, no pair of adjacent vertices can have a common neighbor. For this purpose, we shall need to introduce the following definitions.

A set $S \subseteq V(G)$ is a 3-packing if the distance between u and v is greater than three for every pair of different vertices $u, v \in S$. The 3-packing number of G, denoted by $\rho_3(G)$, is the maximum cardinality among all 3-packings of G. We also define

$$\mathcal{P}_3(G) = \{S \subseteq V(G) : S \text{ is a 3-packing of } G\}.$$

Theorem 7. *For any triangle-free graph G of order n,*

$$\gamma_{oiR}(G) \leq n - \max_{S \in \mathcal{P}_3(G)} \left\{ \sum_{v \in S} (\deg(v) - 1) \right\}.$$

Proof. Let $S \in \mathcal{P}_3(G)$. As G is triangle-free, it follows that $N(v)$ is an independent set of G for every $v \in V(G)$. Hence, $N(S)$ is an independent set of G, which implies that the function $f(V_0, V_1, V_2)$, defined by $V_2 = S$, $V_0 = N(S)$ and $V_1 = V(G) \setminus N[S]$, is an OIRDF on G. Thus, $\gamma_{oiR}(G) \leq 2|V_2| + |V_1| = 2|S| + (n - |N[S]|) = n - \sum_{v \in S}(\deg(v) - 1)$. Since the inequality holds for every $S \in \mathcal{P}_3(G)$, the result follows. □

Corollary 2. *For any triangle-free graph G of order n and minimum degree δ,*

$$\gamma_{oiR}(G) \leq n - \rho_3(G)(\delta - 1).$$

In [8], the bound $\gamma_{oiR}(G) \leq n - \Delta(G) + 1$ was given for the case of triangle-free graph. Next, we state a result which improve the bound above for the triangle-free graphs G that satisfy the condition $diam(G)(\delta(G) - 1) \geq 4(\Delta(G) - 1)$.

Proposition 3. *Let G be a connected triangle-free graph of order n, minimum degree δ and maximum degree Δ. If $diam(G) \geq 4$, then*

$$\gamma_{oiR}(G) \leq n - \left\lceil \frac{diam(G)}{4} \right\rceil (\delta - 1).$$

Proof. Assume that $diam(G) \geq 4$. Let $P = v_0 v_1 \cdots v_k$ be a diametrical path of G (notice that $k = diam(G)$), and $S = \{v_0, v_4, \ldots, v_{4\lfloor k/4 \rfloor}\}$. It is easy to see that $S \in \mathcal{P}_3(G)$, and so, by Theorem 7 we deduce that $\gamma_{oiR}(G) \leq n - \sum_{v \in S}(\deg(v) - 1) \leq n - \left\lceil \frac{diam(G)}{4} \right\rceil (\delta - 1)$ which completes the proof. □

The bounds given in Corollary 2 and Proposition 3 are tight. For instance, they are achieved for the cycle C_{10}.

3. Rooted Product Graphs

Let G be a graph of order n with vertex set $\{u_1, \ldots, u_n\}$ and H a graph with root $v \in V(H)$. The rooted product graph $G \circ_v H$ is defined as the graph obtained from G and n copies of H, by identifying the vertex u_i of G with the root v in the ith-copy of H, where $i \in \{1, \ldots, n\}$ [12]. If H or G is a trivial graph, then $G \circ_v H$ is equal to G or H, respectively. In this sense, to obtain the rooted product $G \circ_v H$, hereafter we will only consider graphs G and H of orders at least two. Figure 3 shows an example of a rooted product graph.

For every $x \in V(G)$, H_x will denote the copy of H in $G \circ_v H$ containing x. The restriction of any $\gamma_{oiR}(G \circ_v H)$-function f to $V(H_x)$ will be denoted by f_x and the restriction to $V(H_x) \setminus \{x\}$ will be denoted by f_x^-.

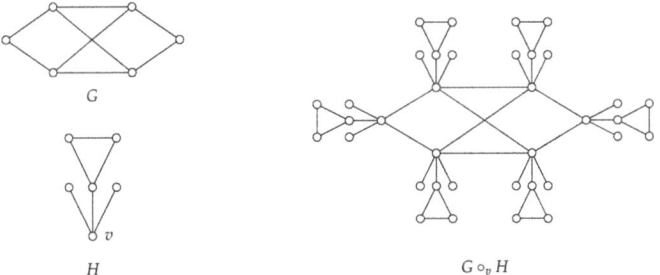

Figure 3. The rooted product graph $G \circ_v H$.

If v is a vertex of a graph H, then the subgraph $H - v$ is the subgraph of H induced by $V(H) \setminus \{v\}$. The following three results will be the main tools to deduce our results.

Lemma 2. *Let H be a graph without isolated vertices. For any $v \in V(H)$,*

$$\gamma_{oiR}(H - v) \geq \gamma_{oiR}(H) - 1.$$

Proof. Let g' be a $\gamma_{oiR}(H - v)$-function. Notice that the function g, defined by $g(v) = 1$ and $g(u) = g'(u)$ whenever $u \in V(H) \setminus \{v\}$, is an OIRDF on H. Hence, $\gamma_{oiR}(H) - 1 \leq w(g) - 1 = w(g') = \gamma_{oiR}(H - v)$, which completes the proof. □

Lemma 3. Let G and H be two graphs without isolated vertices. If G has order n and $v \in V(H)$, then the following statements hold.

(i) If $g(v) = 0$ for some $\gamma_{oiR}(H)$-function g, then $\gamma_{oiR}(G \circ_v H) \leq \alpha(G) + n\gamma_{oiR}(H)$.
(ii) If $g(v) > 0$ for some $\gamma_{oiR}(H)$-function g, then $\gamma_{oiR}(G \circ_v H) \leq n\gamma_{oiR}(H)$.
(iii) If there exists a $\gamma_{oiR}(H - v)$-function g such that $g(x) > 0$ for every $x \in N(v)$, then $\gamma_{oiR}(G \circ_v H) \leq \gamma_{oiR}(G) + n\gamma_{oiR}(H - v)$.

Proof. From any $\gamma_{oiR}(H)$-function g such that $g(v) = 0$ and any $\alpha(G)$-set, we can construct an OIRDF on $G \circ_v H$ of weight $\alpha(G) + n\gamma_{oiR}(H)$. Thus, $\gamma_{oiR}(G \circ_v H) \leq \alpha(G) + n\gamma_{oiR}(H)$ and (i) follows.

Now, if there exists a $\gamma_{oiR}(H)$-function g such that $g(v) > 0$, then from g we can construct an OIRDF on $G \circ_v H$ of weight $nw(g)$. Thus, $\gamma_{oiR}(G \circ_v H) \leq nw(g) = n\gamma_{oiR}(H)$, and (ii) follows.

Finally, if there exists a $\gamma_{oiR}(H - v)$-function g such that $g(x) > 0$ for every $x \in N(v)$, then from g and any $\gamma_{oiR}(G)$-function we can construct an OIRDF on $G \circ_v H$ of weight $\gamma_{oiR}(G) + n\gamma_{oiR}(H - v)$, which completes the proof. □

Lemma 4. Let $f(V_0, V_1, V_2)$ be a $\gamma_{oiR}(G \circ_v H)$-function. The following statements hold for any vertex $x \in V(G)$.

(i) $w(f_x) \geq \gamma_{oiR}(H) - 1$.
(ii) If $w(f_x) = \gamma_{oiR}(H) - 1$, then $x \in V_0$ and $N(x) \cap V(H_x) \subseteq V_1$.

Proof. Let $x \in V(G)$. Observe that $V_0 \cap V(H_x)$ is an independent set of H_x and also, every vertex in $V_0 \cap (V(H_x) \setminus \{x\})$ has a neighbour in $V_2 \cap V(H_x)$. So, it is easy to see that the function g, defined by $g(x) = \max\{1, f(x)\}$ and $g(u) = f(u)$ whenever $u \in V(H_x) \setminus \{x\}$, is an OIRDF on H_x. Hence, $\gamma_{oiR}(H) - 1 = \gamma_{oiR}(H_x) - 1 \leq w(g) - 1 \leq w(f_x)$, which completes the proof of (i).

Now, we suppose that $w(f_x) = \gamma_{oiR}(H) - 1$. If $x \in V_1 \cup V_2$ or $x \in V_0$ and $N(x) \cap V(H_x) \cap V_2 \neq \emptyset$, then f_x is an OIRDF on H_x, which is a contradiction. Hence, $x \in V_0$ and as $V_0 \cap V(H_x)$ is an independent set, we deduce that $N(x) \cap V(H_x) \subseteq V_1$, which completes the proof. □

From Lemma 4 (i) we deduce that any $\gamma_{oiR}(G \circ_v H)$-function f induces three subsets $\mathcal{A}_f, \mathcal{B}_f$ and $\mathcal{C}_f$ of $V(G)$ as follows.

$$\mathcal{A}_f = \{x \in V(G) : w(f_x) > \gamma_{oiR}(H)\},$$
$$\mathcal{B}_f = \{x \in V(G) : w(f_x) = \gamma_{oiR}(H)\},$$
$$\mathcal{C}_f = \{x \in V(G) : w(f_x) = \gamma_{oiR}(H) - 1\}.$$

Next, we state the four possible values of $\gamma_{oiR}(G \circ_v H)$.

Theorem 8. Let G and H be two graphs with no isolated vertex and $|V(G)| = n$. If $v \in V(H)$, then

$$\gamma_{oiR}(G \circ_v H) \in \{\alpha(G) + n\gamma_{oiR}(H), n\gamma_{oiR}(H), \gamma_{oiR}(G) + n(\gamma_{oiR}(H) - 1), \alpha(G) + n(\gamma_{oiR}(H) - 1)\}.$$

Proof. Let $f(V_0, V_1, V_2)$ be a $\gamma_{oiR}(G \circ_v H)$-function. By Lemma 3 (i) and (ii) we deduce the upper bound $\gamma_{oiR}(G \circ_v H) \leq \alpha(G) + n\gamma_{oiR}(H)$. Now, we consider the subsets $\mathcal{A}_f, \mathcal{B}_f, \mathcal{C}_f \subseteq V(G)$ associated to f and distinguish the following cases.

Case 1. $\mathcal{C}_f = \emptyset$. In this case, for any $x \in V(G)$ we have that $w(f_x) \geq \gamma_{oiR}(H)$ and, as a consequence, $\gamma_{oiR}(G \circ_v H) = w(f) \geq n\gamma_{oiR}(H)$. If $\mathcal{A}_f = \emptyset$, then $\gamma_{oiR}(G \circ_v H) = n\gamma_{oiR}(H)$. Hence, assume that $\mathcal{A}_f \neq \emptyset$. This implies that $w(f) > n\gamma_{oiR}(H)$. Moreover, we note that $\mathcal{B}_f \neq \emptyset$ because $\alpha(G) < n$ and

$w(f) \leq \alpha(G) + n\gamma_{oiR}(H)$. Thus, by Lemma 3 (ii) we obtain that $\mathcal{B}_f \subseteq V_0$, and as V_0 is an independent set, we have that $\mathcal{A}_f$ is a vertex cover of G. Therefore,

$$\gamma_{oiR}(G \circ_v H) = \sum_{x \in \mathcal{A}_f} w(f_x) + \sum_{x \in \mathcal{B}_f} w(f_x)$$
$$\geq \sum_{x \in \mathcal{A}_f} (\gamma_{oiR}(H) + 1) + \sum_{x \in \mathcal{B}_f} \gamma_{oiR}(H)$$
$$= |\mathcal{A}_f| + \sum_{x \in V(G)} \gamma_{oiR}(H)$$
$$\geq \alpha(G) + n\gamma_{oiR}(H).$$

Hence, $\gamma_{oiR}(G \circ_v H) = \alpha(G) + n\gamma_{oiR}(H)$.

Case 2. $\mathcal{C}_f \neq \emptyset$. Let $z \in \mathcal{C}_f$. By Lemma 4 (ii) we obtain that $z \in V_0$ and $N(z) \cap V(H_z) \subseteq V_1$. Hence, f_z^- is an OIRDF on $H_z - z$, and so $\gamma_{oiR}(H - v) = \gamma_{oiR}(H_z - z) \leq w(f_z^-) = \gamma_{oiR}(H) - 1$. Thus, Lemma 2 leads to $\gamma_{oiR}(H_z - z) = \gamma_{oiR}(H) - 1$. This implies that f_z^- is a $\gamma_{oiR}(H_z - z)$-function which satisfies Lemma 3 (iii). Therefore, $\gamma_{oiR}(G \circ_v H) \leq \gamma_{oiR}(G) + n(\gamma_{oiR}(H) - 1)$.

Now, observe the following inequality chain.

$$\gamma_{oiR}(G \circ_v H) = \sum_{x \in \mathcal{A}_f \cup \mathcal{B}_f} w(f_x) + \sum_{x \in \mathcal{C}_f} w(f_x) \geq (2|\mathcal{A}_f| + |\mathcal{B}_f|) + n(\gamma_{oiR}(H) - 1). \quad (1)$$

By Lemma 4 (ii) we have that $\mathcal{C}_f \subseteq V_0$, which implies that $\mathcal{A}_f \cup \mathcal{B}_f$ is a vertex cover of G. Thus, Inequality chain (1) leads to $\gamma_{oiR}(G \circ_v H) = w(f) \geq \alpha(G) + n(\gamma_{oiR}(H) - 1)$. Next, we consider the following two subcases.

Subcase 1. There exists a $\gamma_{oiR}(H)$-function g such that $g(v) = 2$. Let D be an $\alpha(G)$-set. From D, g and f_z, we define a function h on $G \circ_v H$ as follows. For every $x \in D$, the restriction of h to $V(H_x)$ is induced from g. Moreover, if $x \in V(G) \setminus D$, then the restriction of h to $V(H_x)$ is induced from f_z. By the construction of g and f_z, it is straightforward to see that h is an OIRDF on $G \circ_v H$. Thus,

$$\gamma_{oiR}(G \circ_v H) \leq \sum_{x \in D} w(h_x) + \sum_{x \in V(G) \setminus D} w(h_x)$$
$$= \sum_{x \in D} w(g) + \sum_{x \in V(G) \setminus D} w(f_z)$$
$$= \sum_{x \in D} \gamma_{oiR}(H) + \sum_{x \in V(G) \setminus D} (\gamma_{oiR}(H) - 1)$$
$$= |D| + \sum_{x \in V(G)} (\gamma_{oiR}(H) - 1)$$
$$= \alpha(G) + n(\gamma_{oiR}(H) - 1).$$

Therefore, $\gamma_{oiR}(G \circ_v H) = \alpha(G) + n(\gamma_{oiR}(H) - 1)$.

Subcase 2. $g(v) \leq 1$ for every $\gamma_{oiR}(H)$-function g. This condition implies that $V_2 \cap \mathcal{B}_f = \emptyset$. Since every vertex $x \in \mathcal{C}_f$ has a neighbour in V_2, and as Lemma 4 (ii) leads to $N(x) \cap V(H_x) \subseteq V_1$, then we deduce that $N(x) \cap V_2 \cap \mathcal{A}_f \neq \emptyset$. Hence, and as $\mathcal{C}_f \subseteq V_0$, the function $f'(V_0', V_1', V_2')$, defined by $V_2' = \mathcal{A}_f$, $V_1' = \mathcal{B}_f$ and $V_0' = \mathcal{C}_f$, is an OIRDF on G. So $\gamma_{oiR}(G) \leq w(f') = 2|\mathcal{A}_f| + |\mathcal{B}_f|$. Therefore, Inequality chain (1) leads to $\gamma_{oiR}(G \circ_v H) \geq \gamma_{oiR}(G) + n(\gamma_{oiR}(H) - 1)$, which implies that $\gamma_{oiR}(G \circ_v H) = \gamma_{oiR}(G) + n(\gamma_{oiR}(H) - 1)$.

Therefore, the proof is complete. □

In order to see that the four possible values of $\gamma_{oiR}(G \circ_v H)$ described in Theorem 8 are realizable, we consider the following example.

Example 1. *Let G be a graph with no isolated vertex. If H is the graph shown in Figure 4, then the resulting values of $\gamma_{oiR}(G \circ_x H)$ for some specific roots $x \in V(H)$ are described below.*

- $\gamma_{oiR}(G \circ_v H) = \alpha(G) + n\gamma_{oiR}(H)$.
- $\gamma_{oiR}(G \circ_w H) = n\gamma_{oiR}(H)$.
- $\gamma_{oiR}(G \circ_{v'} H) = \gamma_{oiR}(G) + n(\gamma_{oiR}(H) - 1)$.
- $\gamma_{oiR}(G \circ_{w'} H) = \alpha(G) + n(\gamma_{oiR}(H) - 1)$.

Now, we characterize the graphs with $\gamma_{oiR}(G \circ_v H) = \alpha(G) + n\gamma_{oiR}(H)$.

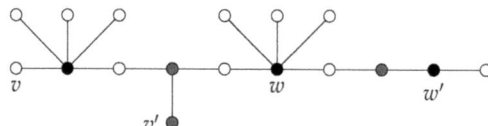

Figure 4. The labels of (gray and black) coloured vertices describe the positive weights of a $\gamma_{oiR}(H)$-function.

Theorem 9. *Let G and H be two graphs with no isolated vertex, let $|V(G)| = n$ and $v \in V(H)$. The following statements are equivalent.*

(i) $\gamma_{oiR}(G \circ_v H) = \alpha(G) + n\gamma_{oiR}(H)$.
(ii) $g(v) = 0$ for every $\gamma_{oiR}(H)$-function g.

Proof. We first assume that (i) holds, i.e., $\gamma_{oiR}(G \circ_v H) = \alpha(G) + n\gamma_{oiR}(H)$. If there exists a $\gamma_{oiR}(H)$-function g such that $g(v) > 0$, then by Lemma 3 (ii) it follows that $\gamma_{oiR}(G \circ_v H) \le n\gamma_{oiR}(H)$, which is a contradiction. Therefore, (ii) holds.

On the other hand, we assume that (ii) holds, i.e., $g(v) = 0$ for every $\gamma_{oiR}(H)$-function g. Let $f(V_0, V_1, V_2)$ be a $\gamma_{oiR}(G \circ_v H)$-function. If $C_f \ne \emptyset$, then by Lemma 4 (ii) we can obtain a $\gamma_{oiR}(H)$-function g such that $g(v) = 1$, which is a contradiction. Hence, $C_f = \emptyset$, and so, by Theorem 8 we deduce that $\gamma_{oiR}(G \circ_v H) \in \{\alpha(G) + n\gamma_{oiR}(H), n\gamma_{oiR}(H)\}$. Now, suppose that $\gamma_{oiR}(G \circ_v H) = n\gamma_{oiR}(H)$. Since $C_f = \emptyset$, it follows that $B_f = V(G)$ and as V_0 is an independent set, there exists $x \in B_f \setminus V_0$. This implies that f_x is a $\gamma_{oiR}(H_x)$-function such that $f_x(x) > 0$, which is a contradiction. Therefore, $\gamma_{oiR}(G \circ_v H) = \alpha(G) + n\gamma_{oiR}(H)$, which completes the proof. □

Next, we characterize the graphs with $\gamma_{oiR}(G \circ_v H) = \alpha(G) + n(\gamma_{oiR}(H) - 1)$.

Theorem 10. *Let G and H be two graphs with no isolated vertex, let $|V(G)| = n$ and $v \in V(H)$. The following statements are equivalent.*

(i) $\gamma_{oiR}(G \circ_v H) = \alpha(G) + n(\gamma_{oiR}(H) - 1)$.
(ii) There exist two $\gamma_{oiR}(H)$-functions g_1 and g_2 such that $g_1(x) = 1$ for every $x \in N[v]$ and $g_2(v) = 2$.

Proof. We first assume that (i) holds, i.e., $\gamma_{oiR}(G \circ_v H) = \alpha(G) + n(\gamma_{oiR}(H) - 1)$. Let $f(V_0, V_1, V_2)$ be a $\gamma_{oiR}(G \circ_v H)$-function. As $\alpha(G) < n$, it follows that $C_f \ne \emptyset$, and so, by Lemma 4 (ii) we can obtain a $\gamma_{oiR}(H)$-function g_1 such that $g_1(x) = 1$ for every $x \in N[v]$. Moreover, if $g(v) \le 1$ for every $\gamma_{oiR}(H)$-function g, then, by proceeding analogously to Subcase 2 in the proof of Theorem 8 we deduce that $\gamma_{oiR}(G \circ_v H) \geq \gamma_{oiR}(G) + n(\gamma_{oiR}(H) - 1)$, which is a contradiction as $\gamma_{oiR}(G) > \alpha(G)$. Therefore, there exists a $\gamma_{oiR}(H)$-function g_2 such that $g_2(v) = 2$, and (ii) follows.

On the other hand, we assume that there exist two $\gamma_{oiR}(H)$-functions g_1 and g_2 such that $g_1(x) = 1$ for every $x \in N[v]$ and $g_2(v) = 2$. Let D be an $\alpha(G)$-set and let g'_1 be a function on H such that $g'_1(v) = 0$ and $g'_1(x) = g_1(x)$ whenever $x \in V(H) \setminus \{v\}$. From D, g'_1 and g_2, we define a function h on $G \circ_v H$ as follows. For every $x \in D$, the restriction of h to $V(H_x)$ is induced from g_2. Moreover, if $x \in V(G) \setminus D$, then the restriction of h to $V(H_x)$ is induced from g'_1. Notice that h is an OIRDF on $G \circ_v H$, and so $\gamma_{oiR}(G \circ_v H) \leq \omega(h) = |D|\gamma_{oiR}(H) + |V(G) \setminus D|(\gamma_{oiR}(H) - 1) = \alpha(G) + n(\gamma_{oiR}(H) - 1)$. Therefore, Theorem 8 leads to $\gamma_{oiR}(G \circ_v H) = \alpha(G) + n(\gamma_{oiR}(H) - 1)$, which completes the proof. □

Next we proceed to characterize the graphs with $\gamma_{oiR}(G \circ_v H) = \gamma_{oiR}(G) + n(\gamma_{oiR}(H) - 1)$. Notice that it is excluded the case $\gamma_{oiR}(G) = n$, since then $\gamma_{oiR}(G \circ_v H) = n\gamma_{oiR}(H)$.

Theorem 11. *Let G be a graph of order n with no isolated vertex such that $\gamma_{oiR}(G) < n$ and let H be a graph with no isolated vertex and $v \in V(H)$. The following statements are equivalent.*

(i) $\gamma_{oiR}(G \circ_v H) = \gamma_{oiR}(G) + n(\gamma_{oiR}(H) - 1)$.
(ii) $g(v) \leq 1$ for every $\gamma_{oiR}(H)$-function g and also, there exists a $\gamma_{oiR}(H)$-function g_1 such that $g_1(x) = 1$ for every $x \in N[v]$.

Proof. We first assume that (i) holds, i.e., $\gamma_{oiR}(G \circ_v H) = \gamma_{oiR}(G) + n(\gamma_{oiR}(H) - 1)$. Let $f(V_0, V_1, V_2)$ be a $\gamma_{oiR}(G \circ_v H)$-function. Since $\gamma_{oiR}(G) < n$, it follows that $C_f \neq \emptyset$, and so, by Lemma 4 (ii) we can obtain a $\gamma_{oiR}(H)$-function g_1 such that $g_1(x) = 1$ for every $x \in N[v]$. Moreover, if there exists a $\gamma_{oiR}(H)$-function g_2 such that $g_2(v) = 2$, then by Theorem 10 we deduce that $\gamma_{oiR}(G \circ_v H) = \alpha(G) + n(\gamma_{oiR}(H) - 1)$, which is a contradiction as $\gamma_{oiR}(G) > \alpha(G)$. Therefore, $g(v) \leq 1$ for every $\gamma_{oiR}(H)$-function g, which implies that (ii) follows.

On the other side, we assume that $g(v) \leq 1$ for every $\gamma_{oiR}(H)$-function g and also, that there exists a $\gamma_{oiR}(H)$-function g_1 such that $g_1(x) = 1$ for every $x \in N[v]$. Under these assumptions, observe that the function g_1 restricted to $V(H) \setminus \{v\}$, namely g'_1, is an OIRDF on $H - v$. Hence, $\gamma_{oiR}(H - v) \leq \omega(g'_1) = \omega(g_1) - 1 = \gamma_{oiR}(H) - 1$ and by Lemma 2 we deduce that $\gamma_{oiR}(H - v) = \gamma_{oiR}(H) - 1$. Hence, g'_1 is a $\gamma_{oiR}(H - v)$-function which satisfies Lemma 3 (iii). Therefore, Lemma 3 and Theorem 8 lead to $\gamma_{oiR}(G \circ_v H) \in \{\gamma_{oiR}(G) + n(\gamma_{oiR}(H) - 1), \alpha(G) + n(\gamma_{oiR}(H) - 1)\}$. Finally, as $g(v) \leq 1$ for every $\gamma_{oiR}(H)$-function g, by Theorem 10 we deduce that $\gamma_{oiR}(G \circ_v H) = \gamma_{oiR}(G) + n(\gamma_{oiR}(H) - 1)$, which completes the proof. □

From Theorem 8 we have that there are four possible expressions for $\gamma_{oiR}(G \circ_v H)$. Theorems 9–11 characterize three of these expressions. In the case of the expression $\gamma_{oiR}(G \circ_v H) = n\gamma_{oiR}(H)$, the corresponding characterization can be derived by elimination from the previous results.

Author Contributions: All authors contributed equally to this work. Investigation, A.C.M., S.C.G., A.C.G. and A.M.G.d.R; writing—review and editing, A.C.M., S.C.G., A.C.G. and A.M.G.d.R. All authors have read and agreed to the published version of the manuscript.

Funding: This research received no external funding.

Conflicts of Interest: The authors declare no conflict of interest.

References

1. Haynes, T.W.; Hedetniemi, S.T.; Slater, P.J. *Fundamentals of Domination in Graphs*; Chapman and Hall/CRC Pure and Applied Mathematics Series; Marcel Dekker, Inc.: New York, NY, USA, 1998.
2. Haynes, T.W.; Hedetniemi, S.T.; Slater, P.J. *Domination in Graphs: Advanced Topics*; Chapman and Hall/CRC Pure and Applied Mathematics Series; Marcel Dekker, Inc.: New York, NY, USA, 1998.
3. Gallai, T. Über extreme Punkt-und Kantenmengen. *Ann. Univ. Sci. Budapestinensis Rolando Eötvös Nomin. Sect. Math.* **1959**, *2*, 133–138.
4. Berge, C. *Theory of Graphs and Its Applications*; CRC Press: Methuen, MA, USA; London, UK, 1962.
5. Ore, O. *Theory of Graphs*; American Mathematical Society: Providence, RI, USA, 1962; pp. 206–212.
6. Goddard, W.; Henning, M.A. Independent domination in graphs: A survey and recent results. *Discret. Math.* **2013**, *313*, 839–854. [CrossRef]
7. Cockayne, E.J.; Dreyer, P.A., Jr.; Hedetniemi, S.M.; Hedetniemi, S.T. Roman domination in graphs. *Discret. Math.* **2004**, *278*, 11–22. [CrossRef]
8. Abdollahzadeh Ahangar, H.; Chellali, M.; Samodivkin, V. Outer independent Roman dominating functions in graphs. *Int. J. Comput. Math.* **2017**, *94*, 2547–2557. [CrossRef]
9. Cabrera Martínez, A.; Kuziak, D.; Yero, I.G. A constructive characterization of vertex cover Roman trees. *Discuss. Math. Graph Theory* **2018**, *41*, 267–283. [CrossRef]
10. Dehgardi, N.; Chellali, M. Outer-independent Roman domination number of tree. *IEEE Access* **2018**, *6*, 35544–35550.
11. Sheikholeslami, S.M.; Nazari-Moghaddam, S. On trees with equal Roman domination and outer-independent Roman domination numbers. *Commun. Comb. Optim.* **2019**, *4*, 185–199.
12. Godsil, C.D.; McKay, B.D. A new graph product and its spectrum. *Bull. Aust. Math. Soc.* **1978**, *18*, 21–28. [CrossRef]

Publisher's Note: MDPI stays neutral with regard to jurisdictional claims in published maps and institutional affiliations.

© 2020 by the authors. Licensee MDPI, Basel, Switzerland. This article is an open access article distributed under the terms and conditions of the Creative Commons Attribution (CC BY) license (http://creativecommons.org/licenses/by/4.0/).

Article

The n-Pythagorean Fuzzy Sets

Anna Bryniarska

Department of Computer Science, Opole University of Technology, Proszkowska 76, 45-758 Opole, Poland; a.bryniarska@po.edu.pl

Received: 16 September 2020; Accepted: 23 October 2020; Published: 26 October 2020

Abstract: The following paper presents deductive theories of n-Pythagorean fuzzy sets (n-PFS). N-PFS objects are a generalization of the intuitionistic fuzzy sets (IFSs) and the Yager Pythagorean fuzzy sets (PFSs). Until now, the values of membership and non-membership functions have been described on a one-to-one scale and a quadratic function scale. There is a symmetry between the values of this membership and non-membership functions. The scales of any power functions are used here in order to increase the scope of the decision-making problems. The theory of n-PFS introduces a conceptual apparatus analogous to the classic theory of Zadeh fuzzy sets, consistently striving to correctly define the n-PFS algebra.

Keywords: fuzzy set; n-Pythagorean; n-PFS algebra; triangular norms

1. Introduction

Zadeh [1] introduced the fuzzy set idea, which generalizes the theory of classical sets. In fuzzy sets, there is a membership function μ, which assigns a number from the set $[0,1]$ to each element of the universe. This determines how much this element belongs to this universe, where 0 means no belonging and 1 means full belonging to the set that is under consideration. Other values between 0 and 1 mean the degree of belonging to this set. This membership function is defined to describe the degree of belonging of an element to some class. The membership function in the fuzzy sets replace the characteristic function that is used in crisp sets. Since the work of Zadeh, the fuzzy set theory has been used in different disciplines such as management sciences, engineering, mathematics, social sciences, statistics, signal processing, artificial intelligence, automata theory, and medical and life sciences.

Atanassov studied the intuitionistic fuzzy sets (IFSs) [2,3]. IFSs have values for two functions: the membership function μ and the non-membership function v. Additionally, there is a constraint $0 \leq \mu + v \leq 1$. This is a symmetric relationship between the values and the membership function. In order to create model for imprecise information, the model of Pythagorean fuzzy sets (PFSs) was proposed by Yager [4,5]. This model is different than the IFSs model because it uses the condition $0 \leq \mu^2 + v^2 \leq 1$. Moreover, there is also the Pythagorean fuzzy number (PFN) idea established by Zhang and Xu [6]. In decision-making problems, there are also applications of PFSs proposed by Garg [7,8].

The decision-making problems in the model of Pythagorean fuzzy sets will significantly increase the application range of solving these problems than in the model of intuitionistic fuzzy sets. It is because more pairs (μ, v) satisfy the condition $0 \leq \mu^2 + v^2 \leq 1$ than the condition $0 \leq \mu + v \leq 1$. Is there any other data scale in decision-making problems that will help to extend its applicability even further? Yes, with the condition $0 \leq \mu^n + v^n \leq 1$, for any natural number $n > 2$.

In articles using local deduction regarding IFSs [3,9–12] and concerning PFSs [4,5,13–17], it was noted that there is a series of mathematical and logical inaccuracies. That is why the conceptual apparatus should be generalized and refined by formulating the deductive theory of n-Pythagorean fuzzy sets with the condition $0 \leq \mu^n + v^n \leq 1$, for any natural number $n \rangle 0$. The theory is presented below.

2. Triangular Norms

Definition 1. *The operation* $\bullet_t : [0,1] \times [0,1] \to [0,1]$ *is called a t-norm in the set* $[0,1]$, *or a triangular norm, when it meets the following conditions (for any numbers* $x, y, z \in [0,1]$*):*

1. boundary conditions
$$0 \bullet_t y = 0, y \bullet_t 1 = y, \tag{1}$$

2. monotonicity
$$x \bullet_t y \leq z \bullet_t y, \text{ when } x \leq z, \tag{2}$$

3. commutativity
$$x \bullet_t y = y \bullet_t x, \tag{3}$$

4. associativity
$$x \bullet_t (y \bullet_t z) = (x \bullet_t y) \bullet_t z, \tag{4}$$

Since there is $\sup\{t \in [0,1] : x \bullet_t t \leq y\}$, there can be specified the operation $\to_t : [0,1] \times [0,1] \to [0,1]$, such that for any numbers $x, y \in [0,1]$:

$$x \to_t y = \sup\{t \in [0,1] : x \bullet_t t \leq y\}. \tag{5}$$

This operation is called *t-residuum* in the set $[0,1]$.
The operation $\bullet_s : [0,1] \times [0,1] \to [0,1]$, described by formula for any numbers $x, y \in [0,1]$:

$$x \bullet_s y = 1 - (1-x) \bullet_t (1-y), \tag{6}$$

is called a *s-norm* or *triangular conorm*.

Using the definitions of t-norm and s-norm, after simple calculations we get (where the names of conditions are given analogously to the definition of t-norm):

Theorem 1. *For any numbers* $x, y, z \in [0,1]$:

- boundary conditions
$$0 \bullet_s y = y, y \bullet_s 1 = 1, \tag{7}$$

- monotonicity
$$x \bullet_s y \leq z \bullet_s y, \text{ when } x \leq z, \tag{8}$$

- commutativity
$$x \bullet_s y = y \bullet_s x, \tag{9}$$

- associativity
$$x \bullet_s (y \bullet_s z) = (x \bullet_s y) \bullet_s z, \tag{10}$$

Further, only continuous t-norms and s-norms are considered. The general discussion on the construction of triangular norms, using the results of functional equations, leads to the theorem from paper [18]:

Theorem 2.

1. There is a continuous and strictly decreasing function for each continuous t-norm $f_t : [0,1] \to [0,+\infty)$ such that $f_t(1) = 0, f_t(0) = 1$ and for any $x, y \in [0,1]$:

$$x \bullet_t y = \begin{cases} f_t^{-1}[f_t(x) + f_t(y)], & f_t(x) + f_t(y) \in [0,1] \\ 0, & \text{otherwise} \end{cases} \tag{11}$$

2. There is a continuous and strictly increasing function for each continuous s-norm $f_s : [0,1] \to [0,+\infty)$ such that $f_s(0) = 0$, $f_s(1) = 1$ and for any $x, y \in [0,1]$:

$$x \bullet_s y = \begin{cases} f_s^{-1}[f_s(x) + f_s(y)], & f_s(x) + f_s(y) \in [0,1] \\ 1, & \text{otherwise.} \end{cases} \quad (12)$$

3. For any $x \in [0,1]$

$$f_s(x) = f_t(1-x). \quad (13)$$

Functions f_t, f_s are called *generators* of t-norm and s-norm, respectively.

Example 1. For the t-norm $x \bullet_t y = \min\{x,y\}$ and any $x, y \in [0,1]$, the generator is $f_t(x) = 1 - x$.
For the s-norm $x \bullet_s y = \max\{x,y\}$ and any $x, y \in [0,1]$, the generator is $f_s(x) = x$.

Example 2. For the t-norm $x \bullet_t y = 1 - \min\{1, ((1-x)^p + (1-y)^p)^{1/p}\}$, $p \geq 1$ and any $x, y \in [0,1]$, the generator is $f_t(x) = 1 - x^p$.
For the s-norm $x \bullet_s y = \min\{1, (x^p + y^p)^{1/p}\}$, $p \geq 1$ and any $x, y \in [0,1]$, the generator is $f_s(x) = x^p$.

Theorem 3. Let f_t, f_s be generators of the triangular norms $\bullet_t, \bullet_s$. Then there exist operations $\bullet_p : [0,1] \times [0,1] \to [0,1]$, $\bullet_l : [0,1] \times [0,1] \to [0,1]$ defined by formulas:

$$\lambda \bullet_p x = \begin{cases} f_t^{-1}[\lambda f_t(x)], & \text{for } \lambda f_t(x) \in [0,1] \\ 0, & \text{otherwise.} \end{cases} \quad (14)$$

$$\lambda \bullet_l x = \begin{cases} f_s^{-1}[\lambda f_s(x)], & \text{for } \lambda f_s(x) \in [0,1] \\ 1, & \text{otherwise.} \end{cases} \quad (15)$$

Proof of Theorem 3. Because f_t is a strictly decreasing function, there is only one value $f_t^{-1}[\lambda f_t(x)] \in [0,1]$, when $\lambda f_t(x) \in [0, f_s(0)]$.
It is noted that $\lambda \bullet_p x =_{df} f_t^{-1}[\lambda f_t(x)]$, for $\lambda f_t(x) \in [0, f_t(0)]$, and $\lambda \bullet_p x =_{df} 0$ otherwise.
Similarly, we define the operation $\bullet_l$. □

Definition 2. Operations $\bullet_p, \bullet_l$ specified in Theorem 3 are called *p-norm* (with properties similar to the power functions) and the *l-norm* (with properties similar to the linear function), respectively, and the system $S_{\text{Yager}} = \langle [0,1], \bullet_t, \bullet_s, \bullet_p, \bullet_l, 0, 1 \rangle$ is called the Yager system of the triangular norms.

The following notation agreement is accepted:

$$\lambda \bullet_p x =_{df} x^\lambda, \quad (16)$$

$$\lambda \bullet_l x =_{df} \lambda x. \quad (17)$$

Fact 1. If $x^\lambda = f_t^{-1}[\lambda f_t(x)]$, for $\lambda f_t(x) \in [0, f_t(0)]$, then $f_t(x^\lambda) = \lambda f_t(x)$.
If $\lambda x = f_s^{-1}[\lambda f_s(x)]$, for $\lambda f_s(x) \in [0, f_s(1)]$, then $f_s(\lambda x) = \lambda f_s(x)$.

Theorem 4. In the system $S_{\text{Yager}} = \langle [0,1], \bullet_t, \bullet_s, \bullet_p, \bullet_l, 0, 1 \rangle$ operations $\bullet_p, \bullet_l$ satisfy the following conditions (for any $x, y, \lambda, \lambda_1, \lambda_2 \in [0,1]$):

$$\lambda(x \bullet_s y) = \lambda x \bullet_s \lambda y, \quad (18)$$

$$(x \bullet_t y)^\lambda = x^\lambda \bullet_t y^\lambda, \quad (19)$$

$$(\lambda_1 + \lambda_2)x = \lambda_1 x \bullet_s \lambda_2 x, \quad (20)$$

$$x^{\lambda_1+\lambda_2} = x^{\lambda_1} \bullet_t x^{\lambda_2}. \tag{21}$$

Proof of Theorem 4.

1. For Equation (18): $\lambda(x \bullet_s y) = f_s^{-1}[\lambda f_s(x \bullet_s y)] = f_s^{-1}[\lambda f_s(x) + \lambda f_s(y)] = f_s^{-1}[f_s(\lambda x) + f_s(\lambda y)] = \lambda x \bullet_s \lambda y$, for $f_s(x \bullet_s y) \in [0,1]$, and since $\lambda f_s(x \bullet_s y) \leq f_s(x \bullet_s y)$, so $\lambda f_s(x \bullet_s y) \in [0,1]$.
2. For Equation (19): $(x \bullet_t y)^\lambda = f_t^{-1}[\lambda f_t(x \bullet_t y)] = f_t^{-1}[\lambda f_t(x) + \lambda f_t(y)] = f_t^{-1}[f_t(x^\lambda) + f_t(y^\lambda)] = x^\lambda \bullet_t y^\lambda$, for $f_t(x \bullet_t y) \in [0,1]$, and since $\lambda f_t(x \bullet_t y) \leq f_t(x \bullet_t y)$, so $\lambda f_t(x \bullet_t y) \in [0,1]$.
3. For Equation (20): $(\lambda_1 + \lambda_2)x = f_s^{-1}[(\lambda_1 + \lambda_2)f_s(x)] = f_s^{-1}[\lambda_1 f_s(x) + \lambda_2 f_s(x)] = f_s^{-1}[f_s(\lambda_1 x) + f_s(\lambda_2 x)] = \lambda_1 x \bullet_s \lambda_2 x$, for $(\lambda_1 + \lambda_2)f_s(x) = f_s(\lambda_1 x) + f_s(\lambda_2 x) \in [0,1]$.
4. For Equation (21): $x^{\lambda_1+\lambda_2} = f_t^{-1}[(\lambda_1 + \lambda_2)f_t(x)] = f_t^{-1}[\lambda_1 f_t(x) + \lambda_2 f_t(x)] = f_t^{-1}[f_t(x^{\lambda_1}) + f_t(x^{\lambda_2})] = x^{\lambda_1} \bullet_t x^{\lambda_2}$, for $(\lambda_1 + \lambda_2)f_t(x) = f_t(x^{\lambda_1}) + f_t(x^{\lambda_2}) \in [0,1]$.

□

3. n-Pythagorean Fuzzy Set and Yager Aggregation Operators

Definition 3. *Let F be a set of all fuzzy sets for the nonempty space X. Any function $p : X \to [0,1] \times [0,1]$ defined for any $\mu_p, v_p \in F$ is:*

$$p = \{\langle x, \langle \mu_p(x), v_p(x) \rangle \rangle : x \in X\}. \tag{22}$$

*It is called the **n-Pythagorean fuzzy set** (n-PFS) if the following condition is satisfied (for any natural number $n > 0$):*

$$0 \leq (\mu_p(x))^n + (v_p(x))^n \leq 1, \text{ for any } x \in X. \tag{23}$$

Let **n-PFS** mean set of all n-PFS.

The fuzzy sets μ_p, v_p indicate the membership and non-membership functions. Zhang and Xu [6] considered $p(x) = \langle \mu_p(x), v_p(x) \rangle$ as n-Pythagorean fuzzy number (n-PFN) represented by $p = \langle \mu_p, v_p \rangle$. The notation is used:

$$\text{n-PFN} =_{df} \{\langle \mu, v \rangle \in [0,1] \times [0,1] : 0 \leq \mu^n + v^n \leq 1\}. \tag{24}$$

Fact 2.

$$\text{n-PFN} = \{p(x) : x \in X, p \in \text{n-PFS}\}. \tag{25}$$

When $n = 1$, then the 1-Pythagorean fuzzy sets are the intuitionistic fuzzy sets (IFS), which were studied by Atanassow [2]. Moreover, when $n = 2$, then the 2-Pythagorean fuzzy sets are the PFS of Yager [4].

Simple arithmetic properties of inequalities $0 \leq \mu^{n+1} + v^{n+1} \leq \mu^n + v^n \leq 1$ result from:

Theorem 5. *For any natural number $n > 1$*

$$\text{n-PFN} \subseteq (n+1)\text{-PFN} \subseteq [0,1] \times [0,1]. \tag{26}$$

Thus, entering a power scale for a value of the membership and non-membership functions allows to replace $\langle \mu, v \rangle \in [0,1] \times [0,1]$ such that $\mu + v > 1$, by $\langle \mu^n, v^n \rangle \in$ 1-PFN, for some n. As a result, the aggregation operations on the IFS can be extended to the aggregation operations on the n-PFS.

Theorem 6. *In any system $S_{Yager} = \langle [0,1], \bullet_t, \bullet_s, \bullet_p, \bullet_l, 0, 1 \rangle$, for any $\langle \mu_1, v_1 \rangle, \langle \mu_2, v_2 \rangle \in$ 1-PFN and a number $\lambda \in [0,1]$, the following conditions are satisfied:*

$$\langle \mu_1 \bullet_s \mu_2, v_1 \bullet_t v_2 \rangle \in \text{1-PFN}, \tag{27}$$

$$\langle \mu_1 \bullet_t \mu_2, v_1 \bullet_s v_2 \rangle \in \text{1-PFN}, \tag{28}$$

Furthermore, in some systems S_{Yager} (not in all - see Proof of Theorem 3,4 and Remark 1) there are additional conditions:

$$\langle \lambda \bullet_l \mu_1, \lambda \bullet_p v_1 \rangle \in \textbf{1-PFN}, \tag{29}$$

$$\langle \lambda \bullet_p \mu_1, \lambda \bullet_l v_1 \rangle \in \textbf{1-PFN}. \tag{30}$$

Proof of Theorem 6.

1. For Equation (27): $\langle \mu_1, v_1 \rangle, \langle \mu_2, v_2 \rangle \in \textbf{1-PFN}$ iff $\mu_1 + v_1 \leq 1, \mu_2 + v_2 \leq 1$ if and only if $\mu_1 \leq 1 - v_1, \mu_2 \leq 1 - v_2$.

    Hence and from the monotonicity of the s-norm and its determination by t-norm:

    $\mu_1 \bullet_s \mu_2 \leq (1-v_1) \bullet_s (1-v_2) = 1 - (1-(1-v_1) \bullet_s (1-v_2)) = 1 - v_1 \bullet_t v_2$ iff $\mu_1 \bullet_s \mu_2 + v_1 \bullet_t v_2 \leq 1$ iff $\langle \mu_1 \bullet_s \mu_2, v_1 \bullet_t v_2 \rangle \in \textbf{1-PFN}$.

2. For Equation (28): $\langle \mu_1, v_1 \rangle, \langle \mu_2, v_2 \rangle \in \textbf{1-PFN}$ iff $\langle v_1, \mu_1 \rangle, \langle v_2, \mu_2 \rangle \in \textbf{1-PFN}$.

    Hence, and from point 1, there is $\langle v_1 \bullet_s v_2, \mu_1 \bullet_t \mu_2 \rangle \in \textbf{1-PFN}$, which is equivalent to $\langle \mu_1 \bullet_t \mu_2, v_1 \bullet_s v_2 \rangle \in \textbf{1-PFN}$.

3. For Equation (29): let $\mu \bullet_s v = \max\{\mu, v\}, \mu \bullet_t v = \min\{\mu, v\}$, then $f_s(x) = x, f_t(x) = 1\check{}x$ (see Example 1);

    $\langle \mu_1, v_1 \rangle \in \textbf{1-PFN}$ iff $\mu_1 + v_1 \leq 1$;

    $\lambda \bullet_l \mu_1 = f_s^{-1}[\lambda f_s(\mu_1)] = \lambda \mu_1$ and $\lambda \bullet_p v_1 = f_s^{-1}[\lambda f_s(v_1)] = 1 - \lambda(1-v_1)$;

    $\lambda \bullet_l \mu_1 + \lambda \bullet_p v_1 = \lambda \mu_1 + 1 - \lambda(1-v_1) = 1 - \lambda + \lambda(\mu_1 + v_1)$;

    Since $\lambda(\mu_1 + v_1) \leq \lambda$, so $0 \leq 1 - \lambda + \lambda(\mu_1 + v_1) \leq 1$.

4. For Equation (30): let $\mu \bullet_s v = \max\{\mu, v\}, \mu \bullet_t v = \min\{\mu, v\}$, then $f_s(x) = x, f_t(x) = 1\check{}x$;

    $\langle \mu_1, v_1 \rangle \in \textbf{1-PFN}$ iff $\mu_1 + v_1 \leq 1$ iff $\langle v_1, \mu_1 \rangle \in \textbf{1-PFN}$.

    Hence, and from point 3, there is $\langle \lambda \bullet_l v_1, \lambda \bullet_p \mu_1 \rangle \in \textbf{1-PFN}$, which is equivalent to $\langle \lambda \bullet_p \mu_1, \lambda \bullet_l v_1 \rangle \in \textbf{1-PFN}$.

□

Remark 1. *Assuming generators of triangular norms from the Example 2* $\lambda \bullet_l \mu_1 = \lambda^{1/p} \mu_1, \lambda \bullet_p v_1 = 1 - \lambda^{1/p}(1-v_1)$, for $\mu_1 = 1/2, v_1 = 7/8, \lambda = 1/4$ and $p = 2$, it is obtained that $\langle \lambda \bullet_l \mu_1, \lambda \bullet_p v_1 \rangle = \langle 1/8, 15/16 \rangle \notin \textbf{1-PFN}, (1 < 1/8 + 15/16)$.

Theorem 7. *Let the system* $S_{Yager} = \langle [0,1], \bullet_t, \bullet_s, \bullet_p, \bullet_l, 0, 1 \rangle$ *conditions from Equations (27)–(30) of the Theorem 6 apply. Then, for any natural number* $n > 1$*, for any* $\langle \mu_1, v_1 \rangle, \langle \mu_2, v_2 \rangle \in n\textbf{-PFN}$*, and number* $\lambda \in [0,1]$ *the following conditions are satisfied:*

$$\langle (\mu_1^n \bullet_s \mu_2^n)^{1/n}, (v_1^n \bullet_t v_2^n)^{1/n} \rangle \in n\textbf{-PFN}, \tag{31}$$

$$\langle (\mu_1^n \bullet_t \mu_2^n)^{1/n}, (v_1^n \bullet_s v_2^n)^{1/n} \rangle \in n\textbf{-PFN}, \tag{32}$$

$$\langle (\lambda \bullet_l \mu_1^n)^{1/n}, (\lambda \bullet_p v_1^n)^{1/n} \rangle \in n\textbf{-PFN}, \tag{33}$$

$$\langle (\lambda \bullet_p \mu_1^n)^{1/n}, (\lambda \bullet_l v_1^n)^{1/n} \rangle \in n\textbf{-PFN}. \tag{34}$$

Proof of Theorem 7.

$$\langle \mu_1, v_1 \rangle, \langle \mu_2, v_2 \rangle \in n\textbf{-PFN} \text{ iff } \langle \mu_1^n, v_1^n \rangle, \langle \mu_2^n, v_2^n \rangle \in \textbf{1-PFN}. \tag{35}$$

Then, the conditions of Theorem 6 are satisfied, which are equivalent to the above conditions (31)–(34). □

Definition 4. *In the system* $\mathbf{S_{Yager}} = \langle [0,1], \bullet_t, \bullet_s, \bullet_p, \bullet_l, 0, 1 \rangle$, *the following aggregation operators are defined: the **Yager operators on n-PFN**: for any* $\langle \mu_1, v_1 \rangle, \langle \mu_2, v_2 \rangle \in $ *n-PFN and number* $\lambda \in [0,1]$.
When conditions (27)–(30) of the Theorem 6 are satisfied:

$$\langle \mu_1, v_1 \rangle \oplus \langle \mu_2, v_2 \rangle = \langle (\mu_1^n \bullet_s \mu_2^n)^{1/n}, (v_1^n \bullet_t v_2^n)^{1/n} \rangle, \quad (36)$$

$$\langle \mu_1, v_1 \rangle \oplus \langle \mu_2, v_2 \rangle = \langle (\mu_1^n \bullet_t \mu_2^n)^{1/n}, (v_1^n \bullet_s v_2^n)^{1/n} \rangle, \quad (37)$$

$$\lambda \langle \mu_1, v_1 \rangle = \langle (\lambda \bullet_l \mu_1^n)^{1/n}, (\lambda \bullet_p v_1^n)^{1/n} \rangle, \quad (38)$$

$$\langle \mu_1, v_1 \rangle \lambda = \langle (\lambda \bullet_p \mu_1^n)^{1/n}, (\lambda \bullet_l v_1^n)^{1/n} \rangle. \quad (39)$$

when $\langle (\lambda \bullet_l \mu_1^n)^{1/n}, (\lambda \bullet_p v_1^n)^{1/n} \rangle \notin $ *n-PFN*, *then:*

$$\lambda \langle \mu_1, v_1 \rangle = \langle 1, 0 \rangle, \quad (40)$$

or when $\langle (\lambda \bullet_p \mu_1^n)^{1/n}, (\lambda \bullet_l v_1^n)^{1/n} \rangle \notin $ *n-PFN*, *then:*

$$\langle \mu_1, v_1 \rangle^\lambda = \langle 0, 1 \rangle. \quad (41)$$

Hence, in any system $\mathbf{S_{Yager}} = \langle [0,1], \bullet_t, \bullet_s, \bullet_p, \bullet_l, 0, 1 \rangle$, using the Theorem 4, there is:

Theorem 8. *For any* $\langle \mu_1, v_1 \rangle, \langle \mu_2, v_2 \rangle, \langle \mu_3, v_3 \rangle \in $ *n-PFN and number* $\lambda \in [0,1]$:

$$\langle \mu_1, v_1 \rangle \oplus \langle 0, 1 \rangle = \langle \mu_1, v_1 \rangle, \langle \mu_1, v_1 \rangle \oplus \langle 1, 0 \rangle = \langle 1, 0 \rangle, \quad (42)$$

$$\langle \mu_1, v_1 \rangle \oplus \langle \mu_2, v_2 \rangle = \langle \mu_2, v_2 \rangle \oplus \langle \mu_1, v_1 \rangle, \quad (43)$$

$$(\langle \mu_1, v_1 \rangle \oplus \langle \mu_2, v_2 \rangle) \oplus \langle \mu_3, v_3 \rangle = \langle \mu_1, v_1 \rangle \oplus (\langle \mu_2, v_2 \rangle \oplus \langle \mu_3, v_3 \rangle), \quad (44)$$

$$\langle \mu_1, v_1 \rangle \oplus \langle 1, 0 \rangle = \langle \mu_1, v_1 \rangle, \langle \mu_1, v_1 \rangle \otimes \langle 0, 1 \rangle = \langle 0, 1 \rangle, \quad (45)$$

$$\langle \mu_1, v_1 \rangle \oplus \langle \mu_2, v_2 \rangle = \langle \mu_2, v_2 \rangle \oplus \langle \mu_1, v_1 \rangle, \quad (46)$$

$$(\langle \mu_1, v_1 \rangle \oplus \langle \mu_2, v_2 \rangle) \oplus \langle \mu_3, v_3 \rangle = \langle \mu_1, v_1 \rangle \oplus (\langle \mu_2, v_2 \rangle \oplus \langle \mu_3, v_3 \rangle), \quad (47)$$

$$\lambda(\langle \mu_1, v_1 \rangle \oplus \langle \mu_2, v_2 \rangle) = \lambda \langle \mu_1, v_1 \rangle \oplus \lambda \langle \mu_2, v_2 \rangle, \quad (48)$$

$$(\langle \mu_1, v_1 \rangle \oplus \langle \mu_2, v_2 \rangle)^\lambda = \langle \mu_1, v_1 \rangle^\lambda \oplus \langle \mu_2, v_2 \rangle^\lambda, \quad (49)$$

$$(\lambda_1 + \lambda_2) \langle \mu_1, v_1 \rangle = \lambda_1 \langle \mu_1, v_1 \rangle \oplus \lambda_2 \langle \mu_1, v_1 \rangle, \quad (50)$$

$$\langle \mu_1, v_1 \rangle^{\lambda_1 + \lambda_2} = \langle \mu_1, v_1 \rangle^{\lambda_1} \oplus \langle \mu_1, v_1 \rangle^{\lambda_2}. \quad (51)$$

4. Triangular Norms in the n-PFN and the n-PFS Algebra

Definition 5. *For any* $\langle \mu_1, v_1 \rangle, \langle \mu_2, v_2 \rangle \in $ *n-PFN*:

$$\langle \mu_1, v_1 \rangle \leq_n \langle \mu_2, v_2 \rangle \text{ iff } \mu_1 \leq \mu_2, v_1 \geq v_2. \quad (52)$$

The results of operations maximum and minimum for any $A \subseteq $ *n-PFN are described for the relation* $\leq_n$ *and are denoted by:* $\max_n A, \min_n A$.

Fact 3. *For any* $x, y \in [0,1]$:

1. $\langle 0, 1 \rangle \leq_n \langle 0, x \rangle \leq_n \langle 0, 0 \rangle \leq_n \langle y, 0 \rangle \leq_n \langle 1, 0 \rangle$.
2. $\langle 0, 1 \rangle \leq_n \langle x, y \rangle \leq_n \langle 1, 0 \rangle$, *when* $\langle x, y \rangle \in $ *n-PFN*.
3. $\langle 0, 1 \rangle = \min_n $ *n-PFN*, $\langle 1, 0 \rangle = \max_n $ *n-PFN*.

Fact 4. There is:
$$n\text{-PFN} = \{\langle x, y\rangle \in [0,1] \times [0,1] : \langle 0,1\rangle \leq_n \langle x,y\rangle \leq_n \langle 1,0\rangle\}. \tag{53}$$

Definition 6. There are:

1. The operation $\circ_t : n\text{-PFN} \times n\text{-PFN} \to n\text{-PFN}$ is called a **t-norm in the set n-PFN** ordered by the relations $\leq_n$, when for any $\langle \mu_1, v_1\rangle, \langle \mu_2, v_2\rangle, \langle \mu_3, v_3\rangle \in n\text{-PFN}$:

   (a) boundary conditions
   $$\langle 0,1\rangle \circ_t \langle \mu_1, v_1\rangle = \langle 0,1\rangle, \langle \mu_1, v_1\rangle \circ_t \langle 1,0\rangle = \langle \mu_1, v_1\rangle, \tag{54}$$

   (b) monotonicity
   $$\langle \mu_1, v_1\rangle \circ_t \langle \mu_2, v_2\rangle \leq_n \langle \mu_3, v_3\rangle \circ_t \langle \mu_2, v_2\rangle, \text{ when } \langle \mu_1, v_1\rangle \leq_n \langle \mu_3, v_3\rangle, \tag{55}$$

   (c) commutativity
   $$\langle \mu_1, v_1\rangle \circ_t \langle \mu_2, v_2\rangle = \langle \mu_2, v_2\rangle \circ_t \langle \mu_1, v_1\rangle, \tag{56}$$

   (d) associativity
   $$\langle \mu_1, v_1\rangle \circ_t (\langle \mu_2, v_2\rangle \circ_t \langle \mu_3, v_3\rangle) = (\langle \mu_1, v_1\rangle \circ_t \langle \mu_2, v_2\rangle) \circ_t \langle \mu_3, v_3\rangle. \tag{57}$$

2. The operation $\circ_s : n\text{-PFN} \times n\text{-PFN} \to n\text{-PFN}$ is called the **s-norm in the set n-PFN** ordered by the relations $\leq_n$, when for any $\langle \mu_1, v_1\rangle, \langle \mu_2, v_2\rangle, \langle \mu_3, v_3\rangle \in n\text{-PFN}$:

   (a) boundary conditions
   $$\langle 1,0\rangle \circ_s \langle \mu_1, v_1\rangle = \langle 1,0\rangle, \langle \mu_1, v_1\rangle \circ_s \langle 0,1\rangle = \langle \mu_1, v_1\rangle, \tag{58}$$

   (b) monotonicity
   $$\langle \mu_1, v_1\rangle \circ_s \langle \mu_2, v_2\rangle \leq_n \langle \mu_3, v_3\rangle \circ_s \langle \mu_2, v_2\rangle, \text{ when } \langle \mu_1, v_1\rangle \leq_n \langle \mu_3, v_3\rangle, \tag{59}$$

   (c) commutativity
   $$\langle \mu_1, v_1\rangle \circ_s \langle \mu_2, v_2\rangle = \langle \mu_2, v_2\rangle \circ_s \langle \mu_1, v_1\rangle, \tag{60}$$

   (d) associativity
   $$\langle \mu_1, v_1\rangle \circ_s (\langle \mu_2, v_2\rangle \circ_s \langle \mu_3, v_3\rangle) = (\langle \mu_1, v_1\rangle \circ_s \langle \mu_2, v_2\rangle) \circ_s \langle \mu_3, v_3\rangle. \tag{61}$$

Theorem 9. *The Yager operator $\otimes$ on the n-PFN is a t-norm in the set n-PFN and the operator $\oplus$ is a s-norm in the set n-PFN.*

Proof of Theorem 9. Conditions (45)–(47) of the Theorem 8 proof that the operator $\otimes$ satisfies conditions (a),(c), and (d) of the Definition 6 (1) of the t-norm in the set **n-PFN**. It is enough to prove that this operation is monotonous.

For any $\langle \mu_1, v_1\rangle, \langle \mu_2, v_2\rangle, \langle \mu_3, v_3\rangle \in \mathbf{n\text{-PFN}}$
$\langle \mu_1, v_1\rangle \otimes \langle \mu_2, v_2\rangle = \langle (\mu_1^n \bullet_t \mu_2^n)^{1/n}, (v_1^n \bullet_s v_2^n)^{1/n}\rangle,$
$\langle \mu_3, v_3\rangle \otimes \langle \mu_2, v_2\rangle = \langle (\mu_3^n \bullet_t \mu_2^n)^{1/n}, (v_3^n \bullet_s v_2^n)^{1/n}\rangle.$
Let $\langle \mu_1, v_1\rangle \leq_n \langle \mu_3, v_3\rangle$. Then $\mu_1 \leq \mu_3, v_1 \geq v_3$. Hence, and from the monotonicity of the t-norm and s-norm, it is obtained that:
$\mu_1^n \bullet_t \mu_2^n \leq \mu_3^n \bullet_t \mu_2^n$ and $v_3^n \bullet_s v_2^n \leq v_1^n \bullet_s v_2^n$ iff
$(\mu_1^n \bullet_t \mu_2^n)^{1/n} \leq (\mu_3^n \bullet_t \mu_2^n)^{1/n}$ and $(v_1^n \bullet_s v_2^n)^{1/n} \geq (v_3^n \bullet_s v_2^n)^{1/n}$ iff
$\langle (\mu_1^n \bullet_t \mu_2^n)^{1/n}, (v_1^n \bullet_s v_2^n)^{1/n}\rangle \leq_n \langle (\mu_3^n \bullet_t \mu_2^n)^{1/n}, (v_3^n \bullet_s v_2^n)^{1/n}\rangle$ iff
$\langle \mu_1, v_1\rangle \otimes \langle \mu_2, v_2\rangle \leq_n \langle \mu_3, v_3\rangle \otimes \langle \mu_2, v_2\rangle.$

Proof that the operator $\oplus$ satisfies conditions of the Definition 6 (2) about the s-norm in the set **n-PFN** is analogical. □

Summarizing, for the knowledge of operations and relationships introduced in the **n-PFN**, the following operations and conclusion relationships can be determined for the **n-PFS**:

Definition 7. *For any $p_1, p_2 \in$ n-PFS, and number $\lambda \in [0,1]$:*

$$p_1 \oplus p_2 = \{\langle x, \langle \mu_{p_1}(x), v_{p_1}(x)\rangle \oplus \langle \mu_{p_2}(x), v_{p_2}(x)\rangle\rangle : x \in X\}, \quad (62)$$

$$p_1 \otimes p_2 = \{\langle x, \langle \mu_{p_1}(x), v_{p_1}(x)\rangle \otimes \langle \mu_{p_2}(x), v_{p_2}(x)\rangle\rangle : x \in X\}, \quad (63)$$

$$\lambda p_1 = \{\langle x, \lambda \langle \mu_{p_1}(x), v_{p_1}(x)\rangle\rangle : x \in X\}, \quad (64)$$

$$p_1^\lambda = \{\langle x, \langle \mu_{p_1}(x), v_{p_1}(x)\rangle^\lambda\rangle : x \in X\}, \quad (65)$$

$$\text{supp}(p_1) = \{x \in X : \mu_{p_1}(x) > 0, v_{p_1}(x) > 0\}, \quad (66)$$

$$p_1 \subseteq_n p_2 \text{ iff for any } x \in X, \langle \mu_{p_1}(x), v_{p_1}(x)\rangle \leq_n \langle \mu_{p_2}(x), v_{p_2}(x)\rangle. \quad (67)$$

The system n-PFS, with defined in the Definition 7 operations (t-norm, s-norm, p-norm, l-norm, and support) and inclusion relations, is called the **n-PFS algebra**.

Let $1 =_{df} \langle 0,1\rangle, 0 =_{df} \langle 1,0\rangle$. Then from the Definition 7 and the Theorem 8 there are:

Theorem 10. *In the algebra n-PFS, for any $p_1, p_2, p_3 \in$ n-PFN and the number $\lambda \in [0,1]$:*

$$p_1 \oplus 0 = p_1, \ p_1 \oplus 1 = 1, \quad (68)$$

$$p_1 \oplus p_2 = p_2 \oplus p_1, \quad (69)$$

$$(p_1 \oplus p_2) \oplus p_3 = p_1 \oplus (p_2 \oplus p_3), \quad (70)$$

$$p_1 \otimes 1 = p_1, \ p_1 \otimes 0 = 0, \quad (71)$$

$$p_1 \otimes p_2 = p_2 \otimes p_1, \quad (72)$$

$$(p_1 \otimes p_2) \otimes p_3 = p_1 \otimes (p_2 \otimes p_3), \quad (73)$$

$$\lambda(p_1 \oplus p_2) = \lambda p_1 \oplus \lambda p_2, \quad (74)$$

$$(p_1 \otimes p_2)^\lambda = p_1^\lambda \otimes p_2^\lambda, \quad (75)$$

$$(\lambda_1 + \lambda_2)p_1 = \lambda_1 p_1 \oplus \lambda_2 p_1, \quad (76)$$

$$p_1^{\lambda_1 + \lambda_2} = p_1^{\lambda_1} \otimes p_1^{\lambda_2}. \quad (77)$$

5. Conclusions

This paper presents the elements of deductive theory of the n-PFS, where the membership degree μ and the non-membership degree v determine not only in the square scale, but in any power scale, i.e., $0 \leq \mu^n + v^n \leq 1$. As a result, any local deductions in the n-PFS range can be formulated. It may be interesting to use the described model to create similar models but based on other functional scales, for example for the function of scale $f_n(x) = x/n : 0 \leq \mu/n + v/n \leq 1$ or $f_n(x) = 1 + \log_n(x+1)$, for $x \in [0,1]$, where $f(x) < x : 0 \leq (1 + \log_n(\mu + 1)) + (1 + \log_n(v + 1)) \leq 1$. In the research, the n-PFS theory can be used to describe n-PFS as a system of information granules [19].

Funding: This research received no external funding.

Conflicts of Interest: The authors declare no conflict of interest.

References

1. Zadeh, L.A. Fuzzy sets. *Inf. Control* **1965**, *8*, 338–353. [CrossRef]
2. Atanassov, K.T. Intuitionistic fuzzy sets. *Fuzzy Sets Syst.* **1986**, *20*, 87–96. [CrossRef]
3. Atanassov, K.T. *Intuitionistic Fuzzy Sets*; Springer: Berlin/Heidelberg, Germany, 1999.
4. Yager, R.R. Pythagorean fuzzy subsets. In Proceedings of the 2013 Joint IFSAWorld Congress and NAFIPS Annual Meeting (IFSA/NAFIPS), Edmonton, AB, Canada, 24–28 June 2013; pp. 57–61.
5. Yager, R.R. Pythagorean membership grades in multicriteria decision making. *IEEE Trans. Fuzzy Syst.* **2014**, *22*, 958–965. [CrossRef]
6. Zhang, X.; Xu, Z. Extension of TOPSIS to multiple criteria decision making with Pythagorean fuzzy sets. *Int. J. Intell. Syst.* **2014**, *29*, 1061–1078. [CrossRef]
7. Garg, H. A novel correlation coefficients between Pythagorean fuzzy sets and its applications to decisionmaking processes. *Int. J. Intell. Syst.* **2016**, *31*, 1234–1252. [CrossRef]
8. Garg, H. Confidence levels based Pythagorean fuzzy aggregation operators and its application to decisionmaking process. *Comput. Math. Organ. Theory* **2017**, *23*, 546–571. [CrossRef]
9. Atanassov, K.T. On the Modal Operators Defined Over The Intuitionistic fuzzy Sets. *Notes Intuit. Fuzzy Sets* **2004**, *10*, 7–12.
10. Atanassov, K.T. *On Intuitionistic Fuzzy Sets Theory*; Springer: Berlin/Heidelberg, Germany, 2012.
11. Dencheva, K. Extension of intuitionistic fuzzy modal operators $\otimes$ and $\oplus$. *Proc. Second Int. IEEE Symp. Intell. syst.* **2004**, *3*, 21–22.
12. Yılmaz, S.; Bal, A. Extentsion of intuitionistic fuzzy modal operators diagram with new operators. *Notes Intuit. Fuzzy Sets* **2014**, *20*, 26–35.
13. Akram, M.; Garg, H.; Ilyas, F. Multi-criteria group decision making based on ELECTRE I method in Pythagorean fuzzy information. *Soft Comput.* **2019**, *24*, 3425–3453.
14. Akram, M.; Dudek, W.A.; Dar, J.M. Pythagorean Dombi fuzzy aggregation operators with application in multicriteria decision-making. *Int. J. Intell. Syst.* **2019**, *34*, 3000–3019. [CrossRef]
15. Akram, M.; Dudek, W.A.; Ilyas, F. Group decision-making based on Pythagorean fuzzy TOPSIS method. *Int. J. Intell. Syst.* **2019**, *34*, 1455–1475. [CrossRef]
16. Shahzadi, G.; Akram, M.; Al-KenaniInt, A.N. Decision-Making Approach under Pythagorean Fuzzy Yager Weighted Operators. *Mathematics* **2020**, *8*, 70. [CrossRef]
17. Yager, R.R. Aggregation operators and fuzzy systems modeling. *Fuzzy Sets Syst.* **1994**, *67*, 129–145. [CrossRef]
18. Aczel, J. *Lectures on Functional Equations and Their Applications*; Academic Press: New York, NY, USA, 1966.
19. Bryniarska, A. Certain information granule system as a result of sets approximation by fuzzy context. *Int. J. Approx. Reason.* **2019**, *111*, 1–20. [CrossRef]

Publisher's Note: MDPI stays neutral with regard to jurisdictional claims in published maps and institutional affiliations.

© 2020 by the authors. Licensee MDPI, Basel, Switzerland. This article is an open access article distributed under the terms and conditions of the Creative Commons Attribution (CC BY) license (http://creativecommons.org/licenses/by/4.0/).

Article

An Improved Crow Search Algorithm Applied to the Phase Swapping Problem in Asymmetric Distribution Systems

Brandon Cortés-Caicedo [1], Laura Sofía Avellaneda-Gómez [1], Oscar Danilo Montoya [2,3,*], Lázaro Alvarado-Barrios [4] and César Álvarez-Arroyo [5]

1. Ingeniería Eléctrica, Universidad Distrital Francisco José de Caldas, Bogotá 110231, Colombia; bcortesc@correo.udistrital.edu.co (B.C.-C.); lsavellanedag@correo.udistrital.edu.co (L.S.A.-G.)
2. Facultad de Ingeniería, Universidad Distrital Francisco José de Caldas, Bogotá 110231, Colombia
3. Laboratorio Inteligente de Energía, Universidad Tecnológica de Bolívar, Cartagena 131001, Colombia
4. Department of Engineering, Universidad Loyola Andalucía, 41704 Sevilla, Spain; lalvarado@uloyola.es
5. Department of Electrical Engineering, Universidad de Sevilla, 41092 Sevilla, Spain; cesaralvarez@us.es
* Correspondence: odmontoyag@udistral.edu.co

Citation: Cortés-Caicedo, B.; Avellaneda-Gómez, L.S.; Montoya, O.D.; Alvarado-Barrios, L.; Álvarez-Arroyo, C. An Improved Crow Search Algorithm Applied to the Phase Swapping Problem in Asymmetric Distribution Systems. *Symmetry* **2021**, *13*, 1329. https://doi.org/10.3390/sym13081329

Academic Editors: Juan Alberto Rodríguez Velázquez and Alejandro Estrada-Moreno

Received: 6 July 2021
Accepted: 21 July 2021
Published: 23 July 2021

Publisher's Note: MDPI stays neutral with regard to jurisdictional claims in published maps and institutional affiliations.

Copyright: © 2021 by the authors. Licensee MDPI, Basel, Switzerland. This article is an open access article distributed under the terms and conditions of the Creative Commons Attribution (CC BY) license (https://creativecommons.org/licenses/by/4.0/).

Abstract: This paper discusses the power loss minimization problem in asymmetric distribution systems (ADS) based on phase swapping. This problem is presented using a mixed-integer nonlinear programming model, which is resolved by applying a master–slave methodology. The master stage consists of an improved version of the crow search algorithm. This stage is based on the generation of candidate solutions using a normal Gaussian probability distribution. The master stage is responsible for providing the connection settings for the system loads using integer coding. The slave stage uses a power flow for ADSs based on the three-phase version of the iterative sweep method, which is used to determine the network power losses for each load connection supplied by the master stage. Numerical results on the 8-, 25-, and 37-node test systems show the efficiency of the proposed approach when compared to the classical version of the crow search algorithm, the Chu and Beasley genetic algorithm, and the vortex search algorithm. All simulations were obtained using MATLAB and validated in the DigSILENT power system analysis software.

Keywords: improved crow search algorithm; normal Gaussian distribution; phase swapping problem; power losses; asymmetric distribution grids; vortex search algorithm

1. Introduction

Due to the economic and population growth, the dependence on electrical systems has equally grown to satisfy humanity's basic needs, changing the habits and customs of how individuals live and work [1]. To ensure this, three-phase distribution networks are used, which are responsible for interconnecting transmission and sub-transmission networks with end-users (i.e., residential, industrial, and commercial areas) requiring medium and low voltage [2,3]. These systems generally operate in an asymmetric manner due to the following factors. (i) The configurations on the distribution lines are asymmetrical since the transposition criterion is not applicable due to the short length of the lines [4,5]. (ii) The nature of the loads may be 1φ, 2φ, or 3φ, which generates unbalances in voltages at the nodes and in the line currents [6]. (iii) The arbitrary location of single-phase transformers on the phases of the system causes an unbalance in the currents through the lines [7]. Load unbalances in distribution systems create undesirable scenarios such as the increase of current in any phase system, which produces an increase in power losses through its constituent elements [8]. These power losses can exceed the capacity required to supply the demand, cause equipment to age, and increase investment and operating costs for network operators [7,9].

The importance of reducing power losses in distribution networks has established multiple approaches, such as (i) optimal placement and sizing of distributed generation [10],

(ii) optimal capacitor placement and sizing [11], (iii) optimal network reconfiguration [12], (iv) optimal conductor sizing in distribution networks [13], (v) optimal power system restoration [14], and (vi) optimal phase swapping [15,16]. These strategies can significantly help distribution companies to reduce the number of power losses. However, the first two approaches involve significant investments since they integrate new devices into the distribution network [17]. The third approach requires less investment since few distribution lines need to be constructed to realize the optimal network reconfiguration [18]. The fourth approach also requires high investment as the system conductors need to be renewed [19]. The fifth methodology is more appropriate for operation of the power system after fault isolation. The sixth strategy is the most economical as it requires few teams to reconfigure the system loads without investing in new equipment [20]. Bearing in mind the low phase swapping costs to minimize the power losses in ADSs, a new master–slave optimization strategy is proposed to solve this problem.

In the specialized literature, the balance phase problem, with the minimizing power losses approach, has been solved using different optimization methods, including the Chu and Beasley genetic algorithms [8,16,21–24], particle swarm optimization [9], mixed-integer convex optimization [25], bat optimization algorithm [26], differential evolution algorithm [27], simulated annealing optimizer [28], and vortex search algorithm [15], among others.

The main feature of the optimization methodologies described above is that they employ the master–slave optimization scheme to solve the problem [15]. The master defines the connection of the loads to the nodes. The slave is typically a power flow tool that allows one to revise and exploit the solution space through the power losses calculation [16].

Similar to the metaheuristic optimization methods described above, a master–slave methodology is proposed in this work to solve the phase swapping problem in ADSs. The proposed optimization algorithm corresponds to an improved version of the crow optimization algorithm (CSA) to select the connection of the loads in the master stage, together with the use of the iterative sweep power flow method in its three-phase version in the slave stage. In the master stage, the connection of each load is defined using an integer encoding between 1 and 6, which represents the six possible connection forms for a three-phase charge [8]. The slave stage is responsible for evaluating the power flow to determine the total power losses for the connection set provided in the master stage [15]. Improvements in the classical CSA are carried out in the crow avoidance stage based on a probability criterion [29]. If the probability is higher than the crow knowledge probability (A_p), the new crow position i is provided using the classical CSA exploration proposed in [29]. Likewise, if the possibility is less than the crow knowledge probability (i.e., A_p), the new crow position i is generated through a regular Gaussian distribution (GD) used in the process of evolution of the vortex search algorithm (VSA) [30]. The main benefit of the VSA is that the solution space can be explored and exploited through the use of hyper-spheres derived from the selection of an individual from the current population. It also clarifies that the criterion of evolution in our proposal is applied at each iteration, which implies that this process is of the adaptive type.

The main contributions of our proposal are listed below.

- It proposes an improved approach for the classical CSA using the VSA evolution mechanism to revise and exploit the solution space.
- The interaction between the improved CSA (i.e., ICSA) and the three-phase power flow (TPPF), based on the classical iterative sweep method, allows the application of phase swapping in radial or meshed systems with connected loads, either in Y or Δ.

It is relevant to mention that, upon analyzing the specialized literature, there was no evidence of the CSA application to the phase swapping problem in distribution systems, which corresponds to a research gap that this work intends to fill. In addition, the numerical results obtained in test systems of 8, 25, and 37 nodes prove the quality of the algorithm when compared with classical metaheuristic optimization methodologies.

The structure of the remainder of the document takes the following form: Section 2 presents the optimal phase swapping problem representation in ADSs, Section 3 presents the ICSA incorporated with the TPPF method, Section 4 portrays the electrical networks used in this research, and Section 5 represents the obtained results for the connections set and the grid power losses. Finally, Section 6 states the conclusions drawn from the development of this article.

2. Mathematical Formulation

The optimal phase swapping problem in ADSs is represented by a mixed-integer nonlinear programming (MINLP) model [16]. The binary variables are the decision variables, which correspond to the connections set for each load presented in the system [8]. Additionally, the power flow formulation provides the continuous part of the decision variables. The nonlinear nature of the products appears between the different voltage magnitudes at the nodes and the trigonometric functions [25]. Next, the objective function and the set of constraints representing the phase swapping problem are presented.

2.1. Formulation of the Objective Function

The phase swapping problem has an objective function associated with the minimization of total active power losses of the ADS, as presented in Equation (1):

$$\min z = \sum_{n \in \mathcal{N}} \sum_{m \in \mathcal{N}} \sum_{f \in \mathcal{F}} \sum_{g \in \mathcal{F}} Y_{nfmg} V_{nf} V_{mg} \cos(\delta_{nf} - \delta_{mg} - \theta_{nmfg}), \quad (1)$$

where z defines the value of the objective function. Further, Y_{nfmg} is the admittance magnitude associated with node n in the electrical phase f with node m in the electrical phase g, V_{nf} (V_{mg}) corresponds to the voltage magnitude at node $n(m)$ in the electrical phase $f(g)$, δ_{nf} (δ_{mg}) represents the angle of the voltage at node $n(m)$ in the electrical phase $f(g)$, and θ_{nmfg} represents the admittance angle associated with node n in the electrical phase f with node m in the electrical phase g. It is relevant to mention that $\mathcal{F}$ and $\mathcal{N}$ are the sets containing all phases and nodes, respectively.

Remark 1. *The product between the magnitudes of the voltages and the trigonometric functions makes the objective function nonlinear and nonconvex [16]. The structure of the objective function makes advanced numerical optimization techniques necessary to minimize it efficiently [15]. A master–slave methodology is proposed, as is the case of the improved version of the developed CSA, due to its simplicity in programming terms.*

2.2. Set of Constraints

The phase swapping problem has a set of constraints that corresponds to the different operating limitations in an ADS [15]. These are shown from Equation (2) to Equation (6):

$$P_{nf}^s - \sum_{g \in \mathcal{F}} x_{nfg} P_{ng}^d = V_{nf} \sum_{m \in \mathcal{N}} \sum_{g \in \mathcal{F}} Y_{nfmg} V_{mg} \cos(\delta_{nf} - \delta_{mg} - \theta_{nfmg}), \begin{Bmatrix} \forall f \in \mathcal{F} \\ \forall n \in \mathcal{N} \end{Bmatrix}, \quad (2)$$

$$Q_{nf}^s - \sum_{g \in \mathcal{F}} x_{nfg} Q_{ng}^d = V_{nf} \sum_{m \in \mathcal{N}} \sum_{g \in \mathcal{F}} Y_{nfmg} V_{mg} \sin(\delta_{nf} - \delta_{mg} - \theta_{nfmg}), \begin{Bmatrix} \forall f \in \mathcal{F} \\ \forall n \in \mathcal{N} \end{Bmatrix}, \quad (3)$$

$$\sum_{g \in \mathcal{F}} x_{nfg} = 1, \{\forall f \in \mathcal{F}, \; \forall n \in \mathcal{N}\}, \quad (4)$$

$$\sum_{f \in \mathcal{F}} x_{nfg} = 1, \{\forall g \in \mathcal{F}, \; \forall n \in \mathcal{N}\}, \quad (5)$$

$$V_{\min} \leq V_{nf} \leq V_{\max}, \{\forall g \in \mathcal{F}, \; \forall n \in \mathcal{N}\}, \quad (6)$$

where P_{nf}^s is the variable associated with the active power produced at generator s connected to node n in the electrical phase f, and Q_{nf}^s is the generated reactive power at generator s sited at node n in the electrical phase f. P_{ng}^d indicates the active demanded power at node n in the electrical phase g, while Q_{ng}^d describes the reactive power needed at node n in the electrical phase g. x_{nfg} is a binary variable defining the configuration of demand node n at f in the electrical phase g. Finally, V_{min} and V_{max} correspond to the allowable limits of voltage regulation for all the system nodes.

Remark 2. *Equations (2) and (3) are nonlinear and nonconvex. These features highlight the complexity of the TPPF problem for electrical systems, making it necessary to use numerical methods, such as the iterative sweep method, to solve it.*

Note that Equation (1) defines the form of the objective function for the phase swapping problem formulated as the minimization of power losses under a given demand condition in all network sections of the system. Equations (2) and (3) define the apparent power balance constraints maintained at each phase and node of the ADS. Equations (4) and (5) ensure that loads take a unique connection form by using a matrix of connections (i.e., x_{nfg}) at each node [5]. Finally, the constraint presented in (6) defines the allowable limits of voltage regulation for all nodes of the system [15].

3. Methodology Proposed

To solve the optimal phase swapping problem in ADS with the objective of minimize power losses for a specific demand condition, this paper proposes to use an ICSA [29] as the master stage in conjunction with the iterative swept TPPF as the slave stage [15]. The master stage defines the phase configurations at each system demand node to achieve the most balanced system possible, while the slave stage evaluates the power flow constraints defined in (2) and (3). The section below will describe each component of the proposed master–slave methodology.

3.1. Slave Stage: TPPF Method

The iterative sweep power flow method is a numerical method typically used for single-phase distribution networks [31]. Nevertheless, this method has been adapted for three-phase ADSs with wye (i.e., Y) and delta (i.e., Δ) loads [15]. This method is derived from graph theory, where the topology of the network is represented by an incidence matrix, which relates the nodes and the links of the system [31]. First, Kirchhoff's first law is used to calculate currents in the system nodes, starting from the terminal nodes and until the source node, which corresponds to the implementation of the backward sweep stage. Then, Kirchhoff's second law, which corresponds to the implementation of the forward sweep stage, is used to calculate the voltage drops in the network sections from the slack node to the terminal nodes [31].

One of the most important aspects of the iterative sweep power flow method is that it is derivative-free. Likewise, the matrices involved in the calculations are constant, which implies that the computing times required to obtain a solution are in milliseconds [15].

In order to expose the iterative swept TPPF method developed by [15], in any n−node system, we used the relationship between the nodal voltage and the injected current that is presented using the equivalent between the admittance matrix and the incidence matrix [32], as shown in Equation (7).

$$\begin{bmatrix} \mathbb{I}_{g3\varphi} \\ \mathbb{I}_{d3\varphi} \end{bmatrix} = \begin{bmatrix} A_{g3\varphi} \mathbb{Z}_{r3\varphi}^{-1} A_{g3\varphi}^T & A_{g3\varphi} \mathbb{Z}_{r3\varphi}^{-1} A_{d3\varphi}^T \\ A_{d3\varphi} \mathbb{Z}_{r3\varphi}^{-1} A_{g3\varphi}^T & A_{d3\varphi} \mathbb{Z}_{r3\varphi}^{-1} A_{d3\varphi}^T \end{bmatrix} \begin{bmatrix} \mathbb{V}_{g3\varphi} \\ \mathbb{V}_{d3\varphi} \end{bmatrix}, \quad (7)$$

where $\mathbb{V}_{g3\varphi}$ is the vector containing all the voltages at the slack node that are known for power flow purposes [21]. $\mathbb{V}_{d3\varphi}$ is the vector containing all the unknown variables of interest, i.e., the demand voltages. Further, $\mathbb{I}_{g3\varphi}$ represents the vector with the net current injections at the slack node, $\mathbb{I}_{d3\varphi}$ represents the vector that involves all the currents at the nodes of consumption, and $\mathbb{Z}_{r3\varphi}$ is the matrix that contains all the impedance matrices of the distribution lines present in the system. $\mathbf{A}_{g3\varphi}$ is the component of the incidence matrix that associates the source nodes to each other, while $\mathbf{A}_{d3\varphi}$ is the component of the incidence matrix relating the nodes of consumption to each other.

Remark 3. *The voltage variables and current ones in Equation (7) are organized by nodes and phases according to the three-phase condition.*

From Equation (7), it can be observed that the second row has the nodal voltages at the demand nodes (i.e., $\mathbb{V}_{d3\varphi}$), which are the unknown variables in power flow studies [16]. Equation (7) can then be rewritten as follows: (8).

$$\mathbb{V}_{d3\varphi} = -\mathbb{Y}_{dd3\varphi}^{-1}[\mathbb{Y}_{dg3\varphi}\mathbb{V}_{g3\varphi} - \mathbb{I}_{d3\varphi}], \tag{8}$$

where $\mathbb{Y}_{dg3\varphi} = \mathbf{A}_{d3\varphi}\mathbf{Z}_{r3\varphi}^{-1}\mathbf{A}_{g3\varphi}^T$ and $\mathbb{Y}_{dd3\varphi} = \mathbf{A}_{d3\varphi}\mathbf{Z}_{r3\varphi}^{-1}\mathbf{A}_{d3\varphi}^T$.

Equation (8) allows the determination of all the nodal demand voltages per phase. However, it is necessary to consider the type of load, either Y or Δ, to establish the demand current (i.e., $\mathbb{I}_{d3\varphi}$).

In the case where node m has a constant power load with a Y structure (assuming it is solidly earthed [33]), the demand current can be shown as in Equation (9), as reported in [21].

$$\mathbb{I}_{dm3\varphi} = -\mathbf{diag}^{-1}(\mathbb{V}_{dm3\varphi}^*)\mathbb{S}_{dm3\varphi}^* \tag{9}$$

If node m has a load with a connection Δ, the demand current can be as shown in Equation (10), as reported in [21].

$$\mathbb{I}_{dm3\varphi} = -(\mathbf{diag}^{-1}(\mathbf{M}\mathbb{V}_{dm3\varphi}^*) - \mathbf{diag}^{-1}(\mathbf{M}^T\mathbb{V}_{dm3\varphi}^*)\mathbf{H})\mathbb{S}_{dm3\varphi}^*, \tag{10}$$

where the **H** and **M** matrices are defined as follows:

$$\mathbf{H} = \begin{bmatrix} 0 & 0 & 1 \\ 1 & 0 & 0 \\ 0 & 1 & 0 \end{bmatrix}, \mathbf{M} = \begin{bmatrix} 1 & -1 & 0 \\ 0 & 1 & -1 \\ -1 & 0 & 1 \end{bmatrix}$$

Through a t iteration counter, the solution of Equation (8) is obtained if $\max\left\{||\mathbb{V}_{d3\varphi}^{t+1}| - |\mathbb{V}_{d3\varphi}^t||\right\} \leq \epsilon$, where ϵ is the maximum tolerance suggested as 1×10^{-10} [34].

When solving the TPPF, the main objective is to evaluate the power losses for the phase connection set established in the master stage. For this purpose, Equation (11), as described in [32], is used.

$$\mathbb{P}_{loss} = \mathbf{real}\left\{|\mathbb{J}_{3\varphi}|^T \mathbb{Z}_{r3\varphi}|\mathbb{J}_{3\varphi}|\right\}, \tag{11}$$

where $\mathbb{P}_{loss}$ describes the total system effective power losses and $\mathbb{J}_{3\varphi}$ represents the current per phase flowing through the system branches expressed as shown in Equation (12) through Ohm's Law, as reported in [15].

$$\mathbb{J}_{3\varphi} = \mathbb{Z}_{r3\varphi}^{-1}\mathbb{E}_{3\varphi}, \tag{12}$$

where $\mathbb{E}_{3\varphi}$ represents the voltage drop per phase in the system branches, which can be written in terms of the generation and demand using the three-phase incidence matrix as shown in Equation (13), as reported in [15].

$$\mathbb{E}_{3\varphi} = \mathbf{A}_{g3\varphi}^T\mathbb{V}_{g3\varphi} + \mathbf{A}_{d3\varphi}^T\mathbb{V}_{d3\varphi} \tag{13}$$

Algorithm 1 shows the general implementation of the TPPF method by an iterative sweep for ADSs with connected loads in Y and Δ.

Algorithm 1: Solution of the TPPF problem for ADSs with Y and Δ loads

---

Define the characteristics of the unbalanced three-phase system under study;
Obtain the per-unit equivalent of the system;
Generate the 3φ incidence matrix $\mathbf{A}_{3\varphi}$;
Extract the components $\mathbf{A}_{g3\varphi}$ and $\mathbf{A}_{d3\varphi}$;
Define $\mathbb{Z}_{r3\varphi}$;
Calculate $\mathbb{Y}_{dg3\varphi}$ and $\mathbb{Y}_{dd3\varphi}$;
Select t_{max};
Define ϵ;
Chose the voltages per phase of the Slack node: $\mathbf{V}_{g3\varphi} = \begin{bmatrix} 1\angle 0, & 1\angle -\frac{2\pi}{3}, & 1\angle \frac{2\pi}{3} \end{bmatrix}^T$;
Do $t = 0$;
for $m \geq n - 1$ **do**
$\quad$ Do $\mathbf{V}^t_{dm3\varphi} = \mathbf{V}_{g3\varphi}$;
end
for $t \leq t_{max}$ **do**
$\quad$ Define $m = 1$;
$\quad$ **for** $m \geq n - 1$ **do**
$\quad\quad$ **if** *node load m is connected at Y* **then**
$\quad\quad\quad$ Calculate $\mathbb{I}^t_{dm3\varphi}$ using Equation (9);
$\quad\quad$ **else**
$\quad\quad\quad$ Calculate $\mathbb{I}^t_{dm3\varphi}$ using Equation (10);
$\quad\quad$ **end**
$\quad$ **end**
$\quad$ Calculate the new voltages at the demand nodes $\mathbf{V}^t_{d3\varphi}$ using Equation (8);
$\quad$ **if** $\max\left\{||\mathbf{V}^{t+1}_{d3\varphi}| - |\mathbf{V}^t_{d3\varphi}||\right\} \leq \epsilon$ **then**
$\quad\quad$ Report the nodal voltages as $\mathbf{V}_{3\varphi} = \begin{bmatrix} \mathbf{V}_{g3\varphi}, & \mathbf{V}_{d3\varphi} \end{bmatrix}^T$;
$\quad\quad$ Calculate the voltage drop across the branches of the system using Equation (13);
$\quad\quad$ Calculate the current flowing in the system branches using Equation (12);
$\quad\quad$ Calculate the power losses using Equation (11);
$\quad\quad$ **break**;
$\quad$ **else**
$\quad\quad$ Do $\mathbf{V}^t_{d3\varphi} = \mathbf{V}^{t+1}_{d3\varphi}$;
$\quad$ **end**
end

Remark 4. *The convergence of the matrix iterative sweep method can be demonstrated with the Banach fixed-point theorem, as reported in [31]. So, if the system is far enough from the stress collapse point, it can be guaranteed that the solution of Equations (2) and (3) obtain any combination of nodal loads provided by the master stage.*

3.2. Master Stage: ICSA

The master stage is responsible for providing the nodal connections set for evaluation in the iterative sweep TPPF presented in Algorithm 2. This paper proposes an improved version of the classical CSA modifying the solution space exploration by introducing a normal GD employed by the VSA optimization method [35]. Before explaining the improvements made to the CSA, the encoding of the phase swapping problem in ADSs according to the possible configurations is shown in Table 1.

Table 1. Possible connection types for the system loads [8].

Connection Type	Phases	Sequence
1	UVW	
2	VWU	No vary
3	WUV	
4	UWV	
5	WVU	Vary
6	VUW	

The coding used to represent an individual i in iteration k is presented as follows:

$$X_{i,k} = [6, 2, 4, 5, ..., c, ..., 1]$$

Here, the dimensions are $1 \times (n-1)$ and c is an integer that defines the type of connection (see Table 1) [15]. The ICSA is developed using a probability criterion at each iteration, which will define the methodology to be used to define the new site of crow i. If the probability criterion is greater than the crow knowledge probability (i.e., A_p) at iteration k, the traditional classical CSA search is employed [29]; otherwise, it will use the GD of the VSA to generate the new crow position i.

3.2.1. Classical Approach: CSA

CSA is a metaheuristic optimization technique inspired by the intelligent performance of crows [29]. Crows are considered to be the smartest birds in nature; they possess a brain much larger in relation to the size of their bodies [36]. In groups, crows show notable traits of intelligence. They can learn and remember faces, use tools, communicate using sophisticated manners, and manage their food throughout the seasons because they hide their excess food in certain places (caches) in their environment and retrieve it when necessary [29].

Crows are ambitious birds as they pursue each other for better food reserves, observe where other birds hide their food, and steal it once they have left [37]. If a crow has stolen something, it takes additional cautions such as relocating its hiding places to prevent becoming a future victim [36]. They use their experience in theft to predict another robber's behavior and to determine the quickest course to protect their caches from being stolen [38]. The main bases of CSA are listed below [29]: (i) crows live in flocks, (ii) crows remember the site of their hiding places, (iii) crows pursue others to commit robberies, and (iv) crows shield their caches from being robbed using probability.

The following is an example of how the CSA mechanism works. Crows explore and exploit their environment, which is the solution space. Each environment cache corresponds to a feasible solution, the quality of the food source is the objective function. The best food source in the environment is the solution to the problem, Thus, the CSA seeks to simulate the intelligent behavior of crows to find the optimal solution [29].

The CSA is also based on a population. The population size (i.e., flock) consists of N individuals (number of crows), which belong to a d dimensional solution space, where $d = (n-1)$. The position $X_{i,k}$ of crow i at iteration k is described in Equation (14) and represents a feasible problem solution.

$$X_{i,k} = [x_{i,k}^1, x_{i,k}^2, ..., x_{i,k}^n], \tag{14}$$

where $i = 1, 2, ..., N$ and $k = 1, 2, ..., t_{max}$. Further, t_{max} is the maximum number of iterations of the exploration and exploitation process of the solution space. Each crow (individual) can memorize the position of its hiding place. At iteration k, the position of crow i hiding place is represented as $M_{i,k}$, being the best position that crow i has obtained so far. Of course, the position of its best experience is memorized because crows move in their environment and search for the best food source (i.e., hiding places).

At the start of the optimization process, it is assumed that at iteration k, crow j wants to visit its hideout, $M_{j,k}$. In this iteration, crow i decided to follow crow j to approach crow j's hideout. In this case, two situations can occur.

Situation 1: Search

Crow j does not know that crow i is following it. As a result, crow i approaches crow j's hiding place. In this case, through Equation (15), the new position of crow i is obtained.

$$X_{i,k+1} = X_{i,k} + r_i fl_{i,k}\left(M_{j,k} - X_{i,k}\right), \tag{15}$$

where r_i a random number between 0 and 1, and $fl_{i,k}$ is the flight length of the crow i at each iteration k. According to [29], small values of fl allow a local exploration of the solution space (close to $X_{i,k}$), while large values of fl allow a global solution space exploration (far from $X_{i,k}$).

Situation 2: Evasion

Crow j knows that crow i is following it. As a result, crow j tries to deceive crow i by heading towards another position in the solution space to protect its caches from being stolen. Either way, situations 1 and 2 can be represented as shown in Equation (16).

$$X_{i,k+1} = \begin{cases} X_{i,k} + r_i fl_{i,k}(M_{j,k} - X_{i,k}) & \text{if } r_j \geq A_P \\ \text{random} & \text{otherwise} \end{cases}, \tag{16}$$

where r_j is a random number between 0 and 1, and A_P denotes the probability of crow j's knowledge of crow i.

Once the crows' positions are modified, the memory of each crow is updated based on the objective function values of the new spots. So, if the objective function of the new location is better than the objective function of the memorized position, the crows update their memory to the new area, as shown in Equation (17).

$$M_{i,k+1} = \begin{cases} F(X_{i,k+1}) & \text{if } F(X_{i,k+1}) < F(M_{i,k}) \\ M_{i,k} & \text{otherwise} \end{cases}, \tag{17}$$

where $F(\cdot)$ represents the minimized objective function.

3.2.2. Proposed ICSA

The CSA has confirmed its capability to reach the optimal solution for particular solution space configurations [29,39,40]. Nevertheless, its convergence is not ensured due to the inefficient exploration of its search strategy [37]. Therefore, it presents difficulties when facing high-dimensional problems [37,41]. In the original CSA method [29], there are two elements responsible for the search process: knowledge probability (i.e., A_P) and random motion (i.e., Situation 2: Avoidance) [41].

The value of A_P is entrusted with providing an adequate equilibrium between diversification and intensification [29]. With small A_P values, a local solution space search is obtained, increasing intensification [29]. On the other hand, with large A_P values, a global solution space search is obtained, which increases diversification [29]. Since metaheuristic algorithms require a balance between diversification and intensification to find a globally optimal solution when solving problems with large dimensions [42,43], A_P is taken as an intermediate value, i.e., 50%.

Moreover, the random motion specifically impacts the CSA search mechanism since it resets the candidate solutions, deviating them from the current best solution and delaying the convergence of the problem [37]. In the proposed ICSA, the random motion is reformulated, as shown below.

3.2.3. Improved Approach: VSA for Random Movement

The evasion behavior is simulated by a random motion implementation computed through a uniformly distributed random value [29]. In the proposed ICSA, to have a better solution space diversification, the possibility of generated candidate solutions based on the VSA evolution criteria is added to the classical CSA [15], considering that the main feature of the VSA is the use of a regular GD to generate neighbors around the current best solution named as the center of the hyper-sphere μ [44]. In this paper, the hyper-sphere center is selected as the best solution in the current population during iteration k as $\mu_k = X_{best,k}$.

The set of candidate solutions $C_{v,k}(y)$ is generated using a random GD in the space d around the best solution $X_{best,k}$, as shown in Equation (18).

$$C_{v,k}(y) = p(y|\mu_k, \Sigma) = \frac{1}{\sqrt{(2\pi)^d |\Sigma|}} \exp\left\{-\frac{1}{2}(y-\mu_k)^T \Sigma^{-1}(y-\mu_k)\right\}, \quad (18)$$

where d is the solution space dimension, $y \in \mathbb{R}^{d \times 1}$ corresponds to a vector of random variables with values between zero and one, and $\Sigma \in \mathbb{R}^{d \times d}$ is the matrix of covariance. If, in Σ, the on-diagonal elements (variance) are equally defined and if the off-diagonal components (covariance) are chosen as zero, then the GD will generate hyper-spheres in the d-dimensional space [35]. Equation (19) displays a simple way to calculate Σ, considering equal variance and zero covariances.

$$\Sigma = \sigma^2 I_{d \times d}, \quad (19)$$

where $I_{d \times d}$ is an identity matrix and σ represents the variance of the GD. Note that the standard deviation of the GD can be defined as shown in Equation (20).

$$\sigma_0 = \frac{\max\{y^{\max}\} - \min\{y^{\min}\}}{2}, \quad (20)$$

where $y^{\max}$ and $y^{\min}$ are vectors of dimension $d \times 1$ that define the upper and lower bounds of the decision variables of the optimization problem, respectively. Here, σ_0 can also be considered as the initial radius r_0 of the hyper-sphere [35]. To achieve a proper exploration of the solution space, initially, σ_0 is the largest possible hyper-sphere.

The candidate solutions obtained and contained in the set $C_{v,k}(y)$ must guarantee that the results lie within the bounds of the solution space; so, Equation (21) is employed.

$$C_{v,k}(y) = \begin{cases} C_{v,k}(y) & y^{\min} \leq y \leq y^{\max} \\ y^{\min} + (y^{\max} - y^{\min})\text{rand} & \text{otherwise} \end{cases}, \quad (21)$$

where rand generates random numbers between 0 and 1. Once verified the limits, the best solution obtained in the set $C_{v,k}(y)$ is selected to be the new position of the crow i.

It is necessary to mention that the radius of the hyper-sphere decreases as the iteration process progresses using an inverse incomplete gamma function, as reported in [45] and as shown in Equation (22).

$$r_k = \sigma_0 \gamma^{-1}(y, a_k) \quad (22)$$

The inverse incomplete gamma function for the variable radius calculation can be calculated in MATLAB®, as shown in Equation (23) [35].

$$r_k = \sigma_0 \frac{1}{y} \text{gammaincinv}(y, a_k), \quad (23)$$

where a_k is a parameter defined as $a_k = 1 - \frac{k}{t_{\max}}$. Moreover, the parameter y is chosen as 0.1, as recommended in [35].

Algorithm 2 summarizes the implementation of the CSA, considering the VSA evolution criterion for Situation 2, to solve the phase swapping problem in ADSs.

Remark 5. *For the phase swapping problem solution, when using Algorithm 2, it is recommended to take y^{max} as 6.5 and y^{min} as 0.5. Further, d is the number of system demand nodes, except for the slack node.*

Remark 6. *The size of the solution set $C_{v,k}(y)$ is chosen as 30% of the CSA population size to minimize the number of evaluations needed in the slave stage to determine the power losses.*

Algorithm 2: General ICSA implementation for solving the phase swapping problem in ADSs
---
Read information from the AC distribution system;
Set the initial values $N, AP, fl,$ and t_{max};
Initialize the crows' position $X_{i,0}$ randomly;
Calculate the objective function value for each crow $F(x_{i,0})$ using Algorithm 1;
Select the initial memory value $M_{i,0}$ for each crow i;
Select the initial radius r_0 (or the standard deviation σ_0) of (20);
Set $k = 1$;
while $k \leq iter\ max$ **do**
    for $i = 1: N$ **do**
        Randomly select a crow j for tracking.;
        if $r_j \geq AP$ **then**
            $X_{i,k+1} = X_{i,k} + r_i \cdot fl_{i,k} \cdot (M_{j,k} - X_{i,k})$;
        else
            Determine the center of the hyper-sphere μ_k as $X_{best,k}$;
            Select the radius of the hyper-sphere r_k ;
            Define the individuals' number v as 30% of N;
            Create the set of candidate solutions $C_{v,k}(y)$ using (18);
            Check the lower and upper bounds for each v en $C_{v,k}(y)$ using (21);
            Calculate the objective function value for each crow v in $C_{v,k}(y)$ using Algorithm 1;
            Select $X_{i,k+1} =$ as the individual with the best solution of $C_{v,k}(y)$;
    end
    Verify feasibility of the new positions $X_{i,k+1}$;
    Evaluate the crows' new position $F(X_{i,k+1})$;
    Update the crows' memory $M_{i,k+1}$;
    Update the radius r_{k+1} as shown in en (22);
    $k = k + 1$;
end
Result: Report the best solution $X_{i,t_{max}}$, and its obj. func. value, i.e., $F(X_{i,t_{max}})$.

---

4. Three-Phase Test Feeders

We consider three test systems to validate the proposed ICSA. These test systems correspond to the 8, 25, and 37 node systems with radial topology reported in [15] for the phase swapping study using the VSA. Their main characteristics are presented below.

4.1. 8-Bus Test Feeder

The 8-node test system is a three-phase ADS formed by eight nodes and seven distribution lines, which operates with a nominal voltage of 11 kV at the main node. Figure 1 shows the electrical configuration of the test system. Note that the benchmark active power losses for this system take a value of 13.9925 kW. The electrical parameters for this test feeder can be found in [15].

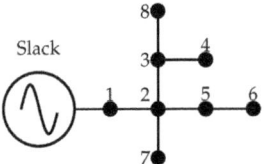

Figure 1. Electrical design of the 8-node system.

4.2. 25-Bus Test Feeder

The 25-node test system is a three-phase ADS formed by twenty five nodes and twenty four distribution lines, which operates with a nominal voltage of 4.16 kV at the main node. Figure 2 displays the grid configuration. Note that the benchmark active power losses for this system take a value of 75.4207 kW. The electrical parameters for this test feeder can be found in [15].

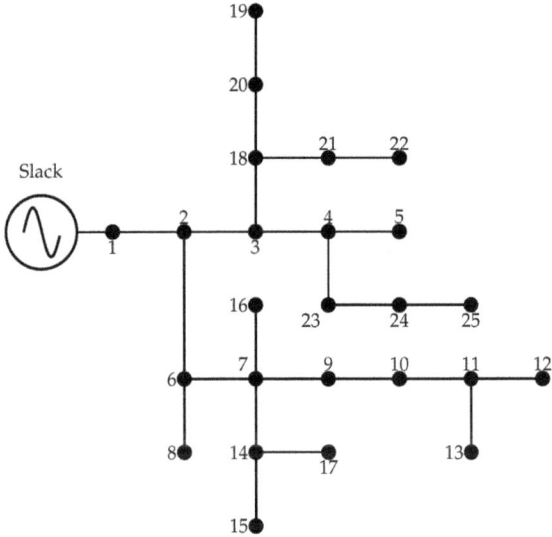

Figure 2. Electrical design of the 25-node system.

4.3. 37-Bus Test Feeder

The 37-node system is part of a current ADS, consisting entirely of subway lines situated in California, USA. It has thirty seven nodes and thirty five distribution lines, and it operates at a nominal voltage of 4.8 kV at the substation. Figure 3 shows the grid electrical design of this test feeder. Note that the benchmark active power losses for this system take a value of 76.1357 kW. The electrical parameters for this test feeder can be found in [15].

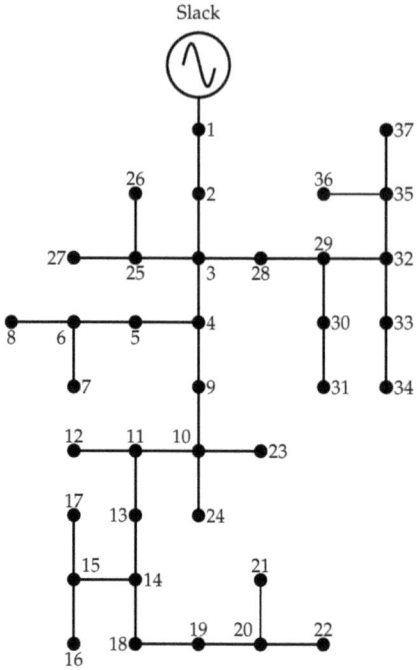

Figure 3. Electrical design of the 37-node system.

5. Numerical Simulations

This section contains the numerical validation of the developed methodology to solve the phase swapping problem in 8-, 25-, and 37-node test systems, considering a given demand condition. In that sense, it uses the information provided by [15], which presents two methodologies to solve the proposed problem. Note that the aim is to compare the results obtained by the proposed procedure with each optimization technique reported in [15].

To solve the MINLP, formulated from (1) through (6), that represents the optimal phase swapping problem for ADSs, MATLAB® V2020a software is used on a laptop computer with an Intel(R) Core(TM) i5-7200U CPU @2.50 Ghz, a RAM of 8.00 GB, and a Windows 10 Home Single Language 64-bit operating system.

To test the efficiency of the proposed algorithm, the ICSA is compared with the Chu and Beasley genetic algorithm (CBGA) [15], the VSA [15], and the CSA. The parameters used for the CSA and ICSA are as provided in Table 2. Furthermore, the parameters were established with ten individuals in the population, six hundred iterations, and one hundred evaluations to calculate the average processing time. The parameters for the CSA are those recommended by the author of the algorithm in [29].

Table 2. Algorithms parameters.

Parameter	CSA	ICSA
fl	2	2
A_p	0.1	0.5

5.1. Results for the 8-Node System

Table 3 shows the results obtained by the proposed ICSA for the 8-node test system as follows: (i) all methodologies allow a reduction of more than 24% in total power losses;

(ii) the solution obtained with the ICSA for the 8-node system equals those reported in [15], which states that the optimal global solution for this system is 10.5968 kW, albeit with a much longer processing time; and (iii) the standard deviation for the ICSA shown in Table 3 is 1.099×10^{-5} kW, which is at least ten times lower than in the VSA [15]. These results indicate that the repeatability of the solutions is close to 100% when solving the phase swapping problem in the 8-node test system, bearing in mind that the solution space has dimensions of 279,936.

Table 3. Performance of the power losses after implementing the phase-swapping plan in the 8-bus test feeder.

Method	Connections	Losses (kW)	Std. (kW)	Reduction (%)	Proc. Time (s)
Benchmark case	{1, 1, 1, 1, 1, 1, 1}	13.9925	-	-	-
CBGA [15]	{6, 1, 5, 1, 4, 4, 1}	10.5869	0.0897	24.34	2.8137
VSA [15]	{6, 1, 5, 1, 2, 1, 1}	10.5869	4×10^{-4}	24.34	6.059
CSA	{5, 3, 4, 5, 6, 6, 3}	10.5869	4.646×10^{-3}	24.34	8.7290
ICSA	{2, 4, 5, 4, 6, 4, 3}	10.5869	1.099×10^{-5}	24.34	47.7428

Figure 4 displays the variations of the phase power losses before and after the application of phase swapping with the ICSA, where the phase losses of a and b increase by 1.025 kW and 1.663 kW, respectively. On the contrary, the power losses in the electrical phase c decrease by about 6.10 kW, which increases the offsets seen in the power losses of phases a and b. Additionally, the power losses per phase are close to the average of the total power losses of approximately 3.50 kW, with differences of less than 0.80 kW. These results indicate that phase swapping by ICSA is an effective way to redistribute the loads in the phases of the system as evenly as possible.

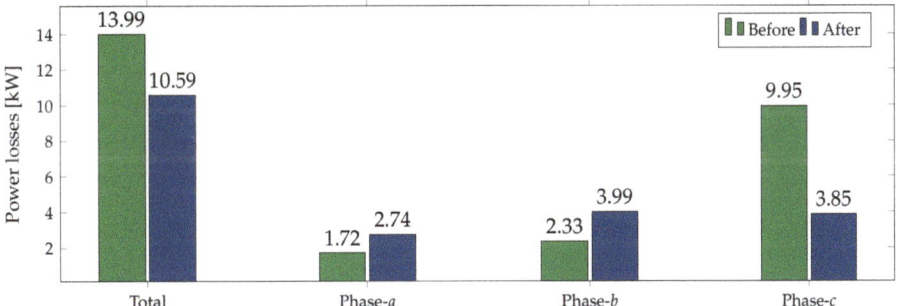

Figure 4. Effect of phase swapping on power losses in the 8-bus test feeder.

5.2. Results for the 25-Node System

Table 4 shows the results obtained by the proposed ICSA for the 25-node test system, where the following is evident: (i) the solution obtained with the ICSA for the 25-node system equals the one reported in [15] for the VSA, which states that the optimal global solution for this system is 72.2888 kW; (ii) the solution obtained with the ICSA for the 25-node system outperforms the solution obtained with the CSA, which shows that the improvement made to the classical CSA explores and exploits the solution space for systems of large dimensions (i.e., 24 nodes in this case) in a better way; and (iii) the standard deviation for the ICSA, shown in Table 4, is 0.0116 kW lower than those reported in [15] and in CSA. This affirms the repeatability properties of the ICSA for solving the phase swapping problem, considering the size of the solution space, i.e., 2.8430×10^{19}.

Table 4. Performance of the power losses after implementing the phase-swapping plan in the 25-bus test feeder.

Method	Connections	Losses (kW)	Std. (kW)	Proc. Time (s)
Benchmark case	{1,1,1,1,1,1,1,1,1,1,1,1,1,1,1,1,1,1,1,1,1,1,1,1}	75.4207	-	-
CBGA [15]	{1,1,3,5,2,1,1,1,2,6,5,1,5,3,6,6,3,3,1,3,5,2,4,3}	72.2919	0.0366	18.6683
VSA [15]	{1,2,4,5,6,1,2,3,1,5,4,3,3,5,5,2,3,3,5,4,2,2,2,3}	72.2888	0.0233	36.6900
CSA	{4,4,4,2,6,5,2,5,3,6,3,1,4,2,2,1,2,3,3,5,2,5,3,4}	72.3296	0.0225	29.4161
ICSA	{4,2,4,5,6,3,2,3,1,5,4,3,3,5,5,2,3,3,5,4,2,2,2,3}	72.2888	0.0116	134.8935

Figure 5 represents the variations of the phase power losses before and after the application of phase swapping with the ICSA, where the phase losses of b increase by 11.3776 kW. On the other hand, the phase power losses of a and c approximately decrease by 11.2156 kW and 3.294 kW, respectively, which offsets the increase seen in the electrical phase b power losses. Additionally, the phase power losses are close to the average of the total power losses, approximately 24 kW, with differences of less than 3.60 kW. These results indicate that phase swapping by ICSA is an effective way to redistribute the loads in the phases of the system as evenly as possible.

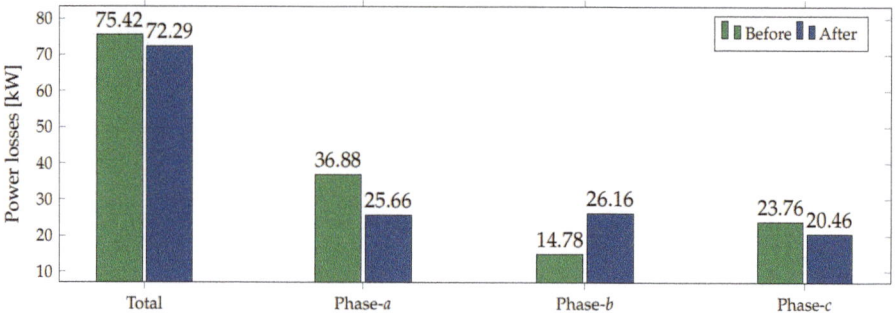

Figure 5. Effect of phase swapping on power losses in the 25-bus test feeder.

Figure 6 shows a comparison of the percentage reductions of the total active power losses of the different methods displayed in Table 4, in contrast to the initial power losses. All methodologies allow a cutback of more than 4%. ICSA obtains a reduction of 4.15%, akin to the VSA reported in [15].

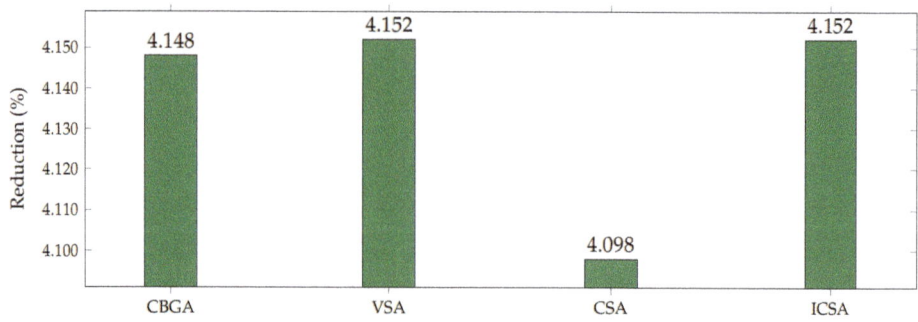

Figure 6. Reduction of the power losses in the 25-bus system.

5.3. Results for the 37-Node System

Table 5 shows the results obtained by the proposed ICSA for the 37-node test system, where the following is evident: (i) the solution obtained with the ICSA for the 37-node system improves the one reported in [15] for the VSA, which indicates that the optimal solution for this system is 61.4781 kW; (ii) the solution obtained with the ICSA for the 37-node system outperforms the solution obtained with the CSA, which shows that the improvement made to the classical CSA explores and exploits the solution space for systems of large dimensions (i.e., 37-nodes) in a better way; and (iii) the standard deviation for the ICSA shown in Table 5 is 0.1344 kW, which is lower than those reported in [15] and the one obtained by the CSA. This affirms the repeatability properties of the ICSA for solving the phase swapping problem, considering the size of the solution space, i.e., 6.1887×10^{28}.

Table 5. Performance of the power losses after implementing the phase-swapping plan in the 37-bus test feeder.

Method	Connections	Losses (kW)	Std. (kW)	Proc. Time (s)
Benchmark case	{1,1,1,1,1,1,1,1,1,1,1,1,1,1,1,1,1,1,1,1,1,1,1,1,1,1,1,1,1,1,1,1,1,1,1,1}	76.1357	-	-
CBGA [15]	{4,1,1,6,4,4,6,4,1,1,6,5,2,1,2,3,1,5,1,4,3,2,6,5,3,2,1,6,5,2,1,4,1,2,3}	61.5785	0.4274	14.1816
VSA [15]	{4,1,1,5,3,4,2,3,1,1,3,2,2,1,3,5,2,3,1,3,6,1,2,3,3,2,1,1,2,4,1,4,1,2,4}	61.4801	0.3286	50.0262
CSA	{4,5,4,4,4,4,5,4,3,3,4,4,1,4,5,3,3,2,3,3,4,3,4,3,3,4,4,4,5,4,4,4,3,2,3}	61.6565	0.2975	89.1698
ICSA	{4,4,4,3,5,3,3,5,6,5,3,2,4,5,3,5,5,3,5,5,6,5,2,6,6,4,6,5,4,4,3,2,5,4,2}	61.4781	0.1344	238.3614

Figure 7 portrays the variations of the phase power losses before and after the application of phase swapping with the ICSA, where the phase b losses increase by 9.9184 kW. On the contrary, the phase power losses in a and c decrease by approximately 6.7771 kW and 17.7989 kW, respectively, offsetting the increase seen in the electrical phase b power losses. Additionally, the phase power losses are close to the average total power losses of approximately 20.50 kW, with differences of less than 1.40 kW. These results indicate that phase swapping by ICSA is an effective way to redistribute the loads in the phases of the system as evenly as possible.

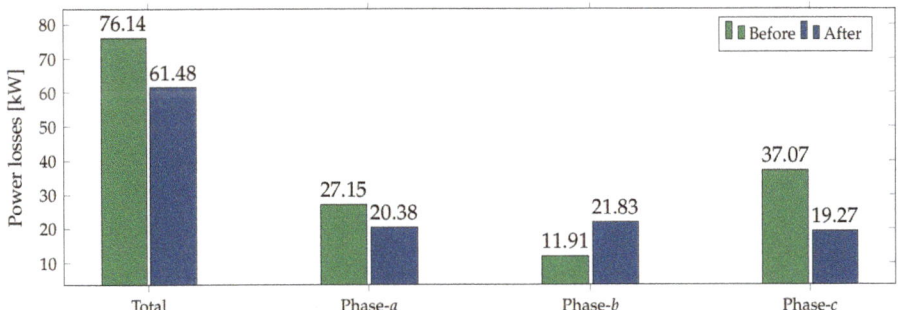

Figure 7. Effect of phase swapping on power losses in the 37-bus test feeder.

Figure 8 compares the reductions in the total active power losses percentage of the different methods presented in Table 5 to the initial power losses. All methodologies allow a decrease of more than 19%, where ICSA obtains a reduction of 19.252%, a higher loss than VSA, as reported in [15], which reported a decrease of 19.249%.

In Figure 8, it is possible to observe that the proposed ICSA allows an additional improvement about 0.003% when compared to the power losses reduction with the VSA. This reduction implies a difference of 0.1784 kW in the total power losses minimization for the IEEE 37-bus system. Even if this power losses value corresponds to a small power losses reduction for this system, this demonstrates that the proposed ICSA finds an optimal solution for the IEEE 37-bus system, which supports the best current literature report obtained by [15] with the VSA method. Thus, this new solution will serve as a reference point for future approaches that can be proposed to solve the phase swapping problem in ADSs.

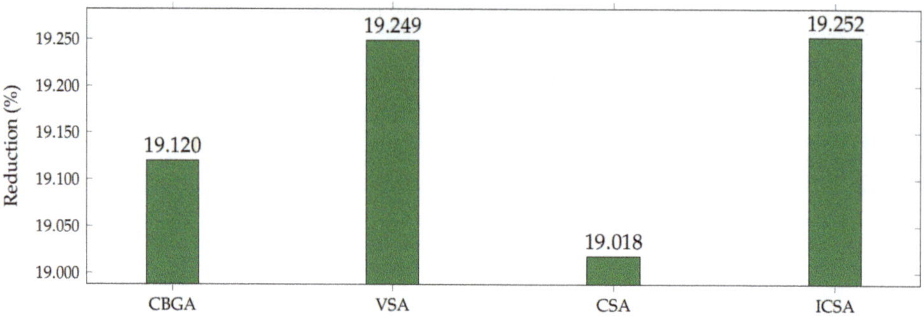

Figure 8. Reduction of the power losses in the 37-bus test feeder.

5.4. Complementary Results

The following can be concluded from the results obtained in Section 5.

- ✓ The optimal solution achieved with the enhanced version of the CSA for each test system is equal to the one reported in [15] for the 8- and 25-node systems. However, it obtained a better solution for the 37-node system. Note that the obtained solutions (power losses) in this research are 10.5869 kW, 72.2888 kW, and 61.4781 kW for the 8-, 25-, and 37-node systems, respectively, while the best solutions reported in [15] are 10.5869 kW, 72.2888 kW, and 61.4801 kW, respectively. It is relevant to highlight that the ICSA requires a longer processing time. Nevertheless, these times do not exceed 6 minutes, which is not significant considering the terms for optimization problems with solution spaces with hundreds of thousands of combinations. It ensures excellent quality solutions and even better ones than the values reported in the literature review in some cases (see results for the 37-node system). However, simulations in tests feeders with large number of nodes will be required to ensure that, in all of the cases, the processing times spent by the proposed ICSA will be compensated with optimal solutions better than the solutions provided by the VSA.
- ✓ The standard deviations reported in Tables 3–5 for the 8-, 25-, and 37-node systems, respectively, for ICSA are lower than those reported in [15]. In addition, a standard deviation of 0.1344 kW for the 37-node test system demonstrates the repeatability properties of the ICSA for solving the phasing problem, considering that the dimensions of the solution space are higher than 1×10^{28}. Regarding metaheuristic optimization methods, the main preoccupation in the literature is associated with the ability of these methods to find the same optimal solution at each simulation. However, this is not possible due to the random nature of the exploration and exploitation criteria inside of each metaheuristic optimizer. Nevertheless, when an optimization methodology exhibits low standard deviations, all the solutions are contained inside of a small hyper-sphere close to the global optimum. This improves the optimization method of the proposed ICSA when compared with a family of metaheuristic optimizers.
- ✓ When comparing the base cases of each test system with the proposed methodologies, as shown in Table 3 and Figures 5 and 7, the reductions in energy losses resulting from ICSA in the 8, 25, and 37-node systems are 24.34%, 4.152%, and 19.252%, respectively. The results show that better results are obtained when applying the proposed methodology for the 37-node system, in contrast to the CSA and the VSA [15]. Likewise, it is observed that the proposed enhancement for the CSA effectively explores and exploits the solution space for large systems higher than 25 nodes in the case of distribution power systems.

✓ The comparison of the effect in the energy losses redistribution at each phase using the classic and improved CSA methods is presented in Table 6.

Table 6. Comparison between the CSA and the proposed ICSA regarding power losses among phases.

Method	Phase a (kW)	Phase b (kW)	Phase b (kW)	Total (kW)
8-bus system				
Benchmark case	1.7158	2.3305	9.9462	13.9925
CSA	3.7617	2.7295	4.0957	10.5869
ICSA	2.7412	3.9930	3.8464	10.5869
25-bus system				
Benchmark case	36.8801	14.7837	23.7570	75.4207
CSA	24.9590	26.2079	21.1627	72.3296
ICSA	25.6645	26.1613	20.4630	72.2888
IEEE 37-bus system				
Benchmark case	27.1532	11.9143	37.0683	76.1357
CSA	19.1177	20.5926	21.1627	61.6565
ICSA	20.3761	21.8327	19.2694	61.4781

With these results, we can make several observations. (i) In the 8-bus system, the phase c presents power losses higher than 4 kW, while the ICSA does not support this value; however, both solutions are indeed optimal since the total power losses is the same for both methods. (ii) In the 25-bus system both methods, i.e., the CSA and its improved version, the phase c has been identified to present higher power losses surpassing 26 kW; however, regarding the final power losses, the ICSA presents better load redistribution, since the amount of power losses is about 0.0408 kW. Finally, (iii) in the 37-bus system, the CSA method presents a difference between phases a and b power losses of 2.0450 kW, while the proposed ICSA has a minor difference between both phases with a value of 1.1067 kW. This directly impacts the final grid results with a general improvement of about 0.1784 kW in favor of the proposed ICSA, which is indeed the best global optimum reported in the current literature for the IEEE 37-bus system.

6. Conclusions and Future Works

This paper presented a master–slave methodology to solve the phase equilibrium problem in ADSs. For this purpose, an ICSA was proposed using the VSA evolution mechanism. The master stage determined the set of phase configurations for the system three-phase charges through the ICSA, using an integer encoding between 1 and 6 representing the load connections in the three-phase system. On the other hand, the slave stage evaluated each of the load connections in the TPPF, which correspond to the extended version of the iterative sweep power flow method for unbalanced three-phase systems.

The numerical results on the 8-, 25-, and 37-node test systems showed that the proposed ICSA, compared to the VSA, achieves identical solutions for the 8-node and 25-node systems. However, for the 37-node system, the ISCA obtains a better optimal solution when compared with the current report employing the VSA method. The solutions obtained by the ICSA are 10.5869 kW, 72.2888 kW, and 61.4781 kW for the 8-, 25-, and 37-node systems, representing a reduction of 24.34%, 4.152%, and 19.252%, respectively. In the same way, the solutions for the VSA are 10.5869 kW, 72.2888 kW, and 61.4801 kW, respectively.

In addition, the proposed methodology has minor standard deviations for solving the phase swapping problem for the 8-, 25-, and IEEE 37-node test systems, which were 1.099×10^{-5} kW, 0.0116 kW, and 0.1344 kW, respectively. This demonstrates better repeatability properties of the improved algorithm, since all the solutions are contained inside small hyper-spheres around the global optimum. Further, the solution space for the phase equilibrium problem potentially increases as a function of the demand nodes. Thus, for the

IEEE 37-node system, there is a solution space bigger than 1×10^{28}, which is by far higher than 100 billion combinations. This result confirmed the effectiveness of the proposed methodology for solving complex MINLP models such as the phase equilibrium problem in ADSs as well as in being better than the optimal solution reported in the current literature using the VSA.

In future work, it is possible to accomplish the following: (i) combine the proposed phase swapping with the optimal placement of distributed generators to reduce power losses in ADSs; (ii) employ typical active and reactive power demand curves to solve the phase swapping problem using the ICSA to reduce power losses; and (iii) use the ICSA to solve the optimal reconfiguration problem in three-phase radial distribution networks.

Author Contributions: Conceptualization, methodology, software, and writing—review and editing, B.C.-C., L.S.A.-G., O.D.M., L.A.-B. and C.Á.-A. All authors have read and agreed to the published version of the manuscript.

Funding: This work was partially supported in part by the Laboratorio de Simulación Hardware-in-the-loop para Sistemas Ciberfísicos de la Universidad Loyola Andalucía.

Institutional Review Board Statement: Not applicable.

Informed Consent Statement: Not applicable.

Data Availability Statement: No new data were created or analyzed in this study. Data sharing is not applicable to this article.

Acknowledgments: This work was supported in part by the Centro de Investigación y Desarrollo Científico de la Universidad Distrital Francisco José de Caldas under grant 1643-12-2020 associated with the project: "Desarrollo de una metodología de optimización para la gestión óptima de recursos energéticos distribuidos en redes de distribución de energía eléctrica." and in part by the Dirección de Investigaciones de la Universidad Tecnológica de Bolívar under grant PS2020002 associated with the project: "Ubicación óptima de bancos de capacitores de paso fijo en redes eléctricas de distribución para reducción de costos y pérdidas de energía: Aplicación de métodos exactos y metaheurísticos."

Conflicts of Interest: The authors declare no conflict of interest.

References

1. Montoya, O.D.; Serra, F.M.; De Angelo, C.H. On the efficiency in electrical networks with ac and dc operation technologies: A comparative study at the distribution stage. *Electronics* **2020**, *9*, 1352. [CrossRef]
2. Montoya, O.D.; Gil-González, W.; Hernández, J.C. Efficient Operative Cost Reduction in Distribution Grids Considering the Optimal Placement and Sizing of D-STATCOMs Using a Discrete-Continuous VSA. *Appl. Sci.* **2021**, *11*, 2175. [CrossRef]
3. Nassar, M.E.; Salama, M. A novel branch-based power flow algorithm for islanded AC microgrids. *Electr. Power Syst. Res.* **2017**, *146*, 51–62. [CrossRef]
4. Aboshady, F.; Thomas, D.W.; Sumner, M. A wideband single end fault location scheme for active untransposed distribution systems. *IEEE Trans. Smart Grid* **2019**, *11*, 2115–2124. [CrossRef]
5. Montoya, O.D.; Arias-Londoño, A.; Grisales-Noreña, L.F.; Barrios, J.Á.; Chamorro, H.R. Optimal Demand Reconfiguration in Three-Phase Distribution Grids Using an MI-Convex Model. *Symmetry* **2021**, *13*, 1124. [CrossRef]
6. Arias, J.; Calle, M.; Turizo, D.; Guerrero, J.; Candelo-Becerra, J.E. Historical load balance in distribution systems using the branch and bound algorithm. *Energies* **2019**, *12*, 1219. [CrossRef]
7. Alvarado-Barrios, L.; Álvarez-Arroyo, C.; Escaño, J.M.; Gonzalez-Longatt, F.M.; Martinez-Ramos, J.L. Two-Level Optimisation and Control Strategy for Unbalanced Active Distribution Systems Management. *IEEE Access* **2020**, *8*, 197992–198009. [CrossRef]
8. Granada Echeverri, M.; Gallego Rendón, R.A.; López Lezama, J.M. Optimal phase balancing planning for loss reduction in distribution systems using a specialized genetic algorithm. *Ing. Cienc.* **2012**, *8*, 121–140. [CrossRef]
9. Hooshmand, R.; Soltani, S. Simultaneous optimization of phase balancing and reconfiguration in distribution networks using BF–NM algorithm. *Int. J. Electr. Power Energy Syst.* **2012**, *41*, 76–86. [CrossRef]
10. Ogunsina, A.A.; Petinrin, M.O.; Petinrin, O.O.; Offornedo, E.N.; Petinrin, J.O.; Asaolu, G.O. Optimal distributed generation location and sizing for loss minimization and voltage profile optimization using ant colony algorithm. *SN Appl. Sci.* **2021**, *3*, 1–10. [CrossRef]
11. Asadi, M.; Shokouhandeh, H.; Rahmani, F.; Hamzehnia, S.M.; Harikandeh, M.N.; Lamouki, H.G.; Asghari, F. Optimal placement and sizing of capacitor banks in harmonic polluted distribution network. In Proceedings of the 2021 IEEE Texas Power and Energy Conference (TPEC), College Station, TX, USA, 2–5 February 2021; pp. 1–6.

12. Jakus, D.; Čađenović, R.; Vasilj, J.; Sarajčev, P. Optimal reconfiguration of distribution networks using hybrid heuristic-genetic algorithm. *Energies* **2020**, *13*, 1544. [CrossRef]
13. Montoya, O.D.; Garces, A.; Castro, C.A. Optimal conductor size selection in radial distribution networks using a mixed-integer non-linear programming formulation. *IEEE Lat. Am. Trans.* **2018**, *16*, 2213–2220. [CrossRef]
14. Łukaszewski, A.; Nogal, Ł.; Robak, S. Weight Calculation Alternative Methods in Prime's Algorithm Dedicated for Power System Restoration Strategies. *Energies* **2020**, *13*, 6063. [CrossRef]
15. Cortés-Caicedo, B.; Avellaneda-Gómez, L.S.; Montoya, O.D.; Alvarado-Barrios, L.; Chamorro, H.R. Application of the Vortex Search Algorithm to the Phase-Balancing Problem in Distribution Systems. *Energies* **2021**, *14*, 1282. [CrossRef]
16. Montoya, O.D.; Molina-Cabrera, A.; Grisales-Noreña, L.F.; Hincapié, R.A.; Granada, M. Improved Genetic Algorithm for Phase-Balancing in Three-Phase Distribution Networks: A Master-Slave Optimization Approach. *Computation* **2021**, *9*, 67. [CrossRef]
17. Saad Al-Sumaiti, A.; Kavousi-Fard, A.; Salama, M.; Pourbehzadi, M.; Reddy, S.; Rasheed, M.B. Economic Assessment of Distributed Generation Technologies: A Feasibility Study and Comparison with the Literature. *Energies* **2020**, *13*, 2764. [CrossRef]
18. Rajaram, R.; Kumar, K.S.; Rajasekar, N. Power system reconfiguration in a radial distribution network for reducing losses and to improve voltage profile using modified plant growth simulation algorithm with Distributed Generation (DG). *Energy Rep.* **2015**, *1*, 116–122. [CrossRef]
19. Ahshan, R. Analysis of Loss Reduction Techniques for Low Voltage Distribution Network. *J. Eng. Res. [TJER]* **2020**, *17*, 100–111.
20. Grigoraş, G.; Neagu, B.C.; Gavrilaş, M.; Triştiu, I.; Bulac, C. Optimal phase load balancing in low voltage distribution networks using a smart meter data-based algorithm. *Mathematics* **2020**, *8*, 549. [CrossRef]
21. Montoya, O.D.; Giraldo, J.S.; Grisales-Noreña, L.F.; Chamorro, H.R.; Alvarado-Barrios, L. Accurate and Efficient Derivative-Free Three-Phase Power Flow Method for Unbalanced Distribution Networks. *Computation* **2021**, *9*, 61. [CrossRef]
22. Taghipour Boroujeni, S.; Mardaneh, M.; Hashemi, Z. A dynamic and heuristic phase balancing method for LV feeders. *Appl. Comput. Intell. Soft Comput.* **2016**, *2016*, doi:10.1155/2016/6928080. [CrossRef]
23. Gandomkar, M. Phase balancing using genetic algorithm. In Proceedings of the 39th International Universities Power Engineering Conference, 2004, UPEC 2004, Bristol, UK, 6–8 September 2004; Volume 1, pp. 377–379.
24. Rios, M.A.; Castaño, J.C.; Garcés, A.; Molina-Cabrera, A. Phase Balancing in Power Distribution Systems: A heuristic approach based on group-theory. In Proceedings of the 2019 IEEE Milan PowerTech, Milan, Italy, 23–27 June 2019; pp. 1–6.
25. Garces, A.; Gil-González, W.; Montoya, O.D.; Chamorro, H.R.; Alvarado-Barrios, L. A Mixed-Integer Quadratic Formulation of the Phase-Balancing Problem in Residential Microgrids. *Appl. Sci.* **2021**, *11*, 1972. [CrossRef]
26. Amon, D.A.; Adeyemi, A. A modified bat algorithm for power loss reduction in electrical distribution system. *Indones. J. Electr. Eng. Comput. Sci. (IJEECS)* **2015**, *14*, 55–61.
27. Sathiskumar, M.; kumar, A.N.; Lakshminarasimman, L.; Thiruvenkadam, S. A self adaptive hybrid differential evolution algorithm for phase balancing of unbalanced distribution system. *Int. J. Electr. Power Energy Syst.* **2012**, *42*, 91–97. [CrossRef]
28. Zhu, J.; Bilbro, G.; Chow, M.Y. Phase balancing using simulated annealing. *IEEE Trans. Power Syst.* **1999**, *14*, 1508–1513. [CrossRef]
29. Askarzadeh, A. A novel metaheuristic method for solving constrained engineering optimization problems: Crow search algorithm. *Comput. Struct.* **2016**, *169*, 1–12. [CrossRef]
30. Dogan, B. A Modified Vortex Search Algorithm for Numerical Function Optimization. *Int. J. Artif. Intell. Appl.* **2016**, *7*, 37–54. [CrossRef]
31. Shen, T.; Li, Y.; Xiang, J. A graph-based power flow method for balanced distribution systems. *Energies* **2018**, *11*, 511. [CrossRef]
32. Herrera-Briñez, M.C.; Montoya, O.D.; Alvarado-Barrios, L.; Chamorro, H.R. The Equivalence between Successive Approximations and Matricial Load Flow Formulations. *Appl. Sci.* **2021**, *11*, 2905. [CrossRef]
33. Łukaszewski, A.; Nogal, Ł. Influence of lightning current surge shape and peak value on grounding parameters. *Bull. Pol. Acad. Sci. Tech. Sci.* **2021**, *69*, e136730. [CrossRef]
34. Gil-González, W.; Montoya, O.D.; Rajagopalan, A.; Grisales-Noreña, L.F.; Hernández, J.C. Optimal Selection and Location of Fixed-Step Capacitor Banks in Distribution Networks Using a Discrete Version of the Vortex Search Algorithm. *Energies* **2020**, *13*, 4914. [CrossRef]
35. Doğan, B.; Ölmez, T. A new metaheuristic for numerical function optimization: Vortex Search algorithm. *Inf. Sci.* **2015**, *293*, 125–145. [CrossRef]
36. Hussien, A.G.; Amin, M.; Wang, M.; Liang, G.; Alsanad, A.; Gumaei, A.; Chen, H. Crow Search Algorithm: Theory, Recent Advances, and Applications. *IEEE Access* **2020**, *8*, 173548–173565. [CrossRef]
37. Jain, M.; Rani, A.; Singh, V. An improved Crow Search Algorithm for high-dimensional problems. *J. Intell. Fuzzy Syst.* **2017**, *33*, 3597–3614. [CrossRef]
38. Clayton, N.; Emery, N. Corvid cognition. *Curr. Biol.* **2005**, *15*, R80–R81. [CrossRef]
39. Askarzadeh, A. Capacitor placement in distribution systems for power loss reduction and voltage improvement: A new methodology. *IET Gener. Transm. Distrib.* **2016**, *10*, 3631–3638. [CrossRef]
40. Rajput, S.; Parashar, M.; Dubey, H.M.; Pandit, M. Optimization of benchmark functions and practical problems using Crow Search Algorithm. In Proceedings of the 2016 Fifth International Conference on Eco-friendly Computing and Communication Systems (ICECCS), Bhopal, India, 8–9 December 2016; pp. 73–78.

41. Díaz, P.; Pérez-Cisneros, M.; Cuevas, E.; Avalos, O.; Gálvez, J.; Hinojosa, S.; Zaldivar, D. An improved crow search algorithm applied to energy problems. *Energies* **2018**, *11*, 571. [CrossRef]
42. Li, Z.; Wang, W.; Yan, Y.; Li, Z. PS–ABC: A hybrid algorithm based on particle swarm and artificial bee colony for high-dimensional optimization problems. *Expert Syst. Appl.* **2015**, *42*, 8881–8895. [CrossRef]
43. Sayed, G.I.; Hassanien, A.E.; Azar, A.T. Feature selection via a novel chaotic crow search algorithm. *Neural Comput. Appl.* **2019**, *31*, 171–188. [CrossRef]
44. Li, P.; Zhao, Y. A quantum-inspired vortex search algorithm with application to function optimization. *Nat. Comput.* **2019**, *18*, 647–674. [CrossRef]
45. Montoya, O.D.; Grisales-Noreña, L.F.; Amin, W.T.; Rojas, L.A.; Campillo, J. Vortex Search Algorithm for Optimal Sizing of Distributed Generators in AC Distribution Networks with Radial Topology. In *Workshop on Engineering Applications*; Springer: Berlin/Heidelberg, Germany, 2019; pp. 235–249. [CrossRef]

Article

Optimal Demand Reconfiguration in Three-Phase Distribution Grids Using an MI-Convex Model

Oscar Danilo Montoya [1,2], Andres Arias-Londoño [3], Luis Fernando Grisales-Noreña [3], José Ángel Barrios [4,*] and Harold R. Chamorro [5]

1. Facultad de Ingeniería, Universidad Distrital Francisco José de Caldas, Bogotá 110231, Colombia; odmontoyag@udistrital.edu.co
2. Laboratorio Inteligente de Energía, Universidad Tecnológica de Bolívar, Cartagena 131001, Colombia
3. Facultad de Ingeniería, Institución Universitaria Pascual Bravo, Campus Robledo, Medellín 050036, Colombia; andres.arias366@pascualbravo.edu.co (A.A.-L.); luis.grisales@pascualbravo.edu.co (L.F.G.-N.)
4. Universidad Politécnica de García, Prolongación 16 de Septiembre, Col. Valles de San José, García 66004, NL, Mexico
5. Department of Electrical Engineering at KTH, Royal Institute of Technology, SE-100 44 Stockholm, Sweden; hr.chamo@ieee.org
* Correspondence: joseangel_barrios@yahoo.com.mx

Citation: Montoya, O.D.; Arias-Londoño, A.; Grisales-Noreña, L.F.; Barrios, J.Á.; Chamorro, H.R. Optimal Demand Reconfiguration in Three-Phase Distribution Grids Using an MI-Convex Model. *Symmetry* **2021**, *13*, 1–15. https://doi.org/10.3390/sym13071124

Academic Editors: Juan Alberto Rodríguez Velázquez and Alejandro Estrada-Moreno

Received: 30 May 2021
Accepted: 23 June 2021
Published: 24 June 2021

Publisher's Note: MDPI stays neutral with regard to jurisdictional claims in published maps and institutional affiliations.

Copyright: © 2021 by the authors. Licensee MDPI, Basel, Switzerland. This article is an open access article distributed under the terms and conditions of the Creative Commons Attribution (CC BY) license (https://creativecommons.org/licenses/by/4.0/).

Abstract: The problem of the optimal load redistribution in electrical three-phase medium-voltage grids is addressed in this research from the point of view of mixed-integer convex optimization. The mathematical formulation of the load redistribution problem is developed in terminals of the distribution node by accumulating all active and reactive power loads per phase. These loads are used to propose an objective function in terms of minimization of the average unbalanced (asymmetry) grade of the network with respect to the ideal mean consumption per-phase. The objective function is defined as the l_1-norm which is a convex function. As the constraints consider the binary nature of the decision variable, each node is conformed by a 3 × 3 matrix where each row and column have to sum 1, and two equations associated with the load redistribution at each phase for each of the network nodes. Numerical results demonstrate the efficiency of the proposed mixed-integer convex model to equilibrate the power consumption per phase in regards with the ideal value in three different test feeders, which are composed of 4, 15, and 37 buses, respectively.

Keywords: load redistribution; leveling power consumption per phase; three-phase asymmetric distribution networks; ideal power consumption; mixed-integer convex optimization

1. Introduction

Most of the electricity users are typically connected to medium- and low-voltage levels, corresponding to three-phase distribution system structures [1]. The main characteristics of these networks are: (i) the radial connection among nodes helps to reduce the investment costs in protective schemes [2]; (ii) the existence of multiple single-, two-, and three-phase loads produce current unbalances that increases the amount of power losses with respect to the perfectly balanced load scenario [3]; and the high grade of active and reactive power imbalances in terminals of the substation causes deterioration of voltage profile in the nodes located at the end of the feeder [4]. The importance of the three-phase distribution networks to supply medium- and low-voltage users shows the need of proposing optimization methodologies to improve their electrical performance when the connection of new loads is required and the consumption at industrial nodes is increased [5]. The most common methodologies to improve the operative performance of the distribution networks are: optimal placement and sizing of reactive power compensators, i.e., capacitor banks and static distribution compensators [6,7]; optimal placement and sizing of disperse generation [8,9]; optimal grid reconfiguration [10]; and optimal phase-balancing [3,11].

Note that the first two methodologies are based on the connection of new devices (shunt generators and compensators) to the network, which implies large amounts of investment to improve the quality of the grid, hardly recovered in short periods of time, i.e., 5 to 15 years [12–14]; the third methodology based on grid reconfiguration involves moderate investments in tie-lines and reconfiguration of protective devices [15]; whereas the phase-balancing method is the most simple strategy to reduce power losses in three-phase distribution networks with minimum investment efforts, since devices are not required to implement the phase-balancing plan and it is only necessary to send few working crews along with the grid infrastructure to interchange the phase connections in the required nodes [3,16,17].

In the current literature can be found multiple optimization strategies, most of them based on evolutionary optimization algorithms to address the problem of the phase-balancing in three-phase networks. Some of these works apply to the following optimization methods: genetic algorithms [18,19]; tabu search algorithm [20]; particle swarm optimization [21]; ant colony optimization [22], and the vortex search algorithm [3], among others. The main characteristic of these evolutionary algorithms is the master–slave optimization strategy, where the master stage is entrusted of defining the connection of the loads using an integer or binary codification, while the slave stage determines the amount of energy losses at each connection provided by the master stage. Even if the master–slave optimization approach is widely accepted in the current literature, its main problem arises with not ensuring the global optimum finding. Recently, authors of [23] have proposed a mixed-integer quadratic programming model that allows to ensure the global optimum finding of the phase-balancing problem in three-phase networks; however, the effectiveness of this methodology was only tested in a small low-voltage microgrid. This fact has reduced the real impact of the convex methodologies in the general operation improvement of the electrical networks.

Based on the aforementioned arguments, this research deals with a problem similar to the phase-balancing approach, which is known as the load redistribution at terminals of the substation, i.e., the problem studied corresponds to leveling the active power demands per phase, making these to be accumulated in the main bus of the network (without considering the effect of the three-phase lines). The main advantage of this optimization problem is that it can be formulated with a mixed-integer convex (MIC) model that allows to ensure the global optimum finding by combining the Branch & and Bound method with the interior point method. The main contributions of this research are listed below:

- The formulation of a MIC to represent the problem of the load redistribution in terminals of the substation that guarantees the global optimum finding with convex optimization tools that deal with integer problems [23];
- The evaluation of each solution provided by the MIC in a three-phase asymmetric power flow method based on its matrix formulation by rotating all the loads connected at the nodes, to find and evaluate the load redistribution configuration with minimum power losses.

The remainder of this document is structured as presented below:

Section 2 presents the proposed mixed-integer convex optimization model to represent the problem of the load redistribution in the terminals of the main substation; Section 3 presents the main characteristics of the optimization methodology based on the combination of the branch and bound (B&B) method with linear programming methods, ensuring the global optimum finding for MIC optimization models; Section 4 presents the main characteristics of the test feeders composed of 4, 15, and 37 nodes used to validate the proposed MIC optimization model using the CVX optimization tool with the MOSEK solver in the MATLAB programming environment; Section 5 describes the main numerical achievements in the three test feeders under study regarding the minimization of the general average grade of unbalance in terminals of the substation and the grid power losses. Finally, Section 6 presents the main concluding remarks derived from this work based on the main numerical results obtained after solving the proposed MIC model with the MOSEK solver.

2. Exact Mathematical Formulation

In general terms, the problem of the load redistribution in electrical three-phase medium-voltage grids is a mixed-integer non-linear programming model due to the presence of the power balance equations [23]; however, here, we propose a mixed-integer convex (MIC) model that allows redistributing the loads at all the nodes by the accumulation in the main substation (i.e., by neglecting the effect of the electrical distribution grid) [11]. The main objective of the MIC is to find the global optimum by combining the (B&B) method with linear programming search methods. Figure 1 presents a schematic model of the load redistribution in a particular node of the network before and after the solution of the proposed MIC model.

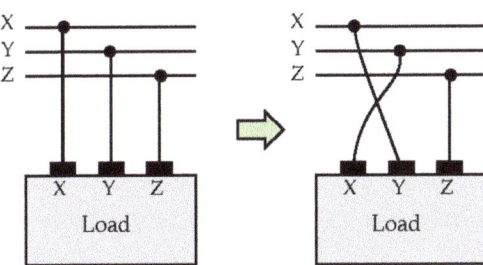

Figure 1. Redistribution of the load in a particular node of the network after solving the proposed MIC model.

Note that the main objective of the proposed MIC model is to reduce the level of asymmetry within all the loads of the network, analyzed at terminal of the main substation, i.e., redistribute all the loads in the nodes to reach the maximum level of balanced (symmetry) in the power consumption.

The complete optimization model for load redistribution in electrical asymmetric networks is fully described below.

2.1. Objective Function

The objective function of the problem is the minimization of general grid unbalance of the network with respect to the ideal consumption per phase. The objective function proposed is defined by Equation (1).

$$\min U_\% = \left(\frac{100}{3P_{ave}}\right) \sum_{f \in \mathcal{F}} \left|P_f - P_{ave}\right|, \tag{1}$$

where $U_\%$ represents the average grid unbalance; P_{ave} corresponds to the average active power consumption per phase, which is calculated as the total active power load divided by three; and P_f is the total active power consumption per phase. Note that the $\mathcal{F}$ represents the set that contains all the phases of the network.

Remark 1. *The main advantage of the objective function defined in (1) comes from the fact that this corresponds to the l_1−norm (i.e., absolute value) which is a convex function. The convexity property is important since it is possible to ensure the global optimum finding of the optimization problem if, and only if, the set of constraints are also convex, or by using the MIC through the combination of the (B&B) method with the Simplex method.*

2.2. Set of Constraints

The problem of the load redistribution is subject to linear constraints which correspond to the load reconfiguration at each phase, the binary nature of the decision variable, and the calculation of the total load per phase, among others. The complete list of constraints is defined as follows:

$$P_{kf}^d = \sum_{g \in \mathcal{G}} x_{kfg} P_{kg}^d, \quad \{\forall k \in \mathcal{N}, \forall f \in \mathcal{F}\}, \tag{2}$$

$$Q_{kf}^d = \sum_{g \in \mathcal{G}} x_{kfg} Q_{kg}^d, \quad \{\forall k \in \mathcal{N}, \forall f \in \mathcal{F}\}, \tag{3}$$

$$P_f = \sum_{k \in \mathcal{N}} P_{kf}^d, \quad \{\forall f \in \mathcal{F}\}, \tag{4}$$

$$Q_f = \sum_{k \in \mathcal{N}} Q_{kf}^d, \quad \{\forall f \in \mathcal{F}\}, \tag{5}$$

$$\sum_{g \in \mathcal{F}} x_{kfg} = 1, \quad \{\forall k \in \mathcal{N}, \forall f \in \mathcal{F}\}, \tag{6}$$

$$\sum_{f \in \mathcal{F}} x_{kfg} = 1, \quad \{\forall k \in \mathcal{N}, \forall g \in \mathcal{F}\}, \tag{7}$$

where P_{kf}^d and Q_{kf}^d are the active and reactive power connected at node k in phase f after the redistribution of the loads; P_{kg}^d and Q_{kg}^d correspond to the active and reactive power connected at the node k in the phase f before the redistribution of the loads; x_{kfg} is the binary variable that determines if the load connected in phase g is reassigned to the phase f in the node k; P_f represents the total active power consumption of the network in the phase f after redistributing all the loads; Q_f defines the total reactive power consumption of the network in the phase f after redistributing all the loads. Note that $\mathcal{N}$ represents the set that contains all the buses of the network.

The set of constraints defined from (2) to (7) are explained as follows: Equations (2) and (3) determine the amount of active and reactive power consumption at each phase and node after redistributing the loads in all the buses and phases of the network. Equations (4) and (5) determine the final equivalent active and reactive power consumption in the terminals of the substation after redistributing all the loads in order to minimize the average unbalance of the network; finally, Equations (6) and (7) ensure that each load is uniquely connected to one phase of the network.

Remark 2. *The general structure of the set of constraints above presented show that the optimization model defined from (1) to (7) is indeed MIC, which implies that its optimal solution is achievable with conventional mixed-integer optimization methods [24].*

Remark 3. *The solution of the optimization model presented in (1) to (7) with conventional optimization techniques such as the (B&B) and interior point methods ensures the global optimum finding based on the mixed-integer convex theory [25]; however, it is not possible that the combination of the variables that produce the optimum value is unique. This behavior is observed in the numerical analysis presented in the Results' section.*

To illustrate the effect of the three-dimensional characteristics of the decision variable x_{kfg}, let us consider the possible three-phase load connections presented in Table 1.

Table 1. Possible load distributions in a three-phase network [3].

Type of Connection	Phases	Sequence	Binary Variable x_{kfg}
1	XYZ	No change	$\begin{bmatrix} 1 & 0 & 0 \\ 0 & 1 & 0 \\ 0 & 0 & 1 \end{bmatrix}$
2	ZXY		$\begin{bmatrix} 0 & 0 & 1 \\ 1 & 0 & 0 \\ 0 & 1 & 0 \end{bmatrix}$
3	YZX		$\begin{bmatrix} 0 & 1 & 0 \\ 0 & 0 & 1 \\ 1 & 0 & 0 \end{bmatrix}$
4	XZY	Change	$\begin{bmatrix} 1 & 0 & 0 \\ 0 & 0 & 1 \\ 0 & 1 & 0 \end{bmatrix}$
5	YXZ		$\begin{bmatrix} 0 & 1 & 0 \\ 1 & 0 & 0 \\ 0 & 0 & 1 \end{bmatrix}$
6	ZYX		$\begin{bmatrix} 0 & 0 & 1 \\ 0 & 1 & 0 \\ 1 & 0 & 0 \end{bmatrix}$

Note that the decision variables in Table 1 in the last column represent the six possible combinations for the connection of the three-phase loads [23], which clearly fulfills the requirements in equality constraints (6) and (7) associated with the uniqueness of the loads per phase.

It is worth mentioning that the exact formulation of the load redistribution problem in three-phase asymmetric networks is indeed a mixed-integer non-linear programming (MINLP) problem due to the power balance equations that relates the power injection at each node with the voltage and angle variables [18]; the main difficulty of the exact MINLP model lies in the non-convexity of the power balance equations that makes impossible to find the global optimum with exact or metaheuristic methods. To reduce this model complexity, here, we propose a reformulation of the load balancing problem in two stages which corresponds to the exact MIC model formulated from (1) to (7) in the first stage that helps with finding the optimal redistribution of the loads in all the nodes. The solution obtained in the first stage is evaluated at the second one to determine the final level of power losses of the grid. The two-level methodology proposed in this research is easily implemented at any optimization software that combines the (B&B) and interior point methods with the main advantage of ensuring the optimal solution as demonstrated in [25,26], some solvers that can deal with this type of problems are available in the GAMS and AMPL software. Numerical results that will be reported in the Results' section were corroborated with the CPLEX software in the GAMS software [27].

3. Methodology of Solution

To efficiently solve the MIC optimization model defined from (1) to (7) it is possible to use any programming language that deals with convex optimization. Here, we adopt the CVX optimization package in the MATLAB programming environment with the MOSEK solver. The main characteristic of the optimization model is that the objective function is defined as the l_1-norm which is a convex function. The illustration of the objective function in a three-dimensional space is presented in Figure 2.

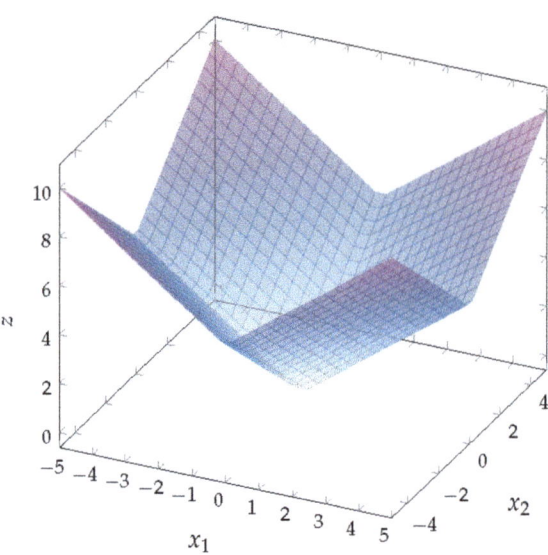

Figure 2. Representation of the objective function in a three-dimensional space $z = |x_1| + |x_2|$.

It is important to mention, that as with the most of the integer optimization models, an MIC can be solved with a modification of the (B&B) method as presented in Figure 3 [28]. Note that at each iteration, it is solved a linear programming model which ensures the optimal solution finding at each nodal exploration [29].

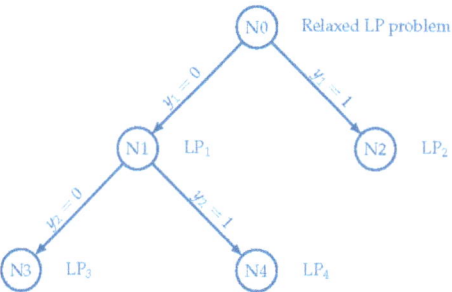

Figure 3. General application of the B&B method for addressing MIC problems.

Remark 4. *Notice that the MIC model defined from (1) to (7) can be rewritten as a mixed-integer quadratic programming problem which is also convex, i.e., it is also possible to find its global optimum by combining the interior point and the (B&B) method [25].*

To verify the efficiency of the optimization model to balance the total power consumption in the substation terminals and its positive effects on the minimization of the power losses, we evaluate the final load reconfiguration in an asymmetrical three-phase power flow method to determine the final power losses and compare with the initial state of the network. The power flow methodology used in this research corresponds to the matrix version of the backward–forward method reported in [3,30].

4. Electric Distribution Grids

The computational validation of the proposed MIC to redistribute loads in three-phase networks considering a simplified model in the substation terminals is made with three different test feeders composed of 4, 15, and 37 buses, respectively. The information of these test feeders is presented below.

4.1. 4-Bus Test Feeder

The 4-bus system is a medium-voltage grid with 4 nodes and 3 lines with a nominal line-to-line voltage of 11.4 kV. The information of the loads and branches are listed in Tables 2 and 3, respectively. This information was obtained from [3].

Table 2. Parametric information of the 4-bus test system (kW and kvar units are used for all powers).

Line	Node i	Node j	Cond.	Length (ft)	P_{ja}	Q_{ja}	P_{jb}	Q_{jb}	P_{jc}	Q_{jc}
1	1	2	1	29,536	500	300	250	100	600	400
2	2	3	2	17,850	0	0	700	350	200	100
3	3	4	3	13,070	750	500	620	540	0	0

Table 3. Impedances' information for the conductors used in the 4-bus system.

Conductor	Impedance Matrix (Ω/mi)		
1	$0.3686 + j0.6852$	$0.0169 + j0.1515$	$0.0155 + j0.1098$
	$0.0169 + j0.1515$	$0.3757 + j0.6715$	$0.0188 + j0.2072$
	$0.0155 + j0.1098$	$0.0188 + j0.2072$	$0.3723 + j0.6782$
2	$0.9775 + j0.8717$	$0.0167 + j0.1697$	$0.0152 + j0.1264$
	$0.0167 + j0.1697$	$0.9844 + j0.8654$	$0.0186 + j0.2275$
	$0.0152 + j0.1264$	$0.0186 + j0.2275$	$0.9810 + j0.8648$
3	$1.9280 + j1.4194$	$0.0161 + j0.1183$	$0.0161 + j0.1183$
	$0.0161 + j0.1183$	$1.9308 + j1.4215$	$0.0161 + j0.1183$
	$0.0161 + j0.1183$	$0.0161 + j0.1183$	$1.9337 + j1.4236$

4.2. 15-Bus Test Feeder

This test feeder is composed by 15 buses and 14 branches, asymmetric three-phase nature network with 13.2 kV of nominal phase voltage at the node of the substation, which corresponds to the typical operating voltage in Colombian power distribution grids. In Figure 4, it is shown the electrical configuration of this test feeder.

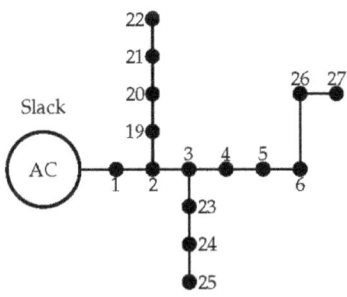

Figure 4. Nodal connections in the 15-bus test feeder.

Tables 3 and 4 present the parametric data of the 15-bus system.

Table 4. Parametric information of the 15-bus test system (kW and kvar units are used for all power values).

Line	Node i	Node j	Cond.	Length (ft)	P_{ja}	Q_{ja}	P_{jb}	Q_{jb}	P_{jc}	Q_{jc}
1	1	2	1	603	0	0	725	300	1100	600
2	2	3	2	776	480	220	720	600	1040	558
3	3	4	3	825	2250	1610	0	0	0	0
4	4	5	3	1182	700	225	0	0	996	765
5	5	6	4	350	0	0	820	700	1220	1050
6	2	7	5	691	2500	1200	0	0	0	0
7	7	8	6	539	0	0	960	540	0	0
8	8	9	6	225	0	0	0	0	2035	1104
9	9	10	6	1050	1519	1250	1259	1200	0	0
10	3	11	3	837	0	0	259	126	1486	1235
11	11	12	4	414	0	0	0	0	1924	1857
12	12	13	5	925	1670	486	0	0	726	509
13	6	14	4	386	0	0	850	752	1450	1100
14	14	15	2	401	486	235	887	722	0	0

4.3. IEEE 37-Bus Test Feeder

The IEEE 37-bus system is a three-phase unbalanced network that is a portion of a real power grid located in California, USA. This grid has 37 nodes with radial connection among them. The line-to-line voltage assigned to the substation bus is 4.8 kV. Note that the electrical configuration of this test feeder was taken from [18] where some variations to the grid topology were included. The single-phase equivalent diagram of the IEEE 37-bus system is presented in Figure 5.

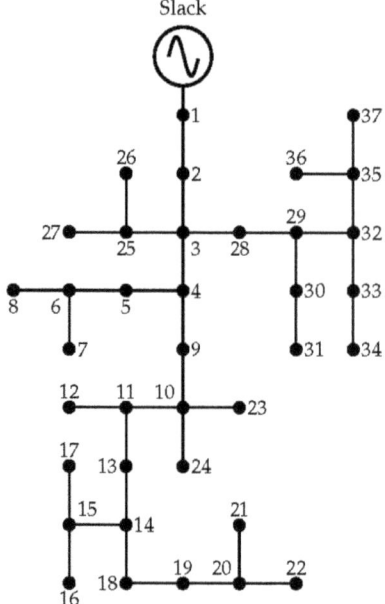

Figure 5. Nodal connection of the IEEE 37-bus system.

The complete parametric information for this test feeder is reported in Tables 5 and 6. Note that this information was taken from [3].

Table 5. Parametric information of the IEEE 37-bus test system (kW and kvar units are used for all power values).

Line	Node i	Node j	Cond.	Length (ft)	P_{ja}	Q_{ja}	P_{jb}	Q_{jb}	P_{jc}	Q_{jc}
1	1	2	1	1850	140	70	140	70	350	175
2	2	3	2	960	0	0	0	0	0	0
3	3	24	4	400	0	0	0	0	0	0
4	3	27	3	360	0	0	0	0	85	40
5	3	4	2	1320	0	0	0	0	0	0
6	4	5	4	240	0	0	0	0	42	21
7	4	9	3	600	0	0	0	0	85	40
8	5	6	3	280	42	21	0	0	0	0
9	6	7	4	200	42	21	42	21	42	21
10	6	8	4	280	42	21	0	0	0	0
11	9	10	3	200	0	0	0	0	0	0
12	10	23	3	600	0	0	85	40	0	0
13	10	11	3	320	0	0	0	0	0	0
14	11	13	3	320	85	40	0	0	0	0
15	11	12	4	320	0	0	0	0	42	21
16	13	14	3	560	0	0	0	0	42	21
17	14	18	3	640	140	70	0	0	0	0
18	14	15	4	520	0	0	0	0	0	0
19	15	16	4	200	0	0	0	0	85	40
20	15	17	4	1280	0	0	42	21	0	0
21	18	19	3	400	126	62	0	0	0	0
22	19	20	3	400	0	0	0	0	0	0
23	20	22	3	400	0	0	0	0	42	21
24	20	21	4	200	0	0	0	0	85	40
25	24	26	4	320	8	4	85	40	0	0
26	24	25	4	240	0	0	0	0	85	40
27	27	28	3	520	0	0	0	0	0	0
28	28	29	4	80	17	8	21	10	0	0
29	28	31	3	800	0	0	0	0	85	40
30	29	30	4	520	85	40	0	0	0	0
31	31	34	4	920	0	0	0	0	0	0
32	31	32	3	600	0	0	0	0	0	0
33	32	33	4	280	0	0	42	21	0	0
34	34	36	4	760	0	0	42	21	0	0
35	34	35	4	120	0	0	140	70	21	10

Table 6. Data of impedance for the conductors used in the IEEE 37-bus system.

Conductor	Impedance Matrix (Ω/mi)		
1	$0.2926 + j0.1973$	$0.0673 - j0.0368$	$0.0337 - j0.0417$
	$0.0673 - j0.0368$	$0.2646 + j0.1900$	$0.0673 - j0.0368$
	$0.0337 - j0.0417$	$0.0673 - j0.0368$	$0.2926 + j0.1973$
2	$0.4751 + j0.2973$	$0.1629 - j0.0326$	$0.1234 - j0.0607$
	$0.1629 - j0.0326$	$0.4488 + j0.2678$	$0.1629 - j0.0326$
	$0.1234 - j0.0607$	$0.1629 - j0.0326$	$0.4751 + j0.2973$
3	$1.2936 + j0.6713$	$0.4871 + j0.2111$	$0.4585 + j0.1521$
	$0.4871 + j0.2111$	$1.3022 + j0.6326$	$0.4871 + j0.2111$
	$0.4585 + j0.1521$	$0.4871 + j0.2111$	$1.2936 + j0.6713$
4	$2.0952 + j0.7758$	$0.5204 + j0.2738$	$0.4926 + j0.2123$
	$0.5204 + j0.2738$	$2.1068 + j0.7398$	$0.5204 + j0.2738$
	$0.4926 + j0.2123$	$0.5204 + j0.2738$	$2.0952 + j0.7758$

5. Computational Validation

The computational validation of the proposed MIC programming model is made in the MATLAB environment with the CVX optimization package and the MOSEK solver. In addition, we evaluate the power losses before and after solving the MIC model using the matrix version of the backward–forward asymmetrical three-phase power flow method [30].

5.1. 4-Bus System

This test feeder presents an initial power losses of 68.6292 kW with an average grid unbalance of 22.47%. After solving the load reconfiguration problem, total grid power losses is 62.5449 kW, i.e., a reduction with respect to the benchmark case is about 8.87%; in addition, the general grid unbalance is reduced until 0.74%. In Figure 6 is presented the comparison between the initial and the final grade of unbalance per phase.

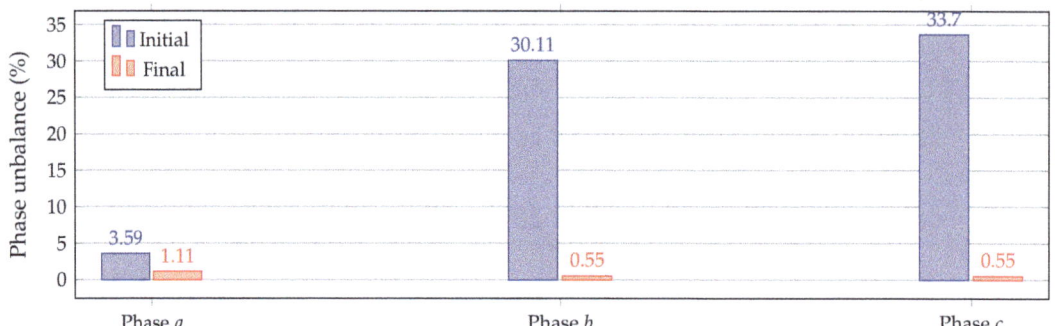

Figure 6. Initial and final unbalances in the 4-bus system.

Note that all the phases are effectively balanced with differences lower than 1.20% with respect to the ideal consumption case, i.e., P_{ave}; in addition, the final active power load at phases a, b, and c in terminals of the substation are 1220 kW, 1200 kW, and 1200 kW, respectively; where the variations with respect to the initial case were 30 kW, 370 kW, and 400 kW, for phases a, b, and c, respectively.

Table 7 presents the final solution regarding load connections after solving the MIC model to redistribute all the loads. The most important result observed in this table corresponds to the existence at least of two possible solutions for the MIC model in the 4-bus system. This happens in this test feeder since two of the phases ends with 1200 kW of total load, which implies that some rotations in the phase connections will exhibit the same final active power losses.

Table 7. Optimal solutions reached by the MIC optimization model in the 4-bus system.

Scenario	Solution	Losses (kW)	Reduction (%)	$U_\%$ (%)
Benchmark case	{1,1,1,1}	68.6292	0.00	22.47
Solution 1	{1,6,4,5}	62.7868	8.51	0.74
Solution 2	{1,2,1,3}	62.5449	8.87	0.74

It is important to mention that the solutions obtained with the MOSEK solver in the CVX environment were validated with the GAMS optimization package with the COUENNE solver. In addition, the average processing time in MATLAB including the power flow evaluations was about 1.83 s.

5.2. 15-Bus System

In this test feeder, previous to the application of the MIC model to redistribute all the loads among the phases of the network, we know that the initial power losses is 134.2472 kW, caused by a general unbalance of 20.48%, which is distributed as 2.68% for phase a, 30.72% pf phase b, and 28.04% for phase c. Once the MIC model defined from (1) to (7) is executed, we observe that the active power losses is uniformly distributed for all the phases with respect to the average value. In Table 8 is presented the load redistribution in the 15-bus system.

Table 8. Comparison between initial and final load distributions per phase (all the values in kW and kVar).

Scenario	P_a	Q_a	P_b	Q_b	P_c	Q_c	$U_\%$ (%)
Benchmark case	9605	5226	6480	4940	11,977	8778	28.04
Solution 1	9354	7112	9354	5794	9354	6038	0.00
Solution 2	9354	6038	9354	5794	9354	7112	0.00

Note that results in Table 8 show that: (i) there are at least two solutions of the optimization model (1)–(7) that present the same objective function performance which in this system is exactly zero; this implies that all the phases have the same active power load consumption per phase, i.e., 9354 kW; (ii) the general unbalance in the case of reactive power for this system in the benchmark case is 26.01% which is reduced to 8.42% after making the load redistribution; this result implies an important effect when redistributing the total consumption per phase in the substation terminals, since the modification of the active power load connection is directly connected with the total reactive power consumption; (iii) the final power losses for solutions 1 and 2 are 117.8982 kW and 115.1107 kW, with reductions respect to the benchmark case of about 12.18%, and 14.25%, respectively; and (iv) note that the main difference between solutions 1 and 2 corresponds to the rotation of the loads connected between phases a and c in all the nodes; this results important since 4 additional solutions can be obtained making possible the 6 load rotations presented in Table 1 with the same objective function of 0.00%, and power losses between 117.8982 kW and 115.1107 kW, respectively.

In regards with the total processing time the MOSEK solver using the MATLAB/CVX environment takes about 174.45 s; which is a quite small processing time taking into account that there are 6^{n-1} possible combinations of the loads, where n is the total number of nodes, i.e., 78,364,164,096, this is, more than 78,000 million of combinations.

5.3. IEEE 37-Bus System

The initial average unbalance in the IEEE 37-bus system is 22.14%, which is distributed in 11.23% for the phase a, 21.98% for the phase b, and 33.21 for the phase c, respectively. Once it is solved the proposed MIC model the final average unbalance in the network is 1.71%. Figure 7 depicts the initial and final unbalances per phase.

Numerical results per phase in Figure 7 show that phases a, b, and c are improved in about 9.16%, 19.41%, and 32.72%, respectively, with respect to the benchmark case of active power; which confirms the efficiency of the proposed approach for optimizing the general average grid unbalance, guaranteeing the global minimum of the problem.

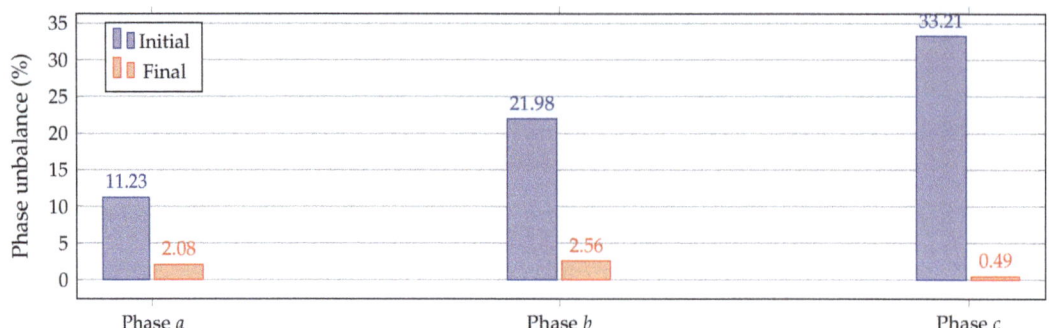

Figure 7. Initial and final unbalances in the IEEE 37-bus system.

In relation with the amount of power losses, the benchmark case presents the initial power losses of 76.1357 kW; however, for this test system after solving the MIC model there are six possible combinations that produce different levels of power losses reduction. Table 9 reports all the possible solutions in regards with power losses obtained by our proposed optimization approach.

Table 9. Optimal solutions reached by the MIC optimization model in the 15-bus system.

Scenario	Solution	Losses (kW)	Reduction (%)	$U_\%$ (%)
Ben. case	$\{1,1,1,1,1,1,1,1,1,1,1,1,1,1,1,1,$ $1,1,1,1,1,1,1,1,1,1,1,1,1,1,1,1\}$	76.1357	0.00	22.14
Sol. 1	$\{1,5,1,1,2,4,1,1,4,1,1,2,5,2,4,3,5,3,$ $6,4,4,6,6,4,4,1,6,4,6,4,2,1,1,1,2,5\}$	66.5829	12.55	1.71
Sol. 2	Rotation of Sol. 1 from XYZ to ZXY in all the nodes	67.2585	11.66	1.71
Sol. 3	Rotation of Sol. 1 from XYZ to YZX in all the nodes	67.3892	11.49	1.71
Sol. 4	Rotation of Sol. 1 from XYZ to XZY in all the nodes	67.4765	11.37	1.71
Sol. 5	Rotation of Sol. 1 from XYZ to YXZ in all the nodes	67.1325	11.83	1.71
Sol. 6	Rotation of Sol. 1 from XYZ to ZYX in all the nodes	66.6432	12.47	1.71

Results in Table 9 allow concluding that: (i) the first solution obtained by the MIC approach presents the best numerical performance regarding grid power losses with a reduction of 12.55% in comparison to the benchmark case; (ii) the worst solution regarding of power losses corresponds to solution 4, which is obtained by rotating solution 1 from XYZ to XZY in all the nodes since the reduction of the power losses in this case decreases until 11.37%; (iii) all the solutions in Table 9 are indeed the global optimum for the optimization model (1)–(7) since the general grid imbalance is 1.71% for all the solution cases; however, the calculation of the final power losses can be considered a decision criterion to select the most attractive solution from the point of view of the grid operator, which, in this context, is solution 1.

Finally, with respect to total processing time, the MOSEK solver using the MAT-LAB/CVX environment takes about 29,879.44 s; which is an acceptable processing time taking into account that there are 6^{n-1} possible combination of the loads, where n is the total number of nodes, i.e., $1.03144247984905 \times 10^{28}$.

5.4. Additional Comments

It is worth mentioning that all the numerical results reported with the CVX tool in the MATLAB environment for the 4-, 15-, and 37-bus systems were confirmed by the CPLEX solver in the GAMS environment with simulation times that do not get over 10 s [27]. These processing times confirm the effectiveness of the MIC model to solve the problem of the load redistribution in asymmetrical three-phase networks by ensuring the global optimum finding in the first stage of the two-level proposed optimization method [26]. The solutions provided in the first stage were evaluated in the triangular-based three-phase power flow which takes less than 10 ms to solve it and determine the final level of power losses in the grid at the second stage [3].

On the other hand, the proposed two-stage optimization methodology to redistribute the load connections in three-phase networks is suitable to be applied in the improvement of the resilience level of the electricity distribution activity [31]. This is in the context of the physical- or cyber-attacks to the distribution network or any electrical disturbance that implies the reconfiguration of the grid topology; since the proposed optimization model can be applied to each possible grid topology to ensure that after clarifying the disturbance the resulting electrical network has the minimum level of unbalance, in other words, minimum power losses.

6. Conclusions

The problem of the load redistribution in three-phase distribution networks was addressed in this research from the point of view of the mathematical optimization, by proposing a mixed-integer convex model that ensures the global optimum finding via (B&B) and linear programming methods. Numerical results in the 4-, 15-, and 37-bus systems demonstrate the effectiveness of the proposed optimization model to reduce the general average grid unbalance of the network by reducing from 22.47% to 0.74% for the 4-bus system, 20.48% to 0.00% for the 15-bus system; and 22.14% to 1.71% for the IEEE-37 bus system, respectively.

In addition, for each test feeder it was observed that when the final load reconfiguration is rotated for all the possible phase combinations, the amount of power losses in the final configuration changes with the same objective function value, which confirms the multi-modal behavior of the load redistribution optimization problem in terminals of the main substation. For the 4-, 15-, and IEEE 37-bus systems the maximum power losses reductions with respect to the benchmark case were 8.87%, 14.25%, and 12.55%, respectively. These reductions demonstrate the strong relationship between the load redistribution problem and the total grid power losses, which can be taken as an advantage of the grid owner to improve the quality of the electricity service at the same time that increases the net profit due to the reduction in the costs of the energy losses.

Regarding the processing times it was observed that for each test feeder the MOSEK solver in the CVX package using the MATLAB environment takes 1.73 s, 174.45 s, and 29,879.44 s, for the 4-, 15-, and IEEE 37-bus systems, respectively, which can be considered as short processing times due to the large dimension of the solution space. This latter can be calculated as 6^{n-1}, being 216, $7.8364164096 \times 10^{10}$, and $1.031442479849054 \times 10^{28}$ for the test feeders mentioned previously. As future work it will be possible to analyze the inclusion in the proposed MIC model of the power balance constraints with a second-order cone representation that will ensure the global optimum finding for the problem of the phase-balancing problem in three-phase networks.

Author Contributions: Conceptualization, O.D.M., A.A.-L., L.F.G.-N., H.R.C. and J.A.B.; Methodology, O.D.M., A.A.-L., L.F.G.-N., H.R.C. and J.A.B.; Investigation, O.D.M., A.A.-L., L.F.G.-N., H.R.C.; and writing—review and editing, O.D.M., A.A.-L., L.F.G.-N., H.R.C. and J.A.B. All authors have read and agreed to the published version of the manuscript.

Funding: This work was supported in part by the Centro de Investigación y Desarrollo Científico de la Universidad Distrital Francisco José de Caldas under grant 1643-12-2020 associated with the project "Desarrollo de una metodología de optimización para la gestión óptima de recursos energéticos distribuidos en redes de distribución de energía eléctrica." and in part by the Dirección de Investigaciones de la Universidad Tecnológica de Bolívar under grant PS2020002 associated with the project "Ubicación óptima de bancos de capacitores de paso fijo en redes eléctricas de distribución para reducción de costos y pérdidas de energía: Aplicación de métodos exactos y metaheurísticos".

Institutional Review Board Statement: Not applicable.

Informed Consent Statement: Not applicable.

Data Availability Statement: No new data were created or analyzed in this study. Data sharing is not applicable to this article.

Conflicts of Interest: The authors declare no conflict of interest.

References

1. Bina, M.T.; Kashefi, A. Three-phase unbalance of distribution systems: Complementary analysis and experimental case study. *Int. J. Electr. Power Energy Syst.* **2011**, *33*, 817–826. [CrossRef]
2. Lavorato, M.; Franco, J.F.; Rider, M.J.; Romero, R. Imposing Radiality Constraints in Distribution System Optimization Problems. *IEEE Trans. Power Syst.* **2012**, *27*, 172–180. [CrossRef]
3. Cortés-Caicedo, B.; Avellaneda-Gómez, L.S.; Montoya, O.D.; Alvarado-Barrios, L.; Chamorro, H.R. Application of the Vortex Search Algorithm to the Phase-Balancing Problem in Distribution Systems. *Energies* **2021**, *14*, 1282. [CrossRef]
4. Caicedo, J.E.; Romero, A.A.; Zini, H.C. Assessment of the harmonic distortion in residential distribution networks: Literature review. *Ing. Investig.* **2017**, *37*, 72–84. [CrossRef]
5. Li, Q.; Ayyanar, R.; Vittal, V. Convex Optimization for DES Planning and Operation in Radial Distribution Systems With High Penetration of Photovoltaic Resources. *IEEE Trans. Sustain. Energy* **2016**, *7*, 985–995. [CrossRef]
6. Montoya, O.D.; Gil-González, W.; Hernández, J.C. Efficient Operative Cost Reduction in Distribution Grids Considering the Optimal Placement and Sizing of D-STATCOMs Using a Discrete-Continuous VSA. *Appl. Sci.* **2021**, *11*, 2175. [CrossRef]
7. Gil-González, W.; Montoya, O.D.; Rajagopalan, A.; Grisales-Noreña, L.F.; Hernández, J.C. Optimal Selection and Location of Fixed-Step Capacitor Banks in Distribution Networks Using a Discrete Version of the Vortex Search Algorithm. *Energies* **2020**, *13*, 4914. [CrossRef]
8. Sadiq, A.; Adamu, S.; Buhari, M. Optimal distributed generation planning in distribution networks: A comparison of transmission network models with FACTS. *Eng. Sci. Technol. Int. J.* **2019**, *22*, 33–46. [CrossRef]
9. Montoya, O.D.; Gil-González, W.; Grisales-Noreña, L. An exact MINLP model for optimal location and sizing of DGs in distribution networks: A general algebraic modeling system approach. *Ain Shams Eng. J.* **2020**, *11*, 409–418. [CrossRef]
10. Dall'Anese, E.; Giannakis, G.B. Convex distribution system reconfiguration using group sparsity. In Proceedings of the 2013 IEEE Power & Energy Society General Meeting, Vancouver, BC, Canada, 21–25 July 2013; doi:10.1109/pesmg.2013.6672702. [CrossRef]
11. Khodr, H.; Zerpa, I.; de Jesu's, P.D.O.; Matos, M. Optimal Phase Balancing in Distribution System Using Mixed-Integer Linear Programming. In Proceedings of the 2006 IEEE/PES Transmission & Distribution Conference and Exposition: Latin America, Caracas, Venezuela, 15–18 August 2006; doi:10.1109/tdcla.2006.311368. [CrossRef]
12. Montoya, O.D.; Fuentes, J.E.; Moya, F.D.; Barrios, J.Á.; Chamorro, H.R. Reduction of Annual Operational Costs in Power Systems through the Optimal Siting and Sizing of STATCOMs. *Appl. Sci.* **2021**, *11*, 4634. [CrossRef]
13. Saif, A.M.; Buccella, C.; Patel, V.; Tinari, M.; Cecati, C. Design and Cost Analysis for STATCOM in Low and Medium Voltage Systems. In Proceedings of the IECON 2018-44th Annual Conference of the IEEE Industrial Electronics Society, Washington, DC, USA, 21–23 October 2018. [CrossRef]
14. Castiblanco-Pérez, C.M.; Toro-Rodríguez, D.E.; Montoya, O.D.; Giral-Ramírez, D.A. Optimal Placement and Sizing of D-STATCOM in Radial and Meshed Distribution Networks Using a Discrete-Continuous Version of the Genetic Algorithm. *Electronics* **2021**, *10*, 1452. [CrossRef]
15. Shojaei, F.; Rastegar, M.; Dabbaghjamanesh, M. Simultaneous placement of tie-lines and distributed generations to optimize distribution system post-outage operation and minimize energy losses. *CSEE J. Power Energy Syst.* **2020**. [CrossRef]
16. Arias, J.; Calle, M.; Turizo, D.; Guerrero, J.; Candelo-Becerra, J. Historical Load Balance in Distribution Systems Using the Branch and Bound Algorithm. *Energies* **2019**, *12*, 1219. [CrossRef]
17. Montoya, O.D.; Molina-Cabrera, A.; Grisales-Noreña, L.F.; Hincapié, R.A.; Granada, M. Improved Genetic Algorithm for Phase-Balancing in Three-Phase Distribution Networks: A Master-Slave Optimization Approach. *Computation* **2021**, *9*, 67. [CrossRef]
18. Granada-Echeverri, M.; Gallego-Rendón, R.A.; López-Lezama, J.M. Optimal Phase Balancing Planning for Loss Reduction in Distribution Systems using a Specialized Genetic Algorithm. *Ing. Y Cienc.* **2012**, *8*, 121–140. [CrossRef]
19. Garcés, A.; Castaño, J.C.; Rios, M.A. Phase Balancing in Power Distribution Grids: A Genetic Algorithm with a Group-Based Codification. In *Energy Systems*; Springer International Publishing: Berlin/Heidelberg, Germany, 2020; pp. 325–342. [CrossRef]
20. Ghasemi, S. Balanced and unbalanced distribution networks reconfiguration considering reliability indices. *Ain Shams Eng. J.* **2018**, *9*, 1567–1579. [CrossRef]

21. Toma, N.; Ivanov, O.; Neagu, B.; Gavrila, M. A PSO Algorithm for Phase Load Balancing in Low Voltage Distribution Networks. In Proceedings of the 2018 International Conference and Exposition on Electrical And Power Engineering (EPE), Iași, Romania, 18–19 October 2018; doi:10.1109/icepe.2018.8559805. [CrossRef]
22. Babu, P.R.; Shenoy, R.; Ramya, N.; Shetty, S. Implementation of ACO technique for load balancing through reconfiguration in electrical distribution system. In Proceedings of the 2014 Annual International Conference on Emerging Research Areas: Magnetics, Machines and Drives (AICERA/iCMMD), Kottayam, Kerala, 24–26 July 2014; doi:10.1109/aicera.2014.6908233. [CrossRef]
23. Garces, A.; Gil-González, W.; Montoya, O.D.; Chamorro, H.R.; Alvarado-Barrios, L. A Mixed-Integer Quadratic Formulation of the Phase-Balancing Problem in Residential Microgrids. *Appl. Sci.* **2021**, *11*, 1972. [CrossRef]
24. Baes, M.; Oertel, T.; Wagner, C.; Weismantel, R. Mirror-Descent Methods in Mixed-Integer Convex Optimization. In *Facets of Combinatorial Optimization*; Springer: Berlin/Heidelberg, Germany, 2013; pp. 101–131. [CrossRef]
25. Benson, H.Y.; Ümit, S. Mixed-Integer Second-Order Cone Programming: A Survey. In *Theory Driven by Influential Applications*; INFORMS: Catonsville, MD, USA, 2013; pp. 13–36. [CrossRef]
26. Wang, J.W.; Wang, H.F.; Zhang, W.J.; Ip, W.H.; Furuta, K. Evacuation Planning Based on the Contraflow Technique With Consideration of Evacuation Priorities and Traffic Setup Time. *IEEE Trans. Intell. Transp. Syst.* **2013**, *14*, 480–485. [CrossRef]
27. Alemany, J.; Kasprzyk, L.; Magnago, F. Effects of binary variables in mixed integer linear programming based unit commitment in large-scale electricity markets. *Electr. Power Syst. Res.* **2018**, *160*, 429–438. [CrossRef]
28. Montoya, O.D.; Gil-González, W.; Molina-Cabrera, A. Exact minimization of the energy losses and the CO_2 emissions in isolated DC distribution networks using PV sources. *DYNA* **2021**, *88*, 178–184. [CrossRef]
29. Aceituno-Cabezas, B.; Dai, H.; Cappelletto, J.; Grieco, J.C.; Fernandez-Lopez, G. A mixed-integer convex optimization framework for robust multilegged robot locomotion planning over challenging terrain. In Proceedings of the 2017 IEEE/RSJ International Conference on Intelligent Robots and Systems (IROS), Vancouver, BC, Canada, 24–28 September 2017; doi:10.1109/iros.2017.8206313. [CrossRef]
30. Shen, T.; Li, Y.; Xiang, J. A Graph-Based Power Flow Method for Balanced Distribution Systems. *Energies* **2018**, *11*, 511. [CrossRef]
31. Zhang, W.; van Luttervelt, C. Toward a resilient manufacturing system. *CIRP Ann.* **2011**, *60*, 469–472. [CrossRef]

Article

From the Quasi-Total Strong Differential to Quasi-Total Italian Domination in Graphs

Abel Cabrera Martínez [†], Alejandro Estrada-Moreno [*,†] and Juan Alberto Rodríguez-Velázquez [†]

Departament d'Enginyeria Informàtica i Matemàtiques, Universitat Rovira i Virgili, Av. Països Catalans 26, 43007 Tarragona, Spain; abel.cabrera@urv.cat (A.C.M.); juanalberto.rodriguez@urv.cat (J.A.R.-V.)
* Correspondence: alejandro.estrada@urv.cat
† These authors contributed equally to this work.

Abstract: This paper is devoted to the study of the quasi-total strong differential of a graph, and it is a contribution to the Special Issue "Theoretical computer science and discrete mathematics" of *Symmetry*. Given a vertex $x \in V(G)$ of a graph G, the neighbourhood of x is denoted by $N(x)$. The neighbourhood of a set $X \subseteq V(G)$ is defined to be $N(X) = \bigcup_{x \in X} N(x)$, while the external neighbourhood of X is defined to be $N_e(X) = N(X) \setminus X$. Now, for every set $X \subseteq V(G)$ and every vertex $x \in X$, the external private neighbourhood of x with respect to X is defined as the set $P_e(x, X) = \{y \in V(G) \setminus X : N(y) \cap X = \{x\}\}$. Let $X_w = \{x \in X : P_e(x, X) \neq \varnothing\}$. The strong differential of X is defined to be $\partial_s(X) = |N_e(X)| - |X_w|$, while the quasi-total strong differential of G is defined to be $\partial_{s^*}(G) = \max\{\partial_s(X) : X \subseteq V(G) \text{ and } X_w \subseteq N(X)\}$. We show that the quasi-total strong differential is closely related to several graph parameters, including the domination number, the total domination number, the 2-domination number, the vertex cover number, the semitotal domination number, the strong differential, and the quasi-total Italian domination number. As a consequence of the study, we show that the problem of finding the quasi-total strong differential of a graph is NP-hard.

Keywords: differentials in graphs; strong differential; quasi-total strong differential; quasi-total Italian domination number

Citation: Cabrera Martínez, A.; Estrada-Moreno, A.; Rodríguez-Velázquez, J.A. From the Quasi-Total Strong Differential to Quasi-Total Italian Domination in Graphs. *Symmetry* **2021**, *13*, 1036. https://doi.org/10.3390/sym13061036

Academic Editor: Markus Meringer

Received: 20 April 2021
Accepted: 3 June 2021
Published: 8 June 2021

Publisher's Note: MDPI stays neutral with regard to jurisdictional claims in published maps and institutional affiliations.

Copyright: © 2021 by the authors. Licensee MDPI, Basel, Switzerland. This article is an open access article distributed under the terms and conditions of the Creative Commons Attribution (CC BY) license (https://creativecommons.org/licenses/by/4.0/).

1. Introduction

Given a graph $G = (V(G), E(G))$, the open neighbourhood of a vertex $x \in V(G)$ is defined to be $N(x) = \{y \in V(G) : xy \in E(G)\}$. The open neighbourhood of a set $X \subseteq V(G)$ is defined by $N(X) = \bigcup_{x \in X} N(x)$, while the external neighbourhood of X, or boundary of X, is defined as $N_e(X) = N(X) \setminus X$.

The differential of a subset $X \subseteq V(G)$ is defined as $\partial(X) = |N_e(X)| - |X|$ and the differential of a graph G is defined as

$$\partial(G) = \max\{\partial(X) : X \subseteq V(G)\}.$$

These concepts were introduced by Hedetniemi about twenty-five years ago in an unpublished paper, and the preliminary results on the topic were developed by Goddard and Henning [1]. The development of the topic was subsequently continued by several authors, including [2–7]. Currently, the study of differentials in graphs and their variants is of great interest because it has been observed that the study of different types of domination can be approached through a variant of the differential which is related to them. Specifically, we are referring to domination parameters that are necessarily defined through the use of functions, such as Roman domination, perfect Roman domination, Italian domination and unique response Roman domination. In each case, the main result linking the domination parameter to the corresponding differential is a Gallai-type theorem, which allows us to study these domination parameters without the use of functions. For instance, the differential is related to the Roman domination number [3], the perfect differential is related to the perfect Roman domination number [5], the strong differential is related to the Italian domination number [8], the 2-packing differential is related to the unique response Roman domination number [9]. Next, we will briefly describe the case of the strong differential and then introduce the study of the quasi-total strong differential. We refer the reader to the corresponding papers for details on the other cases.

For any $x \in X$, the external private neighbourhood of x with respect to X is defined to be

$$P_e(x, X) = \{y \in V(G) \setminus X : N(y) \cap X = \{x\}\}.$$

We define the set $X_w = \{x \in X : P_e(x, X) \neq \emptyset\}$.
The strong differential of a set X is defined to be

$$\partial_s(X) = |N_e(X)| - |X_w|,$$

while the strong differential of G is defined to be

$$\partial_s(G) = \max\{\partial_s(X) : X \subseteq V(G)\}.$$

As shown in [8], the problem of finding the strong differential of a graph is NP-hard, and this parameter is closely related to several graph parameters. In particular, the theory of strong differentials allows us to develop the theory of Italian domination without the use of functions.

In this paper, we study the quasi-total strong differential of G, which is defined as

$$\partial_{s^*}(G) = \max\{\partial_s(X) : X \subseteq V(G) \text{ and } X_w \subseteq N(X)\}.$$

We will show that this novel parameter is perfectly integrated into the theory of domination. In particular, we will show that the quasi-total strong differential is closely related to several graph parameters, including the domination number, the total domination number, the 2-domination number, the vertex cover number, the semitotal domination number, the strong differential, and the quasi-total Italian domination number. As a consequence of the study, we show that the problem of finding the quasi-total strong differential of a graph is NP-hard.

The paper is organised as follows. Section 2 is devoted to establish the main notation, terminology and tools needed to develop the remaining sections. In Section 3 we obtain several bounds on the quasi-total strong differential of a graph and we discuss the tightness of these bounds. In Section 4 we prove a Gallai-type theorem which shows that the theory of quasi-total strong differentials can be applied to develop the theory of Italian domination, provided that the Italian dominating functions fulfil an additional condition. Finally, in Section 5 we show that the problem of finding the quasi-total strong differential of a graph is NP-hard.

2. Notation, Terminology and Basic Tools

Throughout the paper, we will use the notation $G \cong H$ if G and H are isomorphic graphs. Given a set $X \subseteq V(G)$, the subgraph of G induced by X will be denoted by $G[X]$, while (for simplicity) the subgraph induced by $V(G) \setminus X$ will be denoted by $G - X$. The minimum degree, the maximum degree and the order of G will be denoted by $\delta(G)$, $\Delta(G)$ and $n(G)$, respectively.

A leaf of G is a vertex of degree one. A support vertex of G is a vertex which is adjacent to a leaf, while a strong support vertex is a vertex which is adjacent to at least two leaves. The set of leaves, support vertices and strong support vertices of G will be denoted by $\mathcal{L}(G)$, $\mathcal{S}(G)$ and $\mathcal{S}_s(G)$, respectively.

A dominating set of G is a subset $D \subseteq V(G)$ such that $N(v) \cap D \neq \emptyset$ for every $v \in V(G) \setminus D$. Let $\mathcal{D}(G)$ be the set of dominating sets of G. The domination number of G is defined to be,

$$\gamma(G) = \min\{|D| : D \in \mathcal{D}(G)\}.$$

The domination number has been extensively studied. For instance, we cite the following books [10–12].

We define a $\gamma(G)$-set as a set $D \in \mathcal{D}(G)$ with $|D| = \gamma(G)$. The same agreement will be assumed for optimal parameters associated to other characteristic sets of a graph. For instance, a $\partial_{s^*}(G)$-set will be a set $X \subseteq V(G)$ such that $X_w \subseteq N(X)$ and $\partial_s(X) = \partial_{s^*}(G)$.

As described in Figure 1, $X = \{a, b, x, y\}$ is a $\partial_{s^*}(G)$-set while $X' = \{u, v, x, y\}$ is not a $\partial_{s^*}(G)$-set, as $X'_w = \{u, v\} \not\subseteq N(X')$. In contrast, both X and X' are $\partial_s(G)$-sets. Another $\partial_{s^*}(G)$-sets are $Y = \{a, b, u, v, x, y\}$ and $Y' = \{a, b, v, x, y\}$.

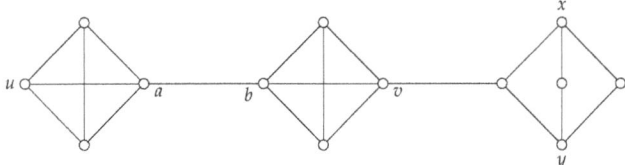

Figure 1. Let $X = \{a, b, x, y\}$ and $X' = \{u, v, x, y\}$. In this case, $X_w = \{a, b\} \subseteq N(X)$ and $\partial_s(X) = \partial_{s^*}(G) = 7$, so that X is a $\partial_{s^*}(G)$-set. In contrast, X' is not a $\partial_{s^*}(G)$-set, although $\partial_s(X') = \partial_{s^*}(G)$.

A total dominating set of G is a subset $D \subseteq V(G)$ such that $N(v) \cap D \neq \emptyset$ for every vertex $v \in V(G)$. Let $\mathcal{D}_t(G)$ be the set of total dominating sets of G. The total domination number of G is defined to be,

$$\gamma_t(G) = \min\{|D| : D \in \mathcal{D}_t(G)\}.$$

The total domination number has been extensively studied. For instance, we cite the book [13].

A k-dominating set of G is a subset $D \subseteq V(G)$ such that $|N(v) \cap D| \geq k$ for every vertex $v \in V(G) \setminus D$. Let $\mathcal{D}_k(G)$ be the set of k-dominating sets of G. The k-domination number of G is defined to be,

$$\gamma_k(G) = \min\{|D| : D \in \mathcal{D}_k(G)\}.$$

For a comprehensive survey on k-domination in graphs, we cite the book [10] published in 2020. In particular, there is a chapter, Multiple Domination, by Hansberg and Volkmann, where they put into context all relevant research results on multiple domination concerning k-domination that have been found up to 2020.

In particular, the following result will be useful in the study of quasi-total strong differentials.

Theorem 1 ([14]). *Let r and k be positive integers. For any graph G with $\delta(G) \geq \frac{r+1}{r}k - 1$,*

$$\gamma_k(G) \leq \frac{r}{r+1} n(G).$$

A semitotal dominating set of a graph G with no isolated vertex, is a dominating set D of G such that every vertex in D is within distance two of another vertex in D. This concept was introduced in 2014 by Goddard et al. in [15]. Let $\mathcal{D}_{t2}(G)$ be the set of semitotal dominating sets of G. The semitotal domination number of G is defined to be

$$\gamma_{t2}(G) = \min\{|D| : D \in \mathcal{D}_{t2}(G)\}.$$

By definition,

$$\gamma(G) \leq \gamma_{t2}(G) \leq \min\{\gamma_t(G), \gamma_2(G)\}.$$

A set $C \subseteq V(G)$ is a vertex cover of G if every edge of G is incident with at least one vertex in C. The vertex cover number of G, denoted by $\beta(G)$, is the minimum cardinality among all vertex covers of G. Recall that the largest cardinality of a set of vertices of G, no two of which are adjacent, is called the independence number of G and it is denoted by $\alpha(G)$. The following well-known result, due to Gallai, states the relationship between the independence number and the vertex cover number of a graph.

Theorem 2 (Gallai's theorem, [16]). *For any graph G,*

$$\alpha(G) + \beta(G) = n(G).$$

The concept of a corona product graph was introduced in 1970 by Frucht and Harary [17]. Given two graphs G_1 and G_2, the corona product graph $G_1 \odot G_2$ is the graph obtained from G_1 and G_2, by taking one copy of G_1 and $n(G_1)$ copies of G_2 and joining by an edge every vertex from the ith-copy of G_2 with the ith-vertex of G_1. Notice that $n(G_1 \odot G_2) = n(G_1)(n(G_2) + 1)$ and $\gamma(G_1 \odot G_2) = n(G_1)$.

The following result will be one of our main tools.

Theorem 3 ([8]). *For any graph G, the following statements hold.*
(i) There exists a $\partial_s(G)$-set which is a dominating set of G.
(ii) $n(G) - \min\{2\gamma(G), \gamma_2(G)\} \leq \partial_s(G) \leq n(G) - \gamma(G) - |\mathcal{S}_s(G)|.$

For the remainder of the paper, definitions will be introduced whenever a concept is needed. In particular, this is the case for concepts, notation and terminology that are used only once or only in a short section.

3. General Results

To begin this section we present some bounds on the quasi-total strong differential of a graph, and then we discuss the tightness of the bounds.

Theorem 4. *For any graph G, the following statements hold.*
(i) $\partial_s(G) - \gamma(G) \leq \partial_{s^*}(G) \leq \partial_s(G).$
(ii) $n(G) - \min\{3\gamma(G), \gamma_2(G)\} \leq \partial_{s^*}(G) \leq n(G) - \gamma(G) - |\mathcal{S}_s(G)|.$

Proof. The inequality $\partial_{s^*}(G) \leq \partial_s(G)$ is straightforward, as for any $\partial_{s^*}(G)$-set X we have $\partial_{s^*}(G) = \partial_s(X) \leq \partial_s(G)$.

We proceed to prove $\partial_{s^*}(G) \geq \partial_s(G) - \gamma(G)$. Let D be a $\partial_s(G)$-set such that $D \in \mathcal{D}(G)$, which exists by Theorem 3. Now, we define $D'' \subseteq V(G)$ as a set of minimum cardinality among all supersets D' of D such that $N(v) \cap D' \neq \emptyset$ for every vertex $v \in D_w$. Since D is a dominating set, $D''_w \subseteq D_w$. Moreover, observe that $|D'' \setminus D| \leq \gamma(G)$, by the minimality of D''. Therefore,

$$\begin{aligned}
\partial_{s^*}(G) &\geq \partial_s(D'') \\
&= |N_e(D'')| - |D''_w| \\
&\geq |N_e(D)| - |D'' \setminus D| - |D''_w| \\
&\geq |N_e(D)| - |D_w| - |D'' \setminus D| \\
&= \partial_s(G) - |D'' \setminus D| \\
&\geq \partial_s(G) - \gamma(G),
\end{aligned}$$

as required.

To prove lower bound $\partial_{s^*}(G) \geq n(G) - \gamma_2(G)$ we only need to observe that for any $\gamma_2(G)$-set S we have $\partial_{s^*}(G) \geq \partial_s(S) = |N_e(S)| - |S_w| = |N_e(S)| = n(G) - |S| = n(G) - \gamma_2(G).$

Finally, to complete the proof of (ii) we only need to combine the previous bounds with Theorem 3. □

Corollary 1. Let G be a graph. If $\partial_s(G) = n(G) - \gamma_2(G)$ or there exists a $\partial_s(G)$-set which is a total dominating set, then $\partial_{s^*}(G) = \partial_s(G)$.

In order to show some classes of graphs with $\partial_{s^*}(G) = \partial_s(G)$ and $\partial_{s^*}(G) = n(G) - \gamma(G) - |S_s(G)|$, we consider the case of corona graphs. It is not difficult to see that if G_1 has no isolated vertex and G_2 is a non trivial graph, then

$$\partial_{s^*}(G_1 \odot G_2) = \partial_s(G_1 \odot G_2) = n(G_1)(n(G_2) - 1).$$

In addition, if G_2 is a graph with at least two isolated vertices, then

$$\partial_{s^*}(G_1 \odot G_2) = n(G_1)(n(G_2) - 1)$$
$$= n(G_1 \odot G_2) - \gamma(G_1 \odot G_2) - |S_s(G_1 \odot G_2)|.$$

Next we discuss some cases where the lower bounds given in Theorem 4 are achieved.

Theorem 5. For any graph G, the following statements are equivalent.
(i) $\partial_{s^*}(G) = \partial_s(G) - \gamma(G)$.
(ii) $\partial_{s^*}(G) = n(G) - 3\gamma(G)$.

Proof. Assume $\partial_{s^*}(G) = \partial_s(G) - \gamma(G)$. By Theorem 3, there exists a set $D \in \mathcal{D}(G)$ which is a $\partial_s(G)$-set. Now, we define $D'' \subseteq V(G)$ as a set of minimum cardinality among all supersets D' of D such that $N(v) \cap D' \neq \emptyset$ for every vertex $v \in D_w$. Obviously, $|D'' \setminus D| \leq |D_w| \leq |D|$. As we have shown in the proof of Theorem 4,

$$\partial_{s^*}(G) \geq \partial_s(G) - |D'' \setminus D| \geq \partial_s(G) - \gamma(G),$$

which implies that $\gamma(G) = |D'' \setminus D|$, and so $\gamma(G) \leq |D_w| \leq |D|$. On the other side, $\partial_s(G) \geq n(G) - 2\gamma(G)$, by Theorem 3. In summary,

$$n(G) - 2\gamma(G) \leq \partial_s(G) = n(G) - |D| - |D_w| \leq n(G) - 2\gamma(G),$$

Therefore, $\partial_s(G) = n(G) - 2\gamma(G)$, and so $\partial_{s^*}(G) = n(G) - 3\gamma(G)$.
Conversely, assume $\partial_{s^*}(G) = n(G) - 3\gamma(G)$. By Theorems 3 and 4 we have

$$n(G) - 3\gamma(G) = \partial_{s^*}(G) \geq \partial_s(G) - \gamma(G) \geq n(G) - 3\gamma(G).$$

Therefore, $\partial_s(G) = n(G) - 2\gamma(G)$ and, as a result, $\partial_{s^*}(G) = \partial_s(G) - \gamma(G)$. □

To continue the study, we need to establish the following lemma.

Lemma 1. For any graph G, there exists a $\partial_{s^*}(G)$-set X which is a dominating set of G and $|P_e(v, X)| \geq 2$ for every $v \in X_w$.

Proof. Let D be a $\partial_{s^*}(G)$-set and $D' = V(G) \setminus N_e(D)$. Since $N_e(D') = N_e(D)$ and $D'_w \subseteq D_w$,

$$\partial_s(D') = |N_e(D')| - |D'_w| \geq |N_e(D)| - |D_w| = \partial_s(D) = \partial_{s^*}(G),$$

which implies that D' is a $\partial_{s^*}(G)$-set, as $D'_w \subseteq N(D')$. Obviously, D' is a dominating set.
Now, let $D_1 \subseteq D'_w$ such that $|P_e(v, D')| = 1$ for every $v \in D_1$ and $|P_e(v, D')| \geq 2$ for every $v \in D'_w \setminus D_1$. Let $X = D' \cup (\bigcup_{v \in D_1} P_e(v, D'))$. Since $|N_e(X)| = |N_e(D')| - |D_1|$ and $|X_w| \leq |D'_w| - |D_1|$,

$$\partial_s(X) = |N_e(X)| - |X_w| \geq |N_e(D')| - |D'_w| = \partial_s(D') = \partial_{s^*}(G).$$

Therefore, X is a $\partial_{s^*}(G)$-set, as $X_w \subseteq N(X)$. Clearly, $|P_e(v, X)| \geq 2$ for every $v \in X_w$. □

We are now able to characterize the graphs with $\partial_{s^*}(G) = n(G) - \gamma(G)$.

Theorem 6. *For any graph G, the following statements are equivalent.*
(i) $\partial_{s^*}(G) = n(G) - \gamma(G)$.
(ii) $\gamma_2(G) = \gamma(G)$.
(iii) $\partial_s(G) = n(G) - \gamma(G)$.

Proof. Assume $\partial_{s^*}(G) = n(G) - \gamma(G)$. By Lemma 1, there exists a set $D \in \mathcal{D}(G)$ which is a $\partial_{s^*}(G)$-set. Hence, $n(G) - \gamma(G) = \partial_{s^*}(G) = |N_e(D)| - |D_w| = n(G) - |D| - |D_w|$, which implies that $|D| + |D_w| = \gamma(G)$. Since $\gamma(G) \leq |D|$, we deduce that $|D| = \gamma(G)$ and $|D_w| = 0$. Therefore, D is a 2-dominating set of G and so, $\gamma_2(G) \leq |D| = \gamma(G) \leq \gamma_2(G)$, which leads to $\gamma_2(G) = \gamma(G)$.

Conversely, from Theorem 4 we deduce that $\gamma_2(G) = \gamma(G)$ implies that $\partial_{s^*}(G) = n(G) - \gamma(G)$.

Finally, the equivalence (ii)⟷(iii) was previously established in [8]. □

By the result above we have that if $\partial_{s^*}(G) = n(G) - \gamma(G)$, then $\partial_{s^*}(G) = n(G) - \gamma_2(G)$. However, the converse does not hold. For instance, as we will see in Corollary 2, if G is a path or a cycle, then $\partial_{s^*}(G) = n(G) - \gamma_2(G) < n(G) - \gamma(G)$.

We next consider some cases of graphs satisfying $\partial_{s^*}(G) = n(G) - \gamma_2(G)$.

Theorem 7. *Let G be a graph. If $\Delta(G) \leq 3$ or G is a claw-free graph, then*

$$\partial_{s^*}(G) = n(G) - \gamma_2(G).$$

Proof. By Lemma 1, there exists $D \in \mathcal{D}(G)$ which is a $\partial_{s^*}(G)$-set and $|P_e(v, D)| \geq 2$ for every $v \in D_w$. Assume that $\Delta(G) \leq 3$. We define a set $D' \subseteq V(G)$ as follows.

$$D' = (D \setminus D_w) \cup \left(\bigcup_{v \in D_w} P_e(v, D) \right).$$

Notice that $N(v) \cap D \neq \emptyset$ and $|N(v) \setminus D| = |P_e(v, D)| = 2$ for every $v \in D_w$. Hence, $D' \in \mathcal{D}(G)$ and $D'_w = \emptyset$, which implies that D' is a 2-dominating set of G and

$$\begin{aligned} n(G) - |D'| &= n(G) - |D| - |D_w| \\ &= |N_e(D)| - |D_w| \\ &= \partial_s(D) \\ &= \partial_{s^*}(G). \end{aligned}$$

Therefore, $\partial_{s^*}(G) = n(G) - |D'| \leq n(G) - \gamma_2(G)$, and we deduce the equality by the lower bound $\partial_{s^*}(G) \geq n(G) - \gamma_2(G)$ given in Theorem 4.

Now, assume that G is a claw-free graph. Observe that in this case $P_e(v, D)$ is a clique for every $v \in D_w$, as $N(v) \cap D \neq \emptyset$. Let $X \subseteq V(G) \setminus D$ such that $|X| = |D_w|$ and $|X \cap P_e(v, D)| = 1$ for every $v \in D_w$. Notice that $X' = D \cup X$ is a 2-dominating set of G. Hence,

$$\partial_{s^*}(G) = \partial_s(D) = |N_e(D)| - |D_w| = n(G) - |D| - |D_w| = n(G) - |X'| \leq n(G) - \gamma_2(G).$$

Therefore, by the lower bound $\partial_{s^*}(G) \geq n(G) - \gamma_2(G)$ given in Theorem 4 we conclude the proof. □

The following result is a direct consequence of Theorem 7 and the well-known equalities $\gamma_2(C_n) = \lceil \frac{n}{2} \rceil$ and $\gamma_2(P_n) = \lceil \frac{n+1}{2} \rceil$ due to Fink and Jacobson [18].

Corollary 2. *For any integer $n \geq 3$,*

$$\partial_{s^*}(P_n) = \left\lfloor \frac{n-1}{2} \right\rfloor \quad \text{and} \quad \partial_{s^*}(C_n) = \left\lfloor \frac{n}{2} \right\rfloor.$$

By Theorems 1 and 4 we derive the following result.

Theorem 8. *Given a graph G, the following statements hold.*
(i) *If $\delta(G) \geq 3$, then $\partial_{s^*}(G) \geq \frac{n(G)}{2}$.*
(ii) *If $\delta(G) = 2$, then $\partial_{s^*}(G) \geq \frac{n(G)}{3}$.*

For instance, for any cubic graph with $\gamma_2(G) = \frac{n(G)}{2}$ we have $\partial_{s^*}(G) = \frac{n(G)}{2}$, and for any corona graph of the form $G \cong G_1 \odot K_2$ we have $\partial_{s^*}(G) = \partial_s(G) = \frac{n(G)}{3}$.

We next discuss the relationship between the quasi-total strong differential and the semitotal domination number.

Theorem 9. *Given a graph G with no isolated vertex, the following statements hold.*
(i) $\partial_{s^*}(G) \leq n(G) - \gamma_{t2}(G)$.
(ii) $\partial_{s^*}(G) = n(G) - \gamma_{t2}(G)$ *if and only if $\gamma_{t2}(G) = \gamma_2(G)$.*
(iii) $\partial_{s^*}(G) = n(G) - \gamma_{t2}(G) - 1$ *if and only if one of the following conditions holds.*
  (a) $\gamma_2(G) = \gamma_{t2}(G) + 1$.
  (b) $\gamma_2(G) \geq \gamma_{t2}(G) + 1$ *and there exist a $\gamma_{t2}(G)$-set D and a vertex $v \in D \cap N(D)$ such that $P_e(v,D) \neq \emptyset$ and D is a 2-dominating set of $G - P_e(v,D)$.*

Proof. By Lemma 1, there exists a dominating set D which is a $\partial_{s^*}(G)$-set. In addition, since G has no isolated vertex, D is also a semitotal dominating set of G, which implies that $|D| \geq \gamma_{t2}(G)$. Hence,

$$\partial_{s^*}(G) = |N_e(D)| - |D_w| \leq n(G) - |D| - |D_w| \leq n(G) - |D| \leq n(G) - \gamma_{t2}(G).$$

Therefore, (i) follows and $\partial_{s^*}(G) = n(G) - \gamma_{t2}(G)$ if and only if D is a 2-dominating set and $|D| = \gamma_{t2}(G)$. Now, since $\gamma_{t2}(G) \leq \gamma_2(G)$, every 2-dominating set of cardinality $\gamma_{t2}(G)$ is a $\gamma_2(G)$-set. Therefore, (ii) follows.

Finally, we proceed to prove (iii). We first assume that $\partial_{s^*}(G) = n(G) - \gamma_{t2}(G) - 1$. By (i) we deduce that $\gamma_{t2}(G) + 1 \leq \gamma_2(G)$. Also, notice that

$$n(G) - \gamma_{t2}(G) - 1 = \partial_{s^*}(G) = \partial_s(D) = n(G) - |D| - |D_w|,$$

which implies that $|D| + |D_w| = \gamma_{t2}(G) + 1$. Since $|D| \geq \gamma_{t2}(G)$, we obtain that $|D_w| \in \{0,1\}$. We distinguish these two cases.

Case 1. $|D_w| = 0$. In this case, we have that D is a 2-dominating set of G of cardinality $\gamma_{t2}(G) + 1$, which implies that $\gamma_{t2}(G) + 1 \leq \gamma_2(G) \leq |D| = \gamma_{t2}(G) + 1$. Therefore, $\gamma_2(G) = \gamma_{t2}(G) + 1$. Conversely, if $\gamma_2(G) = \gamma_{t2}(G) + 1$, then by (i) and Theorem 4 we have that $n(G) - \gamma_{t2}(G) - 1 \leq \partial_{s^*}(G) = n(G) - \gamma_{t2}(G)$, and so (ii) leads to $\partial_{s^*}(G) = n(G) - \gamma_{t2}(G) - 1$.

Case 2. $|D_w| = 1$. If $D_w = \{v\}$, then $v \in D \cap N(D)$ and $P_e(v,D) \neq \emptyset$. In addition, since $|D| + |D_w| = |D| + 1 = \gamma_{t2}(G) + 1$, we have that D is a $\gamma_{t2}(G)$-set and a 2-dominating set of $G - P_e(v,D)$. Therefore, (b) holds. Conversely, assume that (b) holds. Since $\gamma_2(G) \geq \gamma_{t2}(G) + 1$, from (i) and (ii) we conclude that $\partial_{s^*}(G) \leq n(G) - \gamma_{t2}(G) - 1$, and so the $\gamma_{t2}(G)$-set satisfying (b) is a $\partial_{s^*}(G)$-set. □

Next we derive some lower bounds on $\partial_{s^*}(G)$.

Theorem 10. *For any graph G with every component of order at least three,*

$$\partial_{s^*}(G) \geq \left\lceil \frac{1}{2}(n(G) - \gamma(G) + |\mathcal{L}(G)| - 2|\mathcal{S}(G)| - 2|\mathcal{S}_s(G)|) \right\rceil.$$

Proof. Let S be a $\gamma(G)$-set such that $\mathcal{S}(G) \subseteq S$ and $\overline{S} = V(G) \setminus S$.

Now, we define $S'' \subseteq \overline{S}$ as a set of minimum cardinality among all subsets S' of $\overline{S}$ that satisfy the following conditions.

(a) $N(v) \cap \mathcal{L}(G) \cap S' \neq \varnothing$ for every vertex $v \in \mathcal{S}(G) \setminus \mathcal{S}_s(G)$ or $v \in \mathcal{S}(G)$ with $N(v) \subseteq \mathcal{L}(G)$.

(b) $(N(v) \cap S') \setminus \mathcal{L}(G) \neq \varnothing$ for every vertex $v \in \mathcal{S}_s(G)$ such that $N(v) \cap S = \varnothing$ and $N(v) \not\subseteq \mathcal{L}(G)$.

Notice that $|\mathcal{S}(G)| - |\mathcal{S}_s(G)| \leq |S''| \leq |\mathcal{S}(G)|$. Now, let $I \subseteq \overline{S} \setminus S''$ the set of isolated vertices of the graph $G[\overline{S} \setminus S'']$. Hence, by definition of S'' we deduce that $|I| \geq |\mathcal{L}(G)| - |\mathcal{S}(G)| + |\mathcal{S}_s(G)|$.

Now, we define $X'' \subseteq \overline{S} \setminus (I \cup S'')$ as a set of minimum cardinality among all subsets X' of $\overline{S} \setminus (I \cup S'')$ such that $N(v) \cap X' \neq \varnothing$ for every vertex $v \in \overline{S} \setminus (I \cup S'' \cup X')$. It is clear that if $\overline{S} = I \cup S''$, then $X'' = \varnothing$, while if $\overline{S} \setminus (I \cup S'') \neq \varnothing$, then X'' is a $\gamma(G[\overline{S} \setminus (I \cup S'')])$-set. As $G[\overline{S} \setminus (I \cup S'')]$ has no isolated vertex, we have that

$$|X''| \leq \frac{1}{2}(n(G) - (|S| + |I| + |S''|)) \leq \frac{1}{2}(n(G) - \gamma(G) - |\mathcal{L}(G)|).$$

Hence, in any case $|X''| \leq \frac{1}{2}(n(G) - \gamma(G) - |\mathcal{L}(G)|)$ because $|S| + |\mathcal{L}(G)| \leq n(G)$.

Now, let $D = S \cup S'' \cup X''$. Notice that $D \in \mathcal{D}(G)$, $D_w \subseteq \mathcal{S}_s(G)$ and $D_w \subseteq N(D)$. Hence,

$$\begin{aligned}
\partial_{s^*}(G) \geq \partial_s(D) &= |N_e(D)| - |D_w| \\
&= n(G) - |D| - |D_w| \\
&= n(G) - |S| - |S''| - |X''| - |\mathcal{S}_s(G)| \\
&\geq n(G) - \gamma(G) - |\mathcal{S}(G)| - \frac{1}{2}(n(G) - \gamma(G) - |\mathcal{L}(G)|) - |\mathcal{S}_s(G)| \\
&= \frac{1}{2}(n(G) - \gamma(G) + |\mathcal{L}(G)| - 2|\mathcal{S}(G)| - 2|\mathcal{S}_s(G)|).
\end{aligned}$$

Therefore, the result follows. □

The bound above is tight. For instance, it is achieved by the graphs shown in Figure 2.

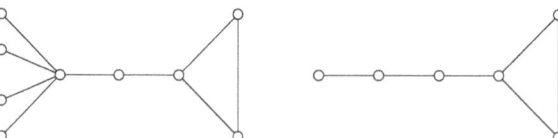

Figure 2. Two graphs achieving the bound given in Theorem 10.

Corollary 3. *For any graph G with $\delta(G) = 2$,*

$$\partial_{s^*}(G) \geq \frac{1}{2}(n(G) - \gamma(G)).$$

The bound above is achieved by any corona graph of the form $G \cong G_1 \odot K_2$, where G_1 is a nontrivial graph. In this case, $\partial_{s^*}(G) = \partial_s(G) = n(G_1) = \frac{1}{2}(n(G) - \gamma(G))$.

Theorem 11. *For any graph G with no isolated vertex,*
$$\partial_{s^*}(G) \geq n(G) - \gamma_t(G) - \gamma(G).$$

Proof. Let S_1 be a $\gamma_t(G)$-set and S_2 a $\gamma(G)$-set. Let $S = S_1 \cup S_2$. As $S_1 \in \mathcal{D}_t(G)$ and $S_2 \in \mathcal{D}(G)$, we deduce that $S_w \subseteq N(S)$ and $S_w \subseteq S_1 \cap S_2$. Hence,

$$\begin{aligned}
\partial_{s^*}(G) &\geq \partial_s(S) \\
&= |N_e(S)| - |S_w| \\
&= n(G) - |S| - |S_w| \\
&\geq n(G) - |S_1| - |S_2| + |S_1 \cap S_2| - |S_w| \\
&\geq n(G) - \gamma_t(G) - \gamma(G),
\end{aligned}$$

as desired. □

The bound above is tight. Figure 3 shows a graph G with $\gamma(G) < \gamma_t(G)$, where $\partial_{s^*}(G) = 5 = n(G) - \gamma_t(G) - \gamma(G)$.

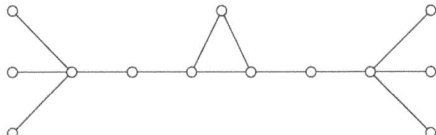

Figure 3. A graph G with $\partial_{s^*}(G) = 5$.

Theorem 12. *For any graph G with every component of order at least three,*
$$\partial_{s^*}(G) \geq n(G) - \beta(G) - |\mathcal{S}(G)| - |\mathcal{S}_s(G)|.$$

Proof. Let S be a $\beta(G)$-set such that $\mathcal{S}(G) \subseteq S$. Now, we define $S' \subseteq \mathcal{L}(G)$ such that $|S'| = |\mathcal{S}(G)|$ and $|N(v) \cap S'| = 1$ for every vertex $v \in \mathcal{S}(G)$. Hence, $S'' = S \cup S'$ is a dominating set, $S''_w \subseteq \mathcal{S}_s(G)$ and $S''_w \subseteq N(S'')$, which implies that

$$\begin{aligned}
\partial_{s^*}(G) &\geq \partial_s(S'') \\
&= |N_e(S'')| - |S''_w| \\
&= n(G) - |S''| - |S''_w| \\
&\geq n(G) - |S| - |\mathcal{S}(G)| - |\mathcal{S}_s(G)|.
\end{aligned}$$

Therefore, the result follows. □

The bound above is tight. For instance, Figure 3 shows a graph G with $\partial_{s^*}(G) = 5 = n(G) - \beta(G) - |\mathcal{S}(G)| - |\mathcal{S}_s(G)| = \alpha(G) - |\mathcal{S}(G)| - |\mathcal{S}_s(G)|$.

Notice that Theorems 2 and 12 lead to the following bound.

Theorem 13. *For any graph G with every component of order at least three,*
$$\partial_{s^*}(G) \geq \alpha(G) - |\mathcal{S}(G)| - |\mathcal{S}_s(G)|.$$

In particular, for graphs of minimum degree at least two we deduce the following result.

Theorem 14. *For any graph G with $\delta(G) \geq 2$, the following statements hold.*
(i) $\partial_{s^*}(G) \geq \alpha(G)$.
(ii) *If $\partial_{s^*}(G) = \alpha(G)$, then $\alpha(G) = n(G) - \gamma_2(G)$.*
(iii) $\partial_{s^*}(G) \geq \gamma(G)$.
(iv) *If $\partial_{s^*}(G) = \gamma(G)$, then $\gamma(G) = n(G) - \gamma_2(G)$.*

Proof. Obviously, (i) is an immediate consequence of Theorem 13 and (iii) is derived from the fact that $\alpha(G) \geq \gamma(G)$.

Now, since $\delta(G) \geq 2$, every vertex cover is a 2-dominating set, which implies that $\gamma_2(G) \leq \beta(G) = n(G) - \alpha(G)$. Thus, by Theorem 4, if $\partial_{s^*}(G) = \alpha(G)$, then

$$\alpha(G) = \partial_{s^*}(G) \geq n(G) - \gamma_2(G) \geq \alpha(G).$$

Therefore, (ii) follows, and by analogy we deduce that (iii) follows. □

The graph shown in Figure 4, on the left, satisfies $\partial_{s^*}(G) = \alpha(G) = n(G) - \gamma_2(G) = 4$. The converse of Theorem 14 (ii) does not hold. For instance, for the right hand side graph shown in Figure 4 we have $\alpha(G) = n(G) - \gamma_2(G) = 3$, while $\partial_{s^*}(G) = 4$.

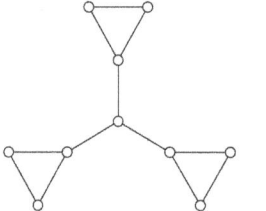 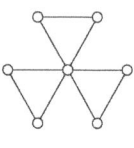

Figure 4. Two graphs with $\partial_{s^*}(G) = 4$.

The graph shown in Figure 5 satisfies $\partial_{s^*}(G) = \gamma(G) = n(G) - \gamma_2(G) = 5$. We would point out that there are several cases of graphs of minimum degree one with $\partial_{s^*}(G) \leq \gamma(G) - 1$.

Next we discuss the trivial bounds on $\partial_{s^*}(G)$ and we characterize the extreme cases.

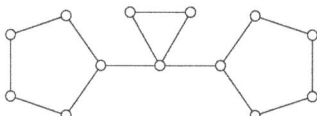

Figure 5. A graph with $\partial_{s^*}(G) = \gamma(G) = n(G) - \gamma_2(G) = 5$.

Proposition 1. *For any graph G of order $n(G) \geq 3$, the following statements hold.*

(i) $\max\{0, \Delta(G) - 2\} \leq \partial_{s^*}(G) \leq n(G) - 2$.
(ii) $\partial_{s^*}(G) = 0$ *if and only if* $\Delta(G) \leq 1$.
(iii) $\partial_{s^*}(G) = 1$ *if and only if* $\Delta(G) \in \{2, 3\}$ *and* $\gamma_2(G) = n(G) - 1$.
(iv) $\partial_{s^*}(G) = n(G) - 2$ *if and only if* $\gamma_2(G) = 2$.
(v) $\partial_{s^*}(G) = n(G) - 3$ *if and only if* $\gamma_2(G) = 3$ *or* $\gamma_2(G) \neq 2$ *and* $\gamma(G) = 1$.

Proof. We first proceed to prove (i). If $\Delta(G) \in \{0, 1\}$, then it is straightforward that $\partial_{s^*}(G) = 0$. We assume that $\Delta(G) \geq 2$. Let $v \in V(G)$ be a vertex of maximum degree, $u \in N(v)$ and $S = \{u\} \cup (V(G) \setminus N(v))$. Notice that either $S_w = \emptyset$ or $S_w = \{v\}$. Hence, $\partial_{s^*}(G) \geq \partial_s(S) = |N_e(S)| - |S_w| \geq \Delta(G) - 2$, as desired. Since $n(G) \geq 3$ every $\partial_{s^*}(G)$-set has cardinality at least two, and so $\partial_{s^*}(G) \leq n(G) - 2$.

We next proceed to prove (ii). if $\partial_{s^*}(G) = 0$, then $\Delta(G) \leq 2$ by (i). Now, if $\Delta(G) = 2$, then for any vertex x of maximum degree we have that $V(G) \setminus \{x\}$ is a 2-dominating set, and so $\partial_{s^*}(G) \geq 1$, which is a contradiction. Therefore, $\Delta(G) \leq 1$. Obviously, if $\Delta(G) \leq 1$, then $\partial_{s^*}(G) = 0$.

Now, we proceed to prove (iii). First, we assume that $\partial_{s^*}(G) = 1$. By (i) and (ii) we deduce that $\Delta(G) \in \{2, 3\}$. Hence, Theorem 7 leads to $\gamma_2(G) = n(G) - 1$. Conversely, if $\Delta(G) \in \{2, 3\}$ and $\gamma_2(G) = n(G) - 1$, then Theorem 7 leads to $\partial_{s^*}(G) = n(G) - \gamma_2(G) = 1$. Therefore, (iii) follows.

To prove the remaining statements, we take a $\partial_{s^*}(G)$-set $D \in \mathcal{D}(G)$, which exists due to Lemma 1.

We next proceed to prove (iv). First, assume that $\partial_{s^*}(G) = n(G) - 2$. In this case, we deduce that $|D| + |D_w| = 2$, which implies that $|D| = 2$ and $D_w = \emptyset$. Therefore, D is a $\gamma_2(G)$-set and so, $\gamma_2(G) = 2$. On the other side, if $\gamma_2(G) = 2$, then by Theorem 4 and (i) we deduce that $\partial_{s^*}(G) = n(G) - 2$.

Finally, we proceed to prove (v). If either $\gamma_2(G) = 3$ or $\gamma_2(G) \neq 2$ and $\gamma(G) = 1$, then by Theorem 4 and the statements (i) and (iv) we deduce that $\partial_{s^*}(G) = n(G) - 3$. Conversely, assume that $\partial_{s^*}(G) = n(G) - 3$. From (iv) we deduce that $\gamma_2(G) \geq 3$. Moreover, we deduce that $|D| + |D_w| = 3$, which implies that either $|D| = 2$ and $|D_w| = 1$ or $|D| = 3$ and $|D_w| = 0$. If $|D| = 2$ and $|D_w| = 1$, then $\gamma(G) = 1$ as $D \in \mathcal{D}(G)$, while if $|D| = 3$ and $|D_w| = 0$, then D is a 2-dominating set, and so $\gamma_2(G) = 3$. □

To conclude this section, we discuss the case of join graphs.

Proposition 2. *For any two graphs G and H we have the following statements.*
(i) $n(G) + n(H) - 4 \leq \partial_{s^*}(G + H) \leq n(G) + n(H) - 2$.
(ii) $\partial_{s^*}(G + H) = n(G) + n(H) - 2$ *if and only if* $\min\{\gamma_2(G), \gamma_2(H)\} = 2$ *or* $\gamma(G) = \gamma(H) = 1$.
(iii) $\partial_{s^*}(G + H) = n(G) + n(H) - 3$ *if and only if one of the following holds.*
- $\min\{\gamma_2(G), \gamma_2(H)\} = 3$ *and* $\max\{\gamma(G), \gamma(H)\} \geq 2$.
- $\min\{\gamma_2(G), \gamma_2(H)\} \geq 3$ *and, in addition,* $\gamma(G) = 2$ *or* $\gamma(H) = 2$.
- $\min\{\gamma(G), \gamma(H)\} = 1$ *and* $\max\{\gamma(G), \gamma(H)\} \geq 2$ *and* $\min\{\gamma_2(G), \gamma_2(H)\} \geq 3$.
(iv) $\partial_{s^*}(G + H) = 4$ *if and only if* $\min\{\gamma(G), \gamma(H)\} \geq 3$ *and* $\min\{\gamma_2(G), \gamma_2(H)\} \geq 4$.

Proof. By Proposition 1 (i) we deduce that $\partial_{s^*}(G + H) \leq n(G) + n(H) - 2$. For any set $D = \{u, v\}$, where $u \in V(G)$ and $v \in V(H)$, we have that $\partial_{s^*}(G + H) \geq |N_e(D)| - |D_w| = n(G) + n(H) - |D| - |D_w| \geq n(G) + n(H) - 4$. Thus, (i) follows. Finally, by (i) and Proposition 1 (iv) and (v), we deduce the remaining statements, which completes the proof. □

4. A Gallai-Type Theorem

A Gallai-type theorem is a result of the form $a(G) + b(G) = n(G)$, where $a(G)$ and $b(G)$ are parameters defined on G. This terminology comes from Theorem 2, which is a well-known result stated by Gallai in 1959. The aim of this section is to identify the parameter $a(G)$ such that $a(G) + \partial_{s^*}(G) = n(G)$. We will show that this invariant, which is associated to a version of the Italian domination, is perfectly integrated into the theory of domination.

Let $f : V(G) \longrightarrow \{0, 1, 2\}$ be a function and $V_i = \{v \in V(G) : f(v) = i\}$ for $i \in \{0, 1, 2\}$. We will identify the function f with these subsets of $V(G)$ induced by f, and write $f(V_0, V_1, V_2)$. The weight of f is defined to be

$$w(f) = f(V(G)) = \sum_{v \in V(G)} f(v) = \sum_i i|V_i|.$$

The theory of Roman domination was introduced by Cockayne et al. [19]. They defined a Roman dominating function on a graph G to be a function $f(V_0, V_1, V_2)$ satisfying the condition that every vertex in V_0 is adjacent to at least one vertex in V_2. Recently, Cabrera García et al. [20] defined a quasi-total Roman dominating function as a Roman dominating function $f(V_0, V_1, V_2)$ such that $N(v) \cap (V_1 \cup V_2) \neq \emptyset$ for every $v \in V_2$.

An Italian dominating function on a graph G is a function $f(V_0, V_1, V_2)$ satisfying that $f(N(v)) = \sum_{u \in N(v)} f(u) \geq 2$ for every $v \in V_0$, i.e., $f(V_0, V_1, V_2)$ is an Italian dominating function if $N(v) \cap V_2 \neq \emptyset$ or $|N(v) \cap V_1| \geq 2$ for every $v \in V_0$. Hence, every Roman dominating function is an Italian dominating function. The concept of Italian domination was introduced by Chellali et al. in [21] under the name Roman {2}-domination. The term

Italian Domination was later introduced by Henning and Klostermeyer [22,23]. The Italian domination number, denoted by $\gamma_I(G)$, is the minimum weight among all dominating functions on G.

The following Gallai-type theorem for the strong differential and the Italian domination number was stated in [8].

Theorem 15 (Gallai-type theorem, [8]). *For any graph G,*
$$\gamma_I(G) + \partial_s(G) = n(G).$$

We say that an Italian dominating function $f(V_0, V_1, V_2)$ is a quasi-total Italian dominating function if $N(v) \cap (V_1 \cup V_2) \neq \emptyset$ for every $v \in V_2$. Clearly, every quasi-total Roman dominating function is a quasi-total Italian dominating function. The quasi-total Italian domination number, denoted by $\gamma_{I*}(G)$, is the minimum weight among all quasi-total dominating functions on G.

Theorem 16 (Gallai-type theorem). *For any graph G,*
$$\gamma_{I*}(G) + \partial_{s*}(G) = n(G).$$

Proof. By Lemma 1, there exists a $\partial_{s*}(G)$-set D which is a dominating set of G. Hence, the function $g(W_0, W_1, W_2)$, defined from $W_1 = D \setminus D_w$ and $W_2 = D_w$, is a quasi-total Italian dominating function on G, which implies that

$$\begin{aligned} \gamma_{I*}(G) &\leq \omega(g) \\ &= 2|D_w| + |D \setminus D_w| \\ &= |D_w| + |D| \\ &= n(G) - (|N_e(D)| - |D_w|) \\ &= n(G) - \partial_{s*}(G). \end{aligned}$$

We proceed to show that $\gamma_{I*}(G) \geq n(G) - \partial_{s*}(G)$. Let $f(V_0, V_1, V_2)$ be a $\gamma_{I*}(G)$-function. It is readily seen that for $D' = V_1 \cup V_2$ we have that $D' \setminus D'_w = V_1$ and $D'_w = V_2$. Thus,

$$\begin{aligned} \partial_{s*}(G) &\geq \partial_s(D') \\ &= |N_e(D')| - |D'_w| \\ &= |V(G) \setminus (V_1 \cup V_2)| - |V_2| \\ &= n(G) - 2|V_2| - |V_1| \\ &= n(G) - \gamma_{I*}(G). \end{aligned}$$

Therefore, the result follows. □

5. Computational Complexity

In this section, we show that the problem of finding the quasi-total strong differential of graph is NP-hard. To this end, we need to establish the following result.

Theorem 17. *For any graph G,*
$$\partial_{s*}(G \odot K_1) = n(G) - \gamma(G).$$

Proof. Given $x \in V(G)$, let x' be the vertex of the copy of K_1 associated to x in $G \odot K_1$, and let $V(G \odot K_1) = V(G) \cup X$, where $X = \bigcup_{x \in V(G)} \{x'\}$.

By Lemma 1, there exists a $\partial_{s^*}(G \odot K_1)$-set A which is a dominating set and $|P_e(v, A)| \geq 2$ for every $v \in A_w$. Hence, $A_w \cap X = \emptyset$. Now, if there exists $x \in V(G) \cap A_w$, then there exists $u \in P_e(x, A) \cap V(G)$ such that $u' \notin A$ and $N(u') \cap A = \emptyset$, which is a contradiction. Hence, $A_w = \emptyset$, which implies that A is a 2-dominating set of $G \odot K_1$. Thus,

$$\partial_{s^*}(G \odot K_1) = \partial_s(A) = n(G \odot K_1) - |A| \leq n(G \odot K_1) - \gamma_2(G \odot K_1).$$

Since Theorem 4 leads to $\partial_{s^*}(G \odot K_1) \geq n(G \odot K_1) - \gamma_2(G \odot K_1)$, we conclude that

$$\partial_{s^*}(G \odot K_1) = 2n(G) - \gamma_2(G \odot K_1).$$

Now, let D be a dominating set of G and $D' = D \cup X$. Since D' is a 2-dominating set of $G \odot K_1$, we have that

$$\gamma_2(G \odot K_1) \leq |D'| = \gamma(G) + n(G).$$

Finally, for any $\gamma_2(G \odot K_1)$-set Y, we have that $X \subseteq Y$ and $Y \cap V(G)$ is a dominating set of G, which implies that

$$\gamma_2(G \odot K_1) = |Y| = |X| + |Y \cap V(G)| \geq |X| + \gamma(G) = n(G) + \gamma(G).$$

Therefore, $\gamma_2(G \odot K_1) = n(G) + \gamma(G)$, and so the result follows. □

A direct consequence of the preceding result is the determination of computational complexity of finding the quasi-total strong differential. Given a graph G and a positive integer t, the domination problem is to decide whether there exists a dominating S in G such that $|S|$ is at most t. It is well known that the domination problem is NP-complete. Hence, the optimization problem of finding $\gamma(G)$ is NP-hard. Therefore, from Theorem 17, we derive the following result.

Corollary 4. *Given a graph G, the problem of finding $\partial_{s^*}(G)$ is NP-hard.*

6. Conclusions and Open Problems

This article is a contribution to the theory differential of graphs. Particularly, we introduce the concept of the quasi-total strong differential of a graph. In our study, we show that the quasi-total strong differential is closely related to several graph parameters, including the domination number, the total domination number, the 2-domination number, the vertex cover number, the semitotal domination number, the strong differential, and the quasi-total Italian domination number. Finally, we proved that the problem of finding the quasi-total strong differential of a graph is NP-hard.

Some open problems have emerged from the study carried out. For instance, we highlight the following.

(a) It would be interesting to obtain some Nordhaus-Gaddum type relations.
(b) We have shown that if $\partial_{s^*}(G) = \alpha(G)$, then $\alpha(G) = n(G) - \gamma_2(G)$. Likewise, we have shown that if $\partial_{s^*}(G) = \gamma(G)$, then $\gamma(G) = n(G) - \gamma_2(G)$. However, the problem of characterizing all graphs such that $\partial_{s^*}(G) = \alpha(G)$ and $\partial_{s^*}(G) = \gamma(G)$ is still an open problem.
(c) Since the optimization problem of finding $\partial_{s^*}(G)$ is NP-hard, it would be interesting to devise polynomial-time algorithm for simple families of graphs or to develop heuristics that allow to estimate as accurately as possible this parameter for any graph.
(d) It would be interesting to investigate the quasi-total strong differential of product graphs, and try to express this invariant in terms of different parameters of the graphs involved in the product.

Author Contributions: All authors contributed equally to this work. The work was organized and led by J.A.R.-V. All authors have read and agreed to the published version of the manuscript.

Funding: This research received no external funding.

Institutional Review Board Statement: Not applicable.

Informed Consent Statement: Not applicable.

Data Availability Statement: Not applicable.

Conflicts of Interest: The authors declare no conflict of interest.

References

1. Goddard, W.; Henning, M.A. Generalised domination and independence in graphs. *Congr. Numer.* **1997**, *123*, 161–171.
2. Bermudo, S. On the differential and Roman domination number of a graph with minimum degree two. *Discret. Appl. Math.* **2017**, *232*, 64–72. [CrossRef]
3. Bermudo, S.; Fernau, H.; Sigarreta, J.M. The differential and the Roman domination number of a graph. *Appl. Anal. Discret. Math.* **2014**, *8*, 155–171. [CrossRef]
4. Bermudo, S.; Hernández-Gómez, J.C.; Rodríguez, J.M.; Sigarreta, J.M. Relations between the differential and parameters in graphs. *Electron. Notes Discret. Math.* **2014**, *46*, 281–288. [CrossRef]
5. Cabrera Martínez, A.; Rodríguez-Velázquez, J.A. On the perfect differential of a graph. *Quaest. Math.* **2021**. [CrossRef]
6. Lewis, J.L.; Haynes, T.W.; Hedetniemi, S.M.; Hedetniemi, S.T.; Slater, P.J. Differentials in graphs. *Util. Math.* **2006**, *69*, 43–54.
7. Pushpam, P.R.L.; Yokesh, D. Differentials in certain classes of graphs. *Tamkang J. Math.* **2010**, *41*, 129–138. [CrossRef]
8. Cabrera Martínez, A.; Rodríguez-Velázquez, J.A. From the strong differential to Italian domination in graphs. *Mediterr. J. Math.* **2021**, accepted.
9. Cabrera Martínez, A.; Puertas, M.L.; Rodríguez-Velazquez, J.A. On the 2-packing differential of a graph. *arXiv* **2021**, arXiv:2105.13438.
10. Haynes, T.W.; Hedetniemi, S.T.; Henning, M.A. Topics in Domination in Graphs. In *Developments in Mathematics*; Springer: Cham, Switzerland, 2020; Volume 64, p. 545. [CrossRef]
11. Haynes, T.W.; Hedetniemi, S.T.; Slater, P.J. *Fundamentals of Domination in Graphs*; Chapman and Hall/CRC Pure and Applied Mathematics Series; Marcel Dekker, Inc.: New York, NY, USA, 1998; p. 446.
12. Haynes, T.; Hedetniemi, S.; Slater, P. *Domination in Graphs: Volume 2: Advanced Topics*; Chapman & Hall/CRC Pure and Applied Mathematics; Taylor & Francis: Boca Raton, FL, USA, 1998.
13. Henning, M.A.; Yeo, A. *Total Domination in Graphs*; Springer Monographs in Mathematics; Springer: New York, NY, USA, 2013; p. 178.
14. Caro, Y.; Roditty, Y. A note on the k-domination number of a graph. *Int. J. Math. Math. Sci.* **1990**, *13*, 205–206. [CrossRef]
15. Goddard, W.; Henning, M.A.; McPillan, C.A. Semitotal domination in graphs. *Util. Math.* **2014**, *94*, 67–81.
16. Gallai, T. Über extreme Punkt- und Kantenmengen. *Ann. Univ. Sci. Budapestinensis Rolando Eötvös Nomin. Sect. Math.* **1959**, *2*, 133–138.
17. Frucht, R.; Harary, F. On the corona of two graphs. *Aequationes Math.* **1970**, *4*, 322–325. [CrossRef]
18. Fink, J.F.; Jacobson, M.S. n-domination in graphs. In *Graph Theory with Applications to Algorithms and Computer Science*; Wiley-Interscience Publication Wiley: New York, NY, USA, 1985; pp. 283–300.
19. Cockayne, E.J.; Dreyer, P.A., Jr.; Hedetniemi, S.M.; Hedetniemi, S.T. Roman domination in graphs. *Discret. Math.* **2004**, *278*, 11–22. [CrossRef]
20. Cabrera García, S.; Cabrera Martínez, A.; Yero, I.G. Quasi-total Roman domination in graphs. *Results Math.* **2019**, *74*, 173. [CrossRef]
21. Chellali, M.; Haynes, T.W.; Hedetniemi, S.T.; McRae, A.A. Roman {2}-domination. *Discret. Appl. Math.* **2016**, *204*, 22–28. [CrossRef]
22. Henning, M.A.; Klostermeyer, W.F. Italian domination in trees. *Discret. Appl. Math.* **2017**, *217*, 557–564. [CrossRef]
23. Klostermeyer, W.F.; MacGillivray, G. Roman, Italian, and 2-domination. *J. Combin. Math. Combin. Comput.* **2019**, *108*, 125–146.

Article

Study of Parameters in the Genetic Algorithm for the Attack on Block Ciphers

Osmani Tito-Corrioso [1,*], Miguel Angel Borges-Trenard [2], Mijail Borges-Quintana [3], Omar Rojas [4] and Guillermo Sosa-Gómez [4,*]

1. Departamento de Matemática, Facultad de Ciencias de la Educación, Universidad de Guantánamo, Av. Che Guevara km 1.5 Carr. Jamaica, Guantánamo 95100, Cuba
2. Doctorate in Mathematics Education, Universidad Antonio Nariño, Bogotá 111321, Colombia; borgestrenard2014@gmail.com
3. Departamento de Matemática, Facultad de Ciencias Naturales y Exactas, Universidad de Oriente, Av. Patricio Lumumba s/n, Santiago de Cuba 90500, Cuba; mijail@uo.edu.cu
4. Facultad de Ciencias Económicas y Empresariales, Universidad Panamericana, Álvaro del Portillo 49, Zapopan, Jalisco 45010, Mexico; orojas@up.edu.mx
* Correspondence: osmanitc@cug.co.cu (O.T.-C.); gsosag@up.edu.mx (G.S.-G.); Tel.: +53-21326113 (O.T.-C.); +52-3313682200 (G.S.-G.)

Abstract: In recent years, the use of Genetic Algorithms (GAs) in symmetric cryptography, in particular in the cryptanalysis of block ciphers, has increased. In this work, the study of certain parameters that intervene in GAs was carried out, such as the time it takes to execute a certain number of iterations, so that a number of generations to be carried out in an available time can be estimated. Accordingly, the size of the set of individuals that constitute admissible solutions for GAs can be chosen. On the other hand, several fitness functions were introduced, and which ones led to better results was analyzed. The experiments were performed with the block ciphers AES(t), for $t \in \{3, 4, 7\}$.

Keywords: genetic algorithm; cryptanalysis; AES(t); optimization; heuristics

1. Introduction

There are several methods and tools that are used as optimization methods and predictive tools. Several heuristic algorithms have been used in the context of cryptography; in [1], the Ant Colony Optimization (ACO) heuristic method was used, and a methodology with S-AES block encryption was tested, using two pairs of plain encrypted texts. In [2], a combination of GA and ACO methods was used for cryptanalysis of stream ciphers. In [3–5], the possibilities of combining and designing these analyzes using machine learning and deep learning tools were shown. In [6–8], the methods of the Artificial Neural Network (ANN), Support Vector Machine (SVM), and Gene-Expression Programming (GEP) were used as predictive tools in other contexts.

The Genetic Algorithm (GA) is an optimization method used in recent years in cryptography for various purposes, mainly to carry out attacks on various encryption types. Some of the research conducted in this direction is mentioned next. In [9], the authors presented a combination of the GA with particle swarm optimization (another heuristic method based on evolutionary techniques); they called their method genetic swarm optimization and applied it to attack the block cipher Data Encryption Standard (DES). Their experimental results showed that better results were obtained by applying their combined method than by using both methods separately. The proposal presented in [10] provided a preliminary exploration of GA's use over a Permutation Substitution Network (SPN) cipher. The purpose of the scan was to determine how to find weak keys. Both works [9,10] used a known plaintext attack, i.e., given a plaintext T and the corresponding ciphertext C, one is interested in finding the key K. In [10], the fitness function evaluates the bitwise difference (Hamming distance) between C and the ciphertext of T, using a candidate for the key, whereas, on the contrary, in [9] the Hamming distance between T and the decryption of the ciphertext of C is measured. In [11], a ciphertext-only attack on simplified DES was shown, obtaining better results than by brute force. The authors used a fitness function that combined the relative frequency of monograms, digrams, and trigrams (for a particular language). Since the key length was very small, they were able to use this kind of function. The approach in [12] was similar to [11]; it used essentially the same fitness function, but with different parameters. It was also more detailed regarding the experiments and compared them concerning brute force and random search. For more details on the area of cryptanalysis using GAs, see [13–15].

As in all evolutionary algorithms, it is always a difficulty in the GA that, as the number of individuals in the space of admissible solutions grows, in this case, the set of keys, it is necessary to perform a greater number of generations in order to obtain the best results. It is clear that the greater the number of generations, the more time the algorithm consumes, so it is important to be able to estimate the time that may be necessary to execute a certain number of desired generations. On the other hand, it is necessary to analyze fitness functions that allow obtaining better results with the fittest individuals obtained.

Symmetry is omnipresent in the universe; in particular, it is present in symmetric cryptography, where the secret key is known for both authorized parts in the communication channel essentially by symmetry. We worked with block ciphers, an important primitive of symmetric cryptographic, where the key space (the population of admissible solutions for the GA in this case) is exponentially big, making it impossible in many cases to fully move in that space.

In the present work, the ideas to divide the key space that were started in [16,17] were followed. Both methodologies for dividing the key space allow the GA search space to be reduced over a subset of individuals. For this case, we studied the behavior of time and the introduction of various fitness functions. The structure of the work is as follows. In Section 2, the general ideas of the GA and two methodologies for partitioning the key space are presented; in Section 3, several parameters of the cryptanalysis for block ciphers using the GA are studied; in Section 3.1, the time it takes to execute a certain number of iterations is analyzed, so that a number of generations to be carried out in an available time can be estimated; and in Section 3.2, other fitness functions are proposed. Finally, Section 4 gives the conclusions.

2. Preliminaries

2.1. The Genetic Algorithm

The GA is a heuristic optimization method. We assume that the reader knows the general ideas of how the GA works; see Algorithm 1. In this section, we briefly describe the GA scheme used in this work.

Algorithm 1 Genetic algorithm.

Input: m (quantity of individuals in the population), F (fitness function), g (number of generations).
Output: the individual with the highest fitness function as the best solution.
1: Randomly generate an initial population P_i with m individuals (possible solutions).
2: Compute the fitness of each individual from P_i with F.
3: **while** the solution is not found, or the g generations are not reached **do**
4:     Select parent pairs in P_i.
5:     Perform the crossover of the selected parents, and generate a pair of offspring.
6:     Mutate each of the resulting descendants.
7:     Compute the fitness of each of the descendants with F and their mutations.
8:     By the tournament method between two, based on the fitness of the parents and descendants, decide what is the new population P_i for the next generation (selecting two individuals at random each time and choosing the one with the highest fitness).
9: **end while**

The individuals from the populations are elements of the key space taken as binary blocks. For **Crossover**, the crossing by two points was used, and the crossover probability was fixed to 0.6. The **Mutate** operation consisted of interchanging the values of the bits of at most three random components of the binary block with a mutation rate of 0.2. The values of 0.6 and 0.2 were fixed for all experiments, and the study of the incidence of the variation of these values in the behavior of the GA was not addressed in this paper. An individual x is better adapted than another y, if it has greater fitness, i.e., if $F(x) > F(y)$. Fitness functions are studied in more detail in Section 3.2. For the specification of the GA to block ciphers, see Section 3 of [16].

2.2. Key Space Partition Methodologies

The methodologies introduced in [16,17] allow GAs to work on a certain subset of the set of admissible solutions as if it were the complete set. The importance of this fact is that it reduces the size of the search space and gives the heuristic method a greater chance of success, assuming that the most suitable individuals are found in the selected subset. Let $\mathbb{F}_2^{k_1}$ be the key space of length $k_1 \in \mathbb{Z}_{>0}$. It is known that $\mathbb{F}_2^{k_1}$ has cardinality 2^{k_1}, and therefore, there is a one-to-one correspondence between $\mathbb{F}_2^{k_1}$ and the range $\left[0, 2^{k_1} - 1\right]$. If an integer k_2 is set, $(1 < k_2 \leq k_1)$, then the key space can be represented by the numbers,

$$q2^{k_2} + r, \qquad (1)$$

where $q \in \left[0, 2^{k_1 - k_2} - 1\right]$ and $r \in \left[0, 2^{k_2} - 1\right]$. In this way, the key space is divided into $2^{k_1 - k_2}$ blocks (determined by the quotient in the division algorithm dividing by 2^{k_2}), and within each block, the corresponding key is determined by its position, which is given by the remainder r. The main idea is to stay in a block (given by q) and move within this block through the elements (given by r) using the GA. Note in this methodology that first q is set to choose a block, and then, r varies to be able to move through the elements of the block; however, the complete key in $\mathbb{F}_2^{k_1}$ is obtained from Expression (1). We refer to this methodology as BBM. For more details on the connection with GAs, see [16].

The following methodology is based on the definition of the quotient group of the keys G_K whose objective is to make a partition of $\mathbb{F}_2^{k_1}$ in equivalence classes. It is known that $\mathbb{F}_2^{k_1}$, as an additive group, is isomorphic to $\mathbb{Z}_{2^{k_1}}$. Let h be the homomorphism defined as follows:

$$\begin{aligned} h : \mathbb{Z}_{2^{k_1}} &\longrightarrow \mathbb{Z}_{2^{k_2}} \\ n &\longrightarrow n \pmod{2^{k_2}}, \end{aligned} \qquad (2)$$

where $k_2 \in \mathbb{Z}_{>0}$ and $0 < k_2 < k_1$. We denote by N the kernel of h, i.e.,

$$N = \{x \in \mathbb{Z}_{2^{k_1}} | h(x) = 0 \in \mathbb{Z}_{2^{k_2}}\}. \qquad (3)$$

Then, by the definition of h, we have that N is composed by the elements of $\mathbb{Z}_{2^{k_1}}$, which are multiples of 2^{k_2}. It is known that N is an invariant subgroup; therefore, the main objective is to calculate the quotient group of $\mathbb{Z}_{2^{k_1}}$ by N, and in this way, the key space will be divided into 2^{k_2} equivalence classes. We denote by G_K the quotient group of $\mathbb{Z}_{2^{k_1}}$ by N ($G_K = \mathbb{Z}_{2^{k_1}}/N$). By Lagrange's theorem, we have that $o(G_K) = o(\mathbb{Z}_{2^{k_1}})/o(N)$, but $o(G_K) = o(\mathbb{Z}_{2^{k_2}}) = 2^{k_2}$, then,

$$o(N) = o(\mathbb{Z}_{2^{k_1}})/o(\mathbb{Z}_{2^{k_2}}) = 2^{k_1 - k_2}. \qquad (4)$$

Now, N can be described, taking into account that its elements are multiples of 2^{k_2}. For this, we take $Q = \{0, 1, 2, \ldots, 2^{k_1-k_2} - 1\}$, then:

$$N = \ <2^{k_2}> \ = \{x \in \mathbb{Z}_{2^{k_1}} | \ \exists q \in Q, x = q\,2^{k_2}\} \qquad (5)$$
$$= \{0, 2^{k_2}, 2*2^{k_2}, 3*2^{k_2}, \ldots, (2^{k_1-k_2} - 1)*2^{k_2}\}.$$

On the other hand,

$$G_K = \{N, 1+N, 2+N, \ldots, (2^{k_2}-2)+N, (2^{k_2}-1)+N\}. \qquad (6)$$

In this way, $\mathbb{Z}_{2^{k_1}}$ is divided into a partition of 2^{k_2} classes given by N. G_K is called the quotient group of keys. Let,

$$E : \{0,1\}^m \times \{0,1\}^n \to \{0,1\}^n, m, n \in \mathbb{N}, m \geq n, \qquad (7)$$

be a block cipher, T a plaintext, K a key, and C the corresponding ciphertext, i.e., $C = E(K, T)$; K' is said to be a consistent key with E, T, and C, if $C = E(K', T)$ (see [16]). The idea here is also to go through, from the total space, the elements that are in a class and then find one (or several) consisting of the keys of that class. To be able to go through the elements of each class, note that $\mathbb{Z}_{2^{k_2}}$ is isomorphic with G_K, and the isomorphism corresponds to each $r \in \mathbb{Z}_{2^{k_2}}$ its equivalence class $r + N$ in G_K; thus, selecting a class is setting an element $r \in \mathbb{Z}_{2^{k_2}}$. On the other hand, the elements of N are of the form $q\,2^{k_2}$ ($q \in Q$); therefore, the elements of the class $r + N$ are of the form,

$$q\,2^{k_2} + r, q \in Q. \qquad (8)$$

Then, the problem of looping through each element of each equivalence class consists of first setting an element of $\mathbb{Z}_{2^{k_2}}$ and then looping through each element of the set Q, to find a key of G_K using Equation (8). The elements of the set Q have block length $k_d = k_1 - k_2$, and each class has 2^{k_d} elements. We refer to this methodology as TBB. Note that the TBB methodology is a kind of dual idea with respect to the BBM methodology, i.e., one first stays in the same class (given by r) and then moves within this class through the elements (given by q) using the GA. In this case, the length of the blocks is 2^{k_d} instead of 2^{k_2}.

The main difficulty in these methodologies is the choice of k_2, since it is the parameter that determines the number of equivalence classes and, therefore, the number of elements within them. If in G_K, k_2 increases, the classes have fewer elements, but there are more classes; on the contrary, if it decreases, so does the number of classes, but the number of elements of each increases. Something similar happens in the first methodology. The operations of the space partitioning and going through the elements of each class are done with the decimal representation and the specific operations of the GA with the binary representation. For more details, see [16,17].

In Figure 1, the relationship of the content by subsections and the attack on block ciphers are shown in a flowchart.

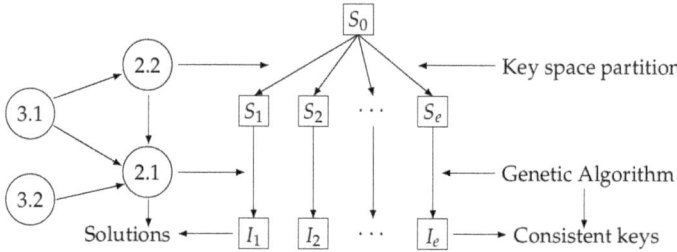

Figure 1. Flowchart of the relationship between content by subsections and the attack on block ciphers.

3. Study of Parameters in the GA

3.1. Time Estimation

In GAs, less complex operations such as mutation and crossing are performed within each class, where the elements have block length $k_2 \leq k_1$ or $k_d \leq k_1$ depending on the way of partitioning the space. However, despite the variation of these two parameters, the calculation of the fitness function, being the function of greater complexity within the GA, is carried out using (8), i.e., with the complete key of length k_1, and not with the part of it found in the class. This means that a variation in the number of elements in a class does not affect the fitness function's cost. Moreover, if all the parameters remain the same, the GA's time in each generation must be quite similar, even if k_2 varies. To check this, experiments were done with a PC with an Intel(R) Core (TM) i3-4160 CPU @ 3.60GHz (four CPUs), and 4GB of RAM. AES(t) encryption was used, a parametric version of AES, where $t \in \{3,4,5,6,7,8\}$ and also AES(8) = AES (see [18,19]). The experiment consisted of executing the GA with the BBM methodology and measuring the time (in minutes) that it took in a generation for different values of k_2 (keeping the other parameters fixed), then verifying if these data were used to forecast the time it would take in n generations. The size of the population was $m = 100$ in all cases.

Tables 1–3 summarize the results corresponding to AES(3), AES(4), and AES(7), respectively. The first column has the different values that were given to k_2. The second column is the average time t_{k_2} that was obtained for a generation in 10 executions of each k_2. The general mean for all the k_2 values is $t_m = 0.0435571$ minutes approximately in Table 1, $t_m = 0.0519393$ in Table 2, and $t_m = 0.1900297$ in Table 3. The third column represents the number of generations (n_g). The real-time that the algorithm takes, t_r, appears in the fourth column. The fifth column is the estimated time, t_e, that should be delayed, the calculation of which is based on:

$$t_e = t_m n_g. \tag{9}$$

Finally, the last column is the error of the prediction, $E_p = |t_r - t_e|$. With these experiments, we wanted to check for the procedure whether if for a specific value of k_2 and having n_g generations, then the approximate time (t) that the GA would take to complete those generations was $t \approx t_e$.

With a generation, or very few, the average time it took for the GA was slightly slower, decreasing and tending to stabilize at a limit as it performed more iterations. This was due to probabilistic functions that intervened in the GA and a set of operations to randomly create an initial population. Therefore, the criterion for calculating the average time t_{k_2} was to let the GA finish executing in a certain number of generations, either because it found the key or because it reached the last iteration without finding it, and then calculate the average. Therefore, calculating t_{k_2} in a few generations or setting the amount to one, would get longer times; however, doing so would be valid if the intention were to go over the top in estimating the time that the algorithm consumed.

Table 1. Time estimation in AES(3).

k_2	1 Gen	n Gen	t_r	t_e	E_p
10	0.0355	2	0.0962	0.0871	0.0091
11	0.0518	20	0.7134	0.8711	0.1577
12	0.0429	27	1.0491	1.1760	0.1269
13	0.042	81	3.3475	3.5281	0.1806
14	0.0429	49	2.0863	2.1343	0.0481
15	0.0454	71	3.1606	3.0926	0.0680
16	0.0444	655	28.9312	28.5299	0.4012

Table 2. Time estimation in AES(4).

k_2	1 Gen	n Gen	t_r	t_e	E_p
10	0.0519	9	0.3739	0.4675	0.0936
11	0.0553	8	0.3838	0.4155	0.0318
12	0.0465	5	0.2756	0.2597	0.0159
13	0.0564	2	0.1303	0.1039	0.0264
14	0.0506	81	4.1554	4.2071	0.0517
15	0.0510	98	4.9621	5.0900	0.1279
16	0.0519	655	34.1330	34.0202	0.1128

In the case of AES(7) (Table 3), we only experimented with the values 17 and 18 of k_2, since considering all the previous (or higher) values would take a considerably longer time (given the greater strength of AES(7)).

Table 3. Time estimation in AES(7).

k_2	1 Gen	n Gen	t_r	t_e	E_p
17	0.1895	373	69.1909	70.8811	1.6902
18	0.1905	932	178.069	177.108	0.9610

Similar results were obtained if more values of k_2 were chosen to calculate t_m. For example, using a PC Laptop with a processor: Intel (R) Celeron (R) CPU N3050 @ 1.60GHz (two CPUs), ~1.6 GHz, and 4 GB of RAM and going through all the values of k_2 from 10 to 48 (AES(3) key length), $t_m = 0.2340212$ was obtained. Now, for $n_g = 215$, we had $t_r = 48.14715$ and $t_e = t_m n_g \approx 50.3145$. In another test: $n_g = 150$, $t_r = 34.9565$, then $t_e \approx 35.1032$. Note that the PC used in this case had different characteristics and less computational capacity than the experiments in Tables 1–3. The interesting thing is that under these conditions, the results were as expected as well.

In a similar way, the GA was executed with the TBB methodology for the search in G_K, for values of k_d equal to those of k_2 and different generations (n_g). It was observed that the time estimates behaved in a similar way to the results presented previously for the BBM methodology. Note that in the AES(t) family of ciphers, the length of the key increases from 48 for AES(3) to 128 for AES(8); however, regardless of the key length, the same behavior was seen in all of them.

Now, we showed with these experiments another application of this study on time estimation. In the GA scheme with the BBM methodology, the total number of generations (iterations) to perform for a given value of k_2 is:

$$g = \left\lfloor \frac{2^{k_2}}{m} \right\rfloor. \tag{10}$$

Taking $n_g = g$, by using t_e, then we can do an a priori estimation for a given value of k_2, of the total time it will take the GA to perform all the generations or a certain desired percent of them. For example, in AES(3), for $k_2 = 16$, in Expression (10), we have $g = 655$; now, since $t_m = 0.0435571$ in Table 1, then the approximate time that the GA will consume to perform 655 generations is $t_e \approx 655 \cdot 0.0435571 \approx 28.5299$, as can be seen in the table. Another example can be seen in Table 2, also for $k_2 = 16$.

On the other hand, supposing we have an available time t_e, to carry out the attack with this model, thus we may use (9) and (10), to compute an approximated value of k_2, which implies doing the corresponding partition of the space and computing the number of generations to perform for this time t_e and the value of k_2. In this sense, doing $n_g = g$ in (9), we have:

$$k_2 \approx \left\lfloor \log_2 \frac{mt_e}{t_m} \right\rfloor. \quad (11)$$

We remark that the above is valid in the TBB methodology, only that k_d is used instead of k_2.

As can be observed, the results on the estimation of time were favorable. In this sense, the following points can be summarized:

1. Taking into account the estimation of time t_e and its observed closeness to the real value t_r, a number of generations to be carried out in an available or desired time can be estimated (using Expression (9)), which can be taken as a starting point for the proper choice of k_2, or k_d in G_K (see Section 2). In this way, it is possible to adapt the size of the search space (to choose a proper value of k_2 using (11)) to the number of generations that it is estimated can be executed in a given time.
2. The time t_{k_2} could be used to perform the time estimation of its own k_2, but as can be seen in the tables, sometimes, it makes predictions with minor errors and other times greater than with t_m. Another drawback is that it cannot be used for other k_2. On the contrary, the main advantage of using t_m is that it can be calculated for some sparse values of k_2 and be used to estimate the time even with values of this parameter whose t_{k_2} has not been calculated.

3.2. Proposal of Other Fitness Functions

In the context of the BBM and TBB methodologies used in this work with the GA, we studied in this section which fitness functions provided a better response, in the sense that consistent keys were obtained as solutions in a greater percentage of occasions. Let E be a block cipher with length n of plaintext and ciphertext, defined as in Expression (7), T a plaintext, K a key, and C the corresponding ciphertext, that is $C = E(K, T)$. Let:

$$D: \{0.1\}^m \times \{0.1\}^n \to \{0.1\}^n, \quad (12)$$

be the function of decryption of E, such that $T = D(K, C)$. Then, the fitness function with which we have been working and based on the Hamming distance d_H, for a certain individual X of the population, is:

$$F_1(X) = \frac{n - d_H(C, E(X, T))}{n}, \quad (13)$$

which measures the closeness between the encrypted texts C and the text obtained from encrypting T with the probable key X (see [16]). A similar function is the one that measures the closeness between plaintexts:

$$F_2(X) = \frac{n - d_H(T, D(X, C))}{n}. \quad (14)$$

Another function that follows the idea of comparing texts in binary with d_H is the weighting of F_1 and F_2. Let $\alpha, \beta \in [0, 1] \subset \mathbb{R}$, such that $\alpha + \beta = 1$, then this function would be defined as follows:

$$F_3(X) = \alpha F_1(X) + \beta F_2(X). \quad (15)$$

It is interesting to note that F_3 is more time consuming than each function separately, but the idea is to be more efficient in searching for the key.

The fitness functions proposed at this point are based on measuring the closeness of the plaintext and ciphertext, but in decimals. Let Y_d be the corresponding conversion to decimals of the binary block Y. The first function is defined as follows,

$$F_4(X) = \frac{2^n - 1 - |C_d - E(X,T)_d|}{2^n - 1}. \tag{16}$$

Note that if the encrypted texts are equal, $C_d = E(X,T)_d$, then $|C_d - E(X,T)_d| = 0$, which implies that $F_4(X) = 1$, i.e., if they are equal, then the fitness function takes the highest value. On the contrary, the greatest difference is the farthest they can be, i.e., $C_d = 2^n - 1$ and $E(X,T)_d = 0$, and therefore, $F_4(X) = 0$. The following is a weighting of the functions F_1 and F_4,

$$F_5(X) = \alpha F_1(X) + \beta F_4(X). \tag{17}$$

Both functions have in common that they measure the closeness between ciphertexts. This is not ambiguous since, for example, if C and $E(X,T)$ differ by two bits, the function F_1 will always have the same value no matter what these two bits are. On the contrary, it is not the same in F_4 if the bits are both more or less significant since the numbers are not the same in their decimal representation. The following function measures the closeness in decimals of plaintexts:

$$F_6(X) = \frac{2^n - 1 - |T_d - D(X,C)_d|}{2^n - 1}. \tag{18}$$

Finally, the functions F_7, F_8, and F_9 are defined with respect to the previous ones as follows,

$$F_7(X) = \alpha F_2(X) + \beta F_6(X), \tag{19}$$
$$F_8(X) = \alpha F_4(X) + \beta F_6(X), \tag{20}$$
$$F_9(X) = \alpha_1 F_1(X) + \alpha_2 F_2(X) + \alpha_3 F_4(X) + \alpha_4 F_6(X), \tag{21}$$

where $\alpha_i \in [0,1] \subset \mathbb{R}, i \in \{1,2,3,4\}$ and $\sum_{i=1}^{4} \alpha_i = 1$. This guarantees that in general, each $F_j(X) \in [0,1] \subset \mathbb{R}, j \in \{1,2,3,4,5,6,7,8,9\}$.

The idea behind the introduction of these functions lies mainly in the fact that there are changes that the Hamming distance does not detect, as opposed to the decimal distance. For example, suppose the key is $a = (1,1,1,1,1,1)_2$, and $b = (0,0,0,0,0,1)_2$ is the possible key, both in binary. It is clear that the Hamming distance is five, and the distance in decimals is 62 since $a = 63$ and $b = 1$; the fitness functions take the values $1 - 5/6 = 0.17$ for the binary version and $1 - 62/63 = 0.016$ for the decimal version. Now, if $b = (0,0,1,0,0,0)_2$, the binary fitness function would still be 0.17 since there are still five different bits; on the other hand, $b = 8$, so the decimal fitness function takes the value $1 - 55/63 = 0.13$. Finally, if we take $b = (1,0,0,0,0,0)_2 = 32$, then the distance in binary remains the same value, but the decimal continues to change, therefore, the fitness function as well, and takes the value 0.49. Therefore, this shows that the change of b, the decimal distance, is always detected, unlike the binary distance, which remains the same for certain changes.

AES(3) encryption attack experiments were carried out for the two methodologies for partitioning the key space to compare these functions. The main idea is to find the key and not do a component percent match analysis between them, where the fitness functions with the Hamming distance would be more useful. A PC with an Inter (R) Core (TM) i3-4160 CPU @ 3.60GHz (four CPUs), and 4 GB of RAM was used. For the results, we took into account the average time it took to find the key, the average number of generations in which it was found, the percentage of failures (in many attacks carried out), and a parameter called efficiency, E_{F_i}, which resulted in a weighting of the three previous criteria.

Definition 1 (Fitness functions' efficiency). *Let $\mu_1, \mu_2, \mu_3 \in [0,1] \subset \mathbb{R}$, $\mu_1 + \mu_2 + \mu_3 = 1$, $t_{F_i}, i = 1, \cdots, k$, the time it takes the GA to find the key with F_i, on an average for g_{F_i} generations,* and p_{F_i} the percent of attempts in that the GA did not find the key with F_i. Then, the efficiency, E_{F_i}, of the fitness function F_i with respect to the other $k-1$ functions, $F_j, j \neq i$, is defined as,

$$E_{F_i} = 1 - \left(\mu_1 \frac{t_{F_i}}{\sum_{\gamma=1}^{k} t_{F_\gamma}} + \mu_2 \frac{g_{F_i}}{\sum_{\gamma=1}^{k} g_{F_\gamma}} + \mu_3 \frac{p_{F_i}}{\sum_{\gamma=1}^{k} p_{F_\gamma}} \right). \tag{22}$$

Note that the number of generations and the failure percentage are inversely proportional to the efficiency E_{F_i} as the higher these parameters, the lower its efficiency fitness function. Table 4 presents the results of the comparison of the different fitness functions for the BBM space partitioning methodology, in this case $k = 9$. We took $\alpha = \beta = 0.5$ and each $\alpha_i = 0.25$. To calculate E_{F_i} the values $\mu_1 = 0.33$, $\mu_2 = 0.33$ and $\mu_3 = 0.34$ were taken for t_{F_i}, g_{F_i}, and p_{F_i}, respectively. Sorting F_i with respect to efficiency, the first five would be F_6, F_8, F_4, F_5, and F_2. It is noteworthy that of the first three that use only the Hamming distance, only F_2 appears.

Table 4. Comparison of fitness functions, with BBM.

F_i	Times	Generations	Failures (%)	E_{F_i}
F_1	5.233	121.2	60	0.8731
F_2	5.402	108.4	50	0.8870
F_3	11.101	117.4	50	0.8584
F_4	4.764	109.2	40	0.8995
F_5	9.451	109.8	30	0.8885
F_6	3.126	63.4	20	0.9433
F_7	12.424	121.3	50	0.8511
F_8	7.054	77.1	10	0.9309
F_9	15.811	87.7	30	0.8682

In the comparison of these functions for the TBB methodology of partitioning the key space and searching in G_K, the experiment results are presented in Table 5. In this case, ordering the functions by their efficiency, the first five would be F_1, F_4, F_5, F_8, and F_6. Again, a single function appears from the first three, in this case F_1, and the others repeat. Note in particular that F_8 (the weight of the functions in decimals) is better than F_3 (the weight of the functions in binary) in each of the parameters measured in both methodologies.

Table 5. Comparison of fitness functions, with TBB.

F_i	Times	Generations	Failures (%)	E_{F_i}
F_1	3.688	83.1	20	0.9278
F_2	5.353	109.1	60	0.8633
F_3	11.403	122.9	40	0.8536
F_4	3.226	67.8	30	0.9240
F_5	7.147	83.4	10	0.9235
F_6	4.871	96.2	40	0.8939
F_7	10.694	113.1	20	0.8840
F_8	8.354	92	20	0.9029
F_9	16.876	95.7	50	0.8270

It is interesting to see what happens if the values of the weights are changed in the functions F_5, F_7, and F_9, which combine the functions with distance in decimals and binary, keeping fixed μ_1, μ_2, and μ_3 for the calculation of E_{F_i}. In this sense, in the following group of experiments, the weights were assigned as follows for each methodology: the values were 0.2 and 0.8; first, in each of these three functions, the subfunctions in binary were favored, from which $\alpha = 0.8$, $\beta = 0.2$ (in F_5, F_7), $\alpha_1 = \alpha_2 = 0.4$, and $\alpha_3 = \alpha_4 = 0.1$ (in F_9; note that this function has two subfunctions with the distance in binary and two in decimals); in this case, we identified the functions as F_{5b}, F_{7b}, and F_{9b}; then, we changed the order of these same weights, and the largest were given to the subfunctions whose distance was in decimals; and we identified the functions for this case as F_{5d}, F_{7d}, and F_{9d}.

For the BBM methodology, the results are presented in Table 6. Note that according to E_{F_i}, the first is F_{7d}, followed by F_{5d} and F_{9d}.

Table 6. Comparison of functions F_5, F_7, and F_9, with BBM.

F_i	Times	Generations	Failures (%)	E_{F_i}
F_{5b}	10.247	115	50	0.772
F_{7b}	9.131	90.6	40	0.814
F_{9b}	20.053	107.4	50	0.728
F_{5d}	7.276	83.3	10	0.891
F_{7d}	5.921	61.3	0	0.933
F_{9d}	13.799	77.5	10	0.862

In Figure 2, these results are compared, according to E_{F_i}, with those of Table 4, also including the values of F_5, F_7, and F_9. Sorting the functions according to their efficiency, the first five are F_{7d}, F_6, F_8, F_{5d}, and F_{9d}.

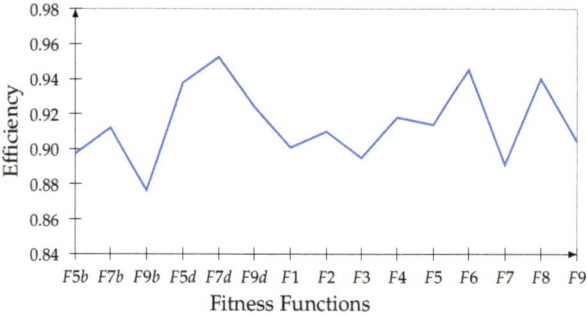

Figure 2. Efficiency of all fitness functions in the BBM methodology.

Notice how the best results prevail in the functions with the distance in decimals. In this sense, F_7 and F_9 (now as F_{7d} and F_{9d}) are incorporated into the first ones and three of those that already were in this group in the above experiments, F_5 (as F_{5d}), F_6, and F_8

In the case of the TBB methodology, the results are presented in Table 7. According to efficiency, the first is F_{7b}, followed by F_{5d} and F_{5b}.

Table 7. Comparison of functions F_5, F_7, and F_9, with TBB.

F_i	Times	Generations	Failures (%)	E_{F_i}
F_{5b}	9.987	111.5	40	0.845
F_{7b}	8.578	86.7	10	0.909
F_{9b}	22.500	119.1	50	0.777
F_{5d}	8.341	96.9	10	0.905
F_{7d}	13.623	141.8	80	0.754
F_{9d}	22.183	114.8	30	0.811

In Figure 3, these results are compared with those of all the functions of Table 5. The first five are now F_1, F_5, F_4, F_{7b}, and F_{5d}; notice how the functions that contain the distance prevail in decimals and this combined with binary. In the experiments, the best global behavior of the functions with the decimal distance is verified, and specifically in the BBM methodology, where the keys are grouped into intervals according to their decimal position in space, contrary to the other methodology, where the keys of each class are positioned throughout the space.

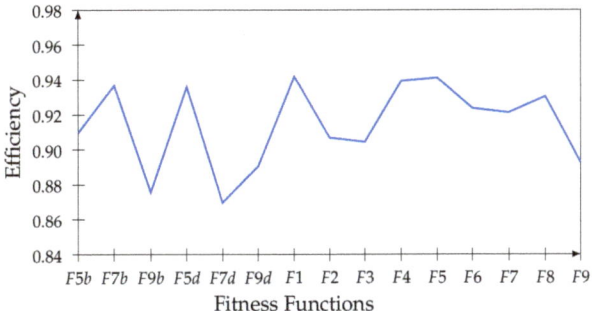

Figure 3. Efficiency of all fitness functions in the TBB methodology.

Note that when comparing Figures 2 and 3, the values of E_{F_i} that are in the tables are not directly compared, but rather, it is necessary to recalculate E_{F_i} taking into account that there are 15 functions. We mean,

$$E_{F_{\delta_i}} = 1 - \left(\mu_1 \frac{tF_{\delta_i}}{\sum\limits_{\gamma=1}^{k} tF_{\delta_\gamma}} + \mu_2 \frac{gF_{\delta_i}}{\sum\limits_{\gamma=1}^{k} gF_{\delta_\gamma}} + \mu_3 \frac{pF_{\delta_i}}{\sum\limits_{\gamma=1}^{k} pF_{\delta_\gamma}} \right), \quad (23)$$

where $\delta_i \in \{1, \cdots, 9, 5b, 5d, 7b, 7d, 9b, 9d\}$, $i = \overline{1, \cdots, k}$, and, $k = 15$.

4. Conclusions

In this article, various aspects of some parameters of the GA for the attack on block ciphers were studied. In the first place, a way of estimating the time that the GA takes in a given number of generations was proposed, having an average of the time that this algorithm takes in one generation. This study is important to jointly evaluate different parameters and make the best decisions according to the computational capacity, available time, and an adequate selection of the size of the search space when using the BBM and TBB methodologies. On the other hand, several fitness functions were proposed with favorable results in the experiments with respect to the fitness functions using only the Hamming distance. In this sense, it was found that the fitness functions that use the decimal distance, in general, are more efficient than those that use only the Hamming distance, especially in the methodology BBM.

As future work, several directions are possible. Similar studies can be carried out with the GA working with other parameters, such as varying the crossover probability and mutation rate and making comparisons regarding the percentage of success of the method. It is also recommended to explore other heuristic techniques and to evaluate the use of whole space partitioning methods so that the methods work closed on the subsets. In the same way, it is also recommended to investigate the combined use with some other tools such as machine learning, deep learning, ANN, SVM, and GEP.

Author Contributions: Conceptualization, O.T.-C. and M.B.-Q.; methodology, O.T.-C., M.B.-Q., M.A.B.-T., and G.S.-G.; software, O.T.-C.; validation, O.T.-C., M.B.-Q., M.A.B.-T., O.R., and G.S.-G.; formal analysis, M.A.B.-T., M.B.-Q., O.R., and G.S.-G.; investigation, O.T.-C., M.B.-Q., M.A.B.-T., and G.S.-G.; writing—original draft preparation, O.T.-C. and M.B.-Q.; writing—review and editing, O.T.-C., M.B.-Q., M.A.B.-T, O.R., and G.S.-G.; visualization, O.T.-C.; supervision, M.A.B.-T., M.B.-Q., O.R., and G.S.-G. All authors read and agreed to the published version of the manuscript.

Funding: This research received no external funding.

Informed Consent Statement: Informed consent was obtained from all subjects involved in the study.

Conflicts of Interest: The authors declare no conflict of interest.

References

1. Grari, H.; Azouaoui, A.; Zine-Dine, K. A cryptanalytic attack of simplified-AES using ant colony optimization. *Int. J. Electr. Comput. Eng.* **2019**, *9*, 4287. [CrossRef]
2. Jawad, R.N.; Ali, F.H. Using Evolving Algorithms to Cryptanalysis Nonlinear Cryptosystems. *Baghdad Sci. J.* **2020**, *17*, 0682. [CrossRef]
3. Blackledge, J.; Mosola, N. Applications of Artificial Intelligence to Cryptography. Transactions on Machine Learning & Artifical Intellengence 6th June 2020. *Trans. Mach. Learn. Artif. Intellengance* **2020**. [CrossRef]
4. Lee, T.R.; Teh, J.S.; Yan, J.L.S.; Jamil, N.; Yeoh, W.Z. A Machine Learning Approach to Predicting Block Cipher Security. In Proceedings of the Cryptology and Information Security Conference, Seoul, Korea, 2–4 December 2020; p. 122.
5. So, J. Deep Learning-Based Cryptanalysis of Lightweight Block Ciphers. *Secur. Commun. Netw.* **2020**, *2020*, 3701067. [CrossRef]
6. You, L.; Yan, K.; Liu, N. Assessing artificial neural network performance for predicting interlayer conditions and layer modulus of multi-layered flexible pavement. *Front. Struct. Civ. Eng.* **2020**, *14*, 487–500. [CrossRef]
7. Qiu, X.; Xu, J.X.; Tao, J.Q.; Yang, Q. Asphalt Pavement Icing Condition Criterion and SVM-Based Prediction Analysis. *J. Highw. Transp. Res. Dev.* **2018**, *12*, 1–9. [CrossRef]
8. Leon, L.P.; Gay, D. Gene expression programming for evaluation of aggregate angularity effects on permanent deformation of asphalt mixtures. *Constr. Build. Mater.* **2019**, *211*, 470–478. [CrossRef]
9. Vimalathithan, R.; Valarmathi, M.L. Cryptanalysis of DES using computational intelligence. *Eur. J. Sci. Res.* **2011**, *55*, 237–244.
10. Brown, J.A.; Houghten, S.; Ombuki-Berman, B. Genetic algorithm cryptanalysis of a substitution permutation network. In Proceedings of the 2009 IEEE Symposium on Computational Intelligence in Cyber Security, Nashville, TN, USA, 30 March–2 April 2009; pp. 115–121. [CrossRef]
11. Garg, P.; Varshney, S.; Bhardwaj, M. Cryptanalysis of Simplified Data Encryption Standard Using Genetic Algorithm. *Am. J. Netw. Commun.* **2015**, *4*, 32. [CrossRef]
12. Al Adwan, F.; Al Shraideh, M.; Saleem Al Saidat, M.R. A genetic algorithm approach for breaking of bimplified data encryption standard. *Int. J. Secur. Appl.* **2015**, *9*, 295–304. [CrossRef]
13. Delman, B. Genetic Algorithms in Cryptography. Master's Thesis, Rochester Institute of Technology, New York, NY, USA, 2004.
14. Baragada, S.R.; Reddy, P.S. A Survey of Cryptanalytic Works Based on Genetic Algorithms-IJETTCS-2013-08-20-024. *Int. J. Emerg. Trends Technol. Comput. Sci. (IJETTCS)* **2013**, *2*, 18–22.
15. Khan, A.H.; Lone, A.H.; Badroo, F.A. The Applicability of Genetic Algorithm in Cryptanalysis: A Survey. *Int. J. Comput. Appl.* **2015**, *130*, 42–46. [CrossRef]
16. Borges-Trenard, M.; Borges-Quintana, M.; Monier-Columbié, L. An application of genetic algorithm to cryptanalysis of block ciphers by partitioning the key space. *J. Discret. Math. Sci. Cryptogr.* **2019**. [CrossRef]
17. Tito, O.; Borges-Trenard, M.A.; Borges-Quintana, M. Ataques a cifrados en bloques mediante búsquedas en grupos cocientes de las claves. *Rev. Cienc. Mat.* **2019**, *33*, 71–74.
18. Monier-Columbié, L. Sobre los Ataques Lineal y Genético a Cifrados en Bloques. Master's Thesis, Universidad de la Habana, Habana, Cuba, 2018.
19. Nakahara, J.; de Freitas, D.S. Mini-ciphers: A reliable testbeb for cryptanalysis? In *Dagstuhl Seminar Proceedings. 09031. Symmetric Cryptography*; Schloss Dagstuhl-Leibniz-Zentrum für Informatik: Wadern, Germany, 2009.

Article

Sensitivity Analysis of Key Formulations of Topology Optimization on an Example of Cantilever Bending Beam

Martin Sotola [1,2], Pavel Marsalek [1,2,*], David Rybansky [1,2], Martin Fusek [1] and Dusan Gabriel [2]

1. Department of Applied Mechanics, Faculty of Mechanical Engineering, VŠB—Technical University of Ostrava, 17. Listopadu 2172/15, 708 00 Ostrava, Czech Republic; martin.sotola@vsb.cz (M.S.); david.rybansky@vsb.cz (D.R.); martin.fusek@vsb.cz (M.F.)
2. Institute of Thermomechanics of the Czech Academy of Sciences, Dolejskova 5, 182 00 Prague, Czech Republic; gabriel@it.cas.cz
* Correspondence: pavel.marsalek@vsb.cz

Citation: Sotola, M.; Marsalek, P.; Rybansky, R.; Fusek, M.; Gabriel, D. Sensitivity Analysis of Key Formulations of Topology Optimization on an Example of Cantilever Bending Beam. *Symmetry* **2021**, *13*, 712. https://doi.org/10.3390/sym13040712

Academic Editors: Sun Young Cho and Aviv Gibali

Received: 11 March 2021
Accepted: 16 April 2021
Published: 18 April 2021

Publisher's Note: MDPI stays neutral with regard to jurisdictional claims in published maps and institutional affiliations.

Copyright: © 2021 by the authors. Licensee MDPI, Basel, Switzerland. This article is an open access article distributed under the terms and conditions of the Creative Commons Attribution (CC BY) license (https://creativecommons.org/licenses/by/4.0/).

Abstract: Topology optimization is a modern method for optimizing the material distribution in a given space, automatically searching for the ideal design of the product. The method aims to maximize the design performance of the system regarding given conditions. In engineering practice, a given space is first described using the finite element method and, subsequently, density-based method with solid isotropic material with penalty. Then, the final shape is found using a gradient-based method, such as the optimality criteria algorithm. However, obtaining the ideal shape is highly dependent on the correct setting of numerical parameters. This paper focuses on the sensitivity analysis of key formulations of topology optimization using the implementation of mathematical programming techniques in MATLAB software. For the purposes of the study, sensitivity analysis of a simple spatial task—cantilever bending—is performed. This paper aims to present the formulations of the optimization problem—in this case, minimization of compliance. It should be noted that this paper does not present any new mathematical formulas but rather provides an introduction into the mathematical theory (including filtering methods and calculating large-size problems using the symmetry of matrices) as well as a step-by step guideline for the minimization of compliance within the density-based topology optimization and search for an optimal shape. The results can be used for complex commercial applications produced by traditional manufacturing processes or by additive manufacturing methods.

Keywords: topology optimization; optimization; filtering; method; penalization; weight factor; FEM; MATLAB; SIMP

1. Introduction

Topology optimization is a calculation of the distribution of materials within a structure without a known pre-defined shape. This distribution calculation yields a "black and white pattern" where black places indicate full material while white places represent voids (i.e., places where material can be removed). Because the distribution is solved over a general region, topology optimization allows us to acquire a unique, innovative, and effective structure. The principle of topology optimization is presented on the example of cantilever bending in Figure 1, where the initial geometry (given space) is depicted on the left and the optimal shape on the right. As apparent from Figure 1, topology optimization plays nowadays an important role in engineering practice. Usually, it allows the designer to reduce the weight of the part without losing too much of its previous properties such as stiffness, natural frequency, etc.

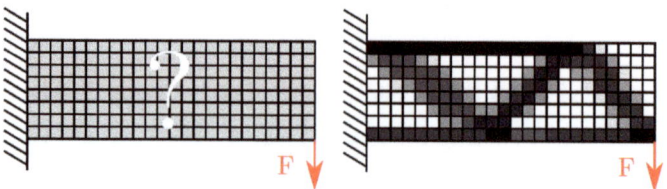

Figure 1. Topology optimization of cantilever bending; the initial geometry (**left**) and the optimal shape in the form of design variables (**right**).

Issues associated with topology optimization are studied by many engineers and researchers. The Finite Elements Method (FEM) is the most widely used technique for the analysis of discretized continuum. Femlab [1], FreeFem++ [2,3] and ToPy [4] are rapidly growing engineering tools supporting topology optimization. The problem of topology optimization was described, e.g., by Bendsøe and Sigmund [5–8]. Many computer tools have been prepared, including tools in the MATLAB platform (MathWorks, Natick, MA, United States of America). Liu et al. [9] described a three-dimensional (3D) topology optimization using MATLAB scripts. They described the necessary steps of optimization and provided scripts for individual steps. Their scripts are freely available and can be modified in accordance with the authors' instructions. It should be noted that although the scripts have great educational value, their practical usage is limited as they are applicable only for simple shapes and cannot work with imported meshes. The same can be said about the paper by Sigmund et al. [7] who investigated two-dimensional optimization and introduced sensitivity filtering. Master thesis by William Hunter [10] worked on 3D topology optimization. The author described in depth the theoretical background as well as its implementation in Python.

Even though topology optimization might seem novel, the first mention of structural topology optimization dates back to 1904 [11]. However, major progress has come only in the last 30 years due to the development of computers and the advancement of new technological processes (in particular, additive manufacturing). Hunar et al. [12] and Pagac et al. [13] illustrated the significance of topology optimization for designing a 3D printed part. Currently, however, topology optimization is perceived by most users as a "black box" producing always correct, i.e., optimal, shapes.

This is, however, largely not true and deeper understanding of the parameters and settings is needed to yield optimal results. A thorough introduction to the problem and a complex guideline for performing sensitivity analysis that would help researchers and engineers with determining correct settings is, however, not available in the current literature. For this reason, we decided to provide such a guideline in this paper.

Hence, the presented paper studies the effect of key formulations of the topology optimization problem on the design performance. In addition, it recommends the values of individual numerical parameters using a cantilever beam problem. Even though it is demonstrated on a simple example, this insight can be used for complex problems of engineering practice. For example, our group [14] described topology optimization of a transtibial bed stump using a custom MATLAB script. Performing sensitivity analysis of the key formulations is recommended for supporting the robustness of the computations in any new problem. Failure to perform so such an analysis may lead to the production of „false-optimal" results. The individual steps of this study were performed in MATLAB as well as in ANSYS Workbench 2019 R3 (ANSYS, Inc., Canonsburg, United States of America, AWB), which helped with the evaluation of the results (assessment of the similarity in resulting values or shapes). The preliminary results of the work are presented in the Master's thesis by Sotola [15].

2. Materials and Methods

The procedure for calculating the optimal shape of the structure can be divided into three stages: (i.) Preparative stage (ii.) Optimization, and (iii.) Postprocessing, see Figure 2. MATLAB scripts were prepared for each of these stages to automatize the procedure.

The first step of the initiation of optimization lies in preparing the finite element analysis. In this step, the boundary conditions and local stiffness matrices of elements are set up. This paper does not describe this stage in depth because it has been already described in many books; for example, Hughes [16] describes the preparation of local and global stiffness matrices. Still, some advanced recommendations are presented in this paper. At this stage, the global stiffness matrix is also assembled and the reference values of the objective function are calculated. Before optimization, it is also necessary to prepare an initial approximation of volume, i.e., the structural elements are assigned a new material model containing individual design variables for each element. Elements are assigned new values of elasticity modulus, which affect the global stiffness matrix and leads to a new value of the objective function with each iteration.

Optimization itself follows, during which new values of design variables are determined. Subsequently, the terminating criterion is queried and if the process is not terminated based on the criterion being met or the maximal number of iterations exceeded, the process is repeated with new design variables.

In the last stage, the results are recalculated and prepared in the *vtk* format, which can be viewed in the open-source software ParaView (Kitware, Inc., Clifton Park, NY, USA). This paper focuses on the first and second stages; we provide the results of the entire process but do not describe the postprocessing in detail.

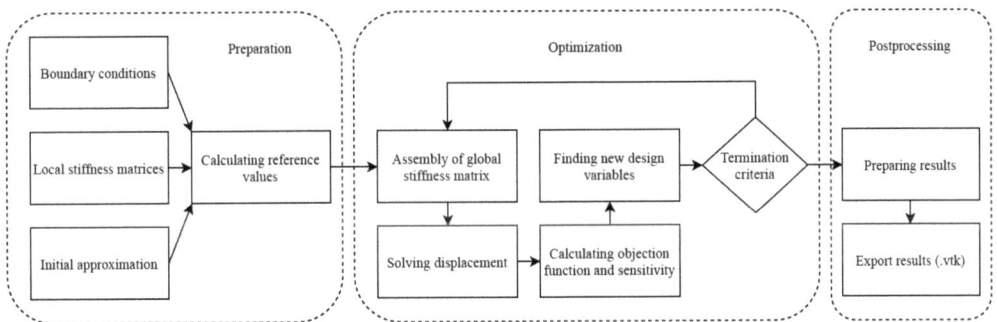

Figure 2. Diagram of the procedure for topology optimization.

2.1. Description of the Optimization Problem

As mentioned already, being able to describe the problem as a mathematical input is the key to the full understanding of topology optimization. Such understanding is needed for proper definition of the design variables, the objective function, and the constraint function (mostly inequality constraints). The most widely used method for solving multivariate optimization is called "Karush–Kuhn–Tucker conditions" [17] (also known as Kuhn–Tucker conditions or just KKT conditions).

2.1.1. The Optimization Problem

The objective of the optimization presented in this paper is to minimize compliance arising during volume reduction defined as volume fraction v_{frac}. The volume fraction is calculated as the ratio of the proposed volume and the original volume. The volume fraction ranges from 0% to 100%.

Several methods can be used for solving the problem of minimization of compliance. For example, the homogenization method [18,19] uses microperforated composites as a base material for shape optimization. Since the number of holes within the domain is not limited, it can be seen as a method of topology optimization. In another approach, the phase-field method, the domain consists of two "phases", the "void" and the fictional "liquid" which interacts with loads [20,21]. One of the newer methods (meaning newly implemented in commercial software) is the Level-set method. Optimization in this method is solved "above" the fixed domain with a fictional velocity [22].

The density-based method is the last of the methods commonly used for solving the topology problem. This method can be viewed as mature and is easy to implement. In this paper, we focused on this method, which uses a continuous design variable ranging from 0% to 100%, see Figure 1. In the literature, the design variable is usually referred to as the density; note, however, that this "density" of the element has no clear physical meaning. In 2D space, the density can be pictured as a variable thickness of sheet metal but in 3D space, it is not easy to assign a tangible meaning to this term. The Solid Isotropic Material with Penalty (SIMP) model is a popular interpolation scheme for definition of material that would be subsequently used in the density-based method. In this model, the elasticity modulus is in a power law relation with the design variable and can be described using the following equation.

$$E_e = \tilde{x}_e^p \cdot E_0, \ \tilde{x}_e \in <0,1>, \tag{1}$$

where E_0 is the elastic modulus of the base material, p is the penalization and $\tilde{x}_e$ is the "modified" filtered design variable. The reasons for the "modification" are described in the following Sections 2.1.2 and 2.1.6.

After preparation of the material model, the objective function of minimizing compliance is defined as

$$\min : c(\tilde{x}_e) = \{f\}^T \cdot \{u(\tilde{x}_e)\}, \tag{2}$$

where c is the deformation energy, $\{f\}$ is the force vector and $\{u\}$ is the displacement. One could argue that the deformation energy of the linear material should be divided by two; however, as it is only a scalar variable, it will not affect the optimization itself. The constraint Equation (or, rather, inequality in this case) is defined as

$$v(\tilde{x}_e) = \{\tilde{x}_e\}^T \cdot \{v_e\} - v_{\text{frac}} \cdot \sum \{v_e\} \leq 0, \tag{3}$$

where $\{v_e\}$ is the vector containing the volume of each element. Displacement is solved from the equilibrium equation

$$\{f\} = [K] \cdot \{u\}, \tag{4}$$

where $[K]$ is the global stiffness matrix. Assembly of stiffness matrices is described in Section 2.1.4. The design variable is defined as

$$0 \leq \tilde{x}_e \leq 1. \tag{5}$$

Equation (2) can be written in a simpler form after applying SIMP

$$\min : c(\tilde{x}_e) = \sum_{e=1}^{n} E_0 \tilde{x}_e^p \{u_e\}^T [k_0] \{u_e\} \tag{6}$$

where $\{u_e\}$ is the vector of element displacement, $[k_0]$ is the stiffness matrix with unit elastic modulus and n is the number of elements. Penalization is introduced in Section 2.1.3.

One can notice that the problem is defined as minimizing the deformation energy, which leads to the higher stiffness of the structure. Using the KKT conditions, the Lagrange multipliers convert the constrained problem to the equivalent unconstrained problem [17,23]. The first step is assembling the Lagrange function

$$L(\tilde{x}_e, \lambda) = c(\tilde{x}_e) + \lambda \cdot v(\tilde{x}_e), \tag{7}$$

where $c(\tilde{x}_e)$ is the objective function, $v(\tilde{x}_e)$ is the constraint function and λ the Lagrange multiplier. It is necessary to solve the derivation of the Lagrange function with respect to the design variables because it is necessary to find a stationary point; at that point, the derivation is equal to 0

$$\nabla L(\tilde{x}_e, \lambda) = \frac{\partial c(\tilde{x}_e)}{\partial \tilde{x}_e} + \lambda \cdot \frac{\partial v(\tilde{x}_e)}{\partial \tilde{x}_e} = 0. \tag{8}$$

The following equation is known as the complementary slackness condition, determining whether the constraint function is active or passive

$$\lambda \cdot v(\tilde{x}_e) = 0. \tag{9}$$

The next condition is that the Lagrange multiplier is not negative

$$\lambda \geq 0. \tag{10}$$

The last condition is the derivation of the Lagrange function with respect to Lagrange multipliers. In simple terms, it determines whether or not the particular point is the KKT point, i.e., the optimum (there are fundamental theorems proving that the solution is automatically optimal, see more in [17]). This condition is not important for solving the problem, it is important for results evaluation. Rearranging the equation and adding the variable B_e leads to the equation

$$B_e = 1 = \frac{-\frac{\partial c(\tilde{x}_e)}{\partial \tilde{x}_e}}{\lambda \frac{\partial v(\tilde{x}_e)}{\partial \tilde{x}_e}}. \tag{11}$$

It is obvious that the optimum of the element is met if $B_e = 1$. Preparing the first derivation of the objective function determined by Equation (2) with respect to the design variables can be tricky as at the first sight, there is no evident "influence" of the design variables. It can be solved numerically but this would require a calculation of the displacement for every individual possible design variable. However, it is possible to calculate derivation using the adjoint method [5], i.e., to add the equilibrium equation into the objective function

$$c(\tilde{x}_e) = \{f\}^T \{u(\tilde{x}_e)\} + \{\eta\}^T ([K]\{u(\tilde{x}_e)\}) - \{f\}), \tag{12}$$

where η is a vector of non-zero variables (also unknown for now). Adding the equilibrium equation does not change the objective function because it is equal to zero. Similarly, the vector η does not change the function. Let us assume that the exterior forces are not independent on the design variables. Then, the first-order derivation of the objective function is

$$\frac{\partial c}{\partial \{\tilde{x}_e\}} = \{f\}^T \frac{\partial \{u\}}{\partial \{\tilde{x}_e\}} + \{\eta\}^T \left(\frac{\partial [K]}{\partial \{\tilde{x}_e\}} \{u\} + [K] \frac{\partial \{u\}}{\partial \{\tilde{x}_e\}} \right). \tag{13}$$

Rearranging the previous Equation (13) leads to factoring out the derivation of the displacement

$$\frac{\partial c}{\partial \{\tilde{x}_e\}} = \left(\{f\}^T + \{\eta\}^T [K] \right) \frac{\partial \{u\}}{\partial \{\tilde{x}_e\}} + \{\eta\}^T \frac{\partial [K]}{\partial \{\tilde{x}_e\}} \{u\}. \tag{14}$$

To get rid of the derivation of displacement, it is critical that the term in brackets (adjoint equation) is equal to zero. This means that our added unknown variable has to be equal to

$$\{\eta\}^T = -\{f\}^T [K]^{-1} = -\{u\}^T. \tag{15}$$

It is apparent that the vector $\{\eta\}$ is already solved. The final form of the derivation of the objective function (with added SIMP model) is

$$\frac{\partial c(\tilde{x}_e)}{\partial \tilde{x}_e} = -E_0\, p\, \tilde{x}_e^{(p-1)} \{u_e\}^T [k_0] \{u_e\}. \tag{16}$$

The first order derivation of the constraint function is defined asIt should be noted that in the literature (for example, [8]), it is possible to find another equation describing the constraint function and its derivation

$$\frac{\partial v(\tilde{x}_e)}{\partial \tilde{x}_e} = \{v_e\}. \tag{17}$$

It should be noted that in the literature (for example, [8]), it is possible to find another equation describing the constraint function and its derivation

$$v(\tilde{x}_e) = \frac{\{\tilde{x}_e\}^T \cdot \{v_e\}}{\sum \{v_e\}} - v_{\text{frac}} \leq 0, \tag{18}$$

$$\frac{\partial v(\tilde{x}_e)}{\partial \tilde{x}_e} = \frac{\{v_e\}}{\sum \{v_e\}}. \tag{19}$$

That form of the equation usually depends on the solver and its settings. For example, "MMA-based" solvers prefer the constraint function defined by Equations (18).

2.1.2. Material Model

In this paper, the density-based method used the Solid Isotropic Material with Penalty (SIMP) material model, which is a power-law relation of the design variables. The SIMP material model is used in solvers such as ANSYS (ANSYS, Inc., Canonsburg, PA, USA), MSC Nastran (MSC Software, Irvine, CA, USA), etc. The elastic modulus of the element is defined as

$$E_e = x_e^p \cdot E_0,\ x_e \in\, <0,1>, \tag{20}$$

where E_0 is the elastic modulus of the base material, p is the penalization and x_e is the "unmodified" design variable. This definition looks very similar to Equation (1); however, here, we use unfiltered designed variable to give a clearer picture of the need for filtering (see below).

During topology optimization, many problems can occur. For example, the so-called checkerboard pattern problem [24–26] is very common. This title describes the distribution of the structural elements in a checkerboard-like arrangement in certain areas of the part. Figure 3 shows the checkerboard patterns on the cantilever beam.

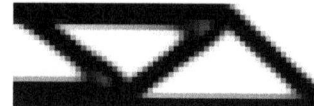

Figure 3. Checkerboard pattern problem (**left**) and its possible solution (**right**).

In our case, i.e., when discussing the problem of minimizing compliance, the computation may consider such a solution to be ideal as the checkerboard pattern creates artificial regions with higher stiffness. However, as obvious from Figure 3, this is not true and, moreover, the part cannot be manufactured. Another common problem is the insufficient number of connected nodes (four or more nodes are needed for the hexahedral element), which causes the formation of possible joints in the structure and, again, makes it impossible to manufacture.

The mesh-dependency of optimization results is a crucial problem [5,26]. As the name suggests, this problem results from the used discretization and its refinement. In context of structure stiffness, the reason is quite simple—increasing the number holes in the structure without changing its volume leads to increase of stiffness. Finer meshes facilitate this operation—they allow us to create a higher numbers of holes and, therefore, are capable of providing different (superior) results than coarser meshes. On the other hand, finer meshes might result in more complex structures that are difficult to manufacture.

The last important problem of topology optimization is the non-uniqueness of solutions, which makes results evaluation trickier [27]. Generally, the problem of optimization can have a single solution or up to an infinite number of solutions depending on whether the problem is convex or not.

One of the ways of solving the above-mentioned problems is to use a suitable filter(s). This solution has not been mathematically confirmed, but many numerical experiments suggest that the results could be considered optimal [7]. The description of each of the filters used in our study is presented in Section 2.1.6.

To prevent numerical difficulties, the modulus of void (passive) elements is introduced into the material model. This modification helps to reduce the risk of having a singular stiffness matrix. The final equation of the material model is

$$E_e = E_{min} + \tilde{x}_e^p \cdot (E_0 - E_{min}),$$
$$\tilde{x}_e = \tilde{x}_e(x_e), \ x_e \in <0,1>, \qquad (21)$$

where E_{min} is the elasticity modulus of passive elements and $\tilde{x}_e$ is the filtered design variable (density).

2.1.3. Penalization

The numerical scheme should lead to a black & white design (or 1-0 design with white to be removed). One of the possible approaches is to ignore the physical importance of elements with intermediate density (grey areas) and consider them "black", leading to their preservation. However, the physical relevance is discussed a lot since many interpolation methods can remove further parts of the grey regions. If the optimization is prematurely terminated, the stiffness (compliance) of the grey areas plays an important role in the evaluation of results. This issue is discussed by Bendsøe [5,6]. As mentioned above, the SIMP material model is suitable for FEM optimization as it assigns elasticity to each element. However, it can be used as the material model only if the penalization meets the following criteria

$$p \geq \max\left\{\frac{2}{1-\mu}, \frac{4}{1+\mu}\right\}, \text{ (in 2D); } p \geq \max\left\{15\frac{1-\mu}{7-5\mu}, \frac{3}{2}\frac{1-\mu}{1-2\mu}\right\}, \text{ (in 3D)} \qquad (22)$$

where μ is the Poisson's ratio. This means that different values of penalization must be calculated for each Poisson's ratio. The resulting values of the penalization for volume elements with the following Poisson's ratio are

$$p\left(\mu = \frac{1}{5}\right) \geq 2, \ p\left(\mu = \frac{1}{3}\right) \geq 3, \ p\left(\mu = \frac{2}{5}\right) \geq 4.5. \qquad (23)$$

In this paper, the base material has a Poisson ratio of $\mu = 0.30$; therefore, the penalization $p = 3$ was chosen for the following topology optimization.

2.1.4. Finite Element Method

From our experience, it is recommended for MATLAB implementation that the stiffness matrices is prepared with unit elastic modulus and saved in memory. The stiffness matrix of the solid element is volume integrated using the stress–strain matrix of the material $[C_0]$ with the unit elasticity modulus

$$[k_0] = \iiint [B]^T [C_0] [B] \, dV, \qquad (24)$$

where $[B]$ is the strain–displacement matrix [28]. The integral is usually solved numerically. Before the assembly of the global stiffness matrix, every local stiffness matrix is multiplied by the corresponding elasticity modulus (i.e., the modulus of the SIMP material model).

$$[k_e(\tilde{x}_e)] = E_e(\tilde{x}_e) \cdot [k_0]. \tag{25}$$

The MATLAB implementation of calculating the stiffness matrix of linear elements is presented by Bhatti [29,30]. With some effort, the procedures were modified to the calculation of quadratic elements. The procedures use full integration of the individual elements using Gauss points ($2 \times 2 \times 2$ scheme for linear elements, $3 \times 3 \times 3$ for quadratic elements).

For testing purposes, meshes were prepared using the AWB software [31]. They were constructed either of tetrahedral (TET) or hexahedral (HEX) elements (i.e., no result with mixed mesh was evaluated in this paper). The calculation of the stiffness matrices of linear elements is fast thanks to the low number of degrees of freedom (DOF); in effect, solving a single load case was swift when using this approach; however, it comes with a disadvantage in the form of locking as linear elements with full integration have a tendency to shear and volume locking [28]. There are numerous ways of fixing this problem; nonetheless, in this paper, the shear locking effects are "neglected" during the optimization but are mentioned in results. Meshes are separated into three groups: coarse, normal, and fine meshes according to the size (length) of individual elements.

2.1.5. The Optimality Criterion Algorithm

The design variable of the elements was updated using an algorithm called the Optimality criterion (OC) [5,32]. The name indirectly refers to the used method, i.e., KKT conditions. The algorithm is also implemented in the AWB software. To find the material distribution of the structure, a fixed updating scheme is proposed as

$$x_e^{new} = \begin{cases} \max(0, x_e - m), & \text{if } x_e \cdot B_e^\eta \leq \max(0, x_e - m) \\ \min(1, x_e + m), & \text{if } x_e \cdot B_e^\eta \geq \min(1, x_e + m) \\ x_e \cdot B_e^\eta & \text{otherwise,} \end{cases} \tag{26}$$

where B_e is constructed using Equation (11). As already mentioned above, the optimum of the element is found if $B_e = 1$. In other words, the design variable increases if $B_e > 1$ and decreases if $B_e < 1$. Changing the move limit m and tuning parameter η can lead to a lower number of iterations. Bendsøe [5] recommended values of $\eta = 0.5$ and $m = 0.2$. New values of design variables depend on the Lagrange multiplier, which has to be solved in the inner loop to ensure that the constraint function is satisfied. This leads to a reduction of the multivariate problem to one-dimensional (1D) optimization, which can be solved by various methods, such as the bisection method, golden section search, or methods using derivation (such as the Newton–Raphson method or secant method) [17,23]. In this work, the bisection method obtained from the paper by Liu [9] was used as the 1D optimization method.

A detailed description of the Optimality criterion method is presented in the dissertation thesis by Munro [32] and the Master's thesis by Hunter [4]. These theses describe the relationships between the Optimality criterion and Sequential approximate optimization (SAO). The SAO method can be solved using the duality principle. The purpose of this method is to find an equivalent subproblem (dual problem) that is easier to solve than the primary problem. The method is used for preparing a scheme (similar to the OC fixed scheme) with only one constraint function, which can be subsequently solved. The primary problem (in this case, minimizing compliance) is then reduced to maximizing the Lagrange multiplier.

2.1.6. Filtering Methods

Density filtering is one of the methods of solving the above-mentioned problems (such as the checkerboard pattern). A common parameter of the filters, weight factor H_{ij}, is defined as

$$H_{ij} = \begin{cases} R - dist(i,j) & \text{if } dist(i,j) \leq R \\ 0 & \text{if } dist(i,j) > R, \end{cases} \qquad (27)$$

where R is the radius of the filter, $dist(i,j)$ is the operator calculating the distance between the center of an element i and the center of an element j. This type of weight factor is called linear. If the distance between elements is greater than the radius, the weight factor is equal to zero; if the distance is equal to zero, the weight factor is equal to the size of the radius. In Figure 4, 2D examples are presented; in 2D, the radius defines a circular neighborhood. In a 3D problem, the radius forms a sphere.

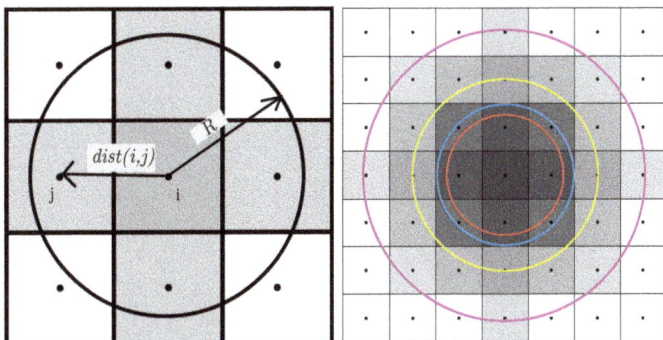

Figure 4. 2D Neighborhood of element (**left**), examples of the radius dependence on the element size ES in 2D; R= 1.2 ES for the red circle, R= 1.5 ES for the blue circle, R= 2.0 ES for the yellow circle and R= 3.0 ES for the purple circle.

An alternative approach is to use the normal distribution (Gauss function) [33,34]. Compared to the linear function, the Gauss function is smoother but in reality, there might not be a real benefit in using this alternative [8]. The weight factor is defined as

$$H_{ij} = e^{-\frac{1}{2}\left(\frac{3 \cdot dist(i,j)}{R}\right)^2}. \qquad (28)$$

In Figure 5, both functions are displayed. Due to the different characters of values of the weight factor, the linear function had to be normalized by the radius to have values ranging from 0 to 1, the same as the Gauss function. Furthermore, the distance was normalized to the radius (i.e., divided by the radius to ensure independence on it).

Preparing a matrix of weight factors can be challenging if the mesh is imported (if the mesh is not imported but created by additional scripts, the weight factor can be calculated during the mesh preparation). The following should be pointed out: Firstly, the preparation should use a function for creating sparse matrices because the matrix of weight factors is mainly sparse. Secondly, if the distance between the i-th and j-th element is constant, there is no need to create a "full" matrix but only the upper triangle of the matrix $[H_{tri}]$. Non-diagonal components of the matrix $[H_{tri}]$ are multiplied by 2. The following equation uses "symmetry" for composing the full matrix

$$[H] = \frac{1}{2} \cdot \left([H_{tri}] + [H_{tri}]^T\right), \qquad (29)$$

where T is the operator of the transpose. Lastly, with the growing amount of elements, the process of preparation is becoming more time-consuming even though calculating only "half" of the matrix. This basic approach is, up to 10 k elements, fast. However, the time of solution is growing exponentially for meshes with over 10k elements. In such cases, it is recommended to invest time into finding an appropriate method for speeding up the preparation. This could be done by dividing the mesh into mutually overlapping subzones with less than 10 k elements to ensure fast calculation. These subzones should be solved individually using the parallel toolbox (package). Using this approach, it should be possible to prepare the weight factors of complex meshes in a reasonable time.

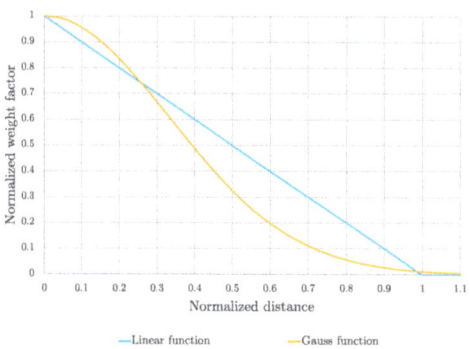

Figure 5. Functions of the weight factor.

Three filters are analyzed in this paper: (i) Density filter, (ii) Sensitivity filter, and (iii) Greyscale filter (G). The Density filter (D) and Sensitivity filter (S) use weight factors while the greyscale filter is an addition to the OC scheme.

Density Filter

The fundamental function of density filtering is

$$\tilde{x}_i = \frac{\sum_{j=1}^{n} H_{ij} x_j}{\sum_{j=1}^{n} H_{ij}}, \quad i = 1, \ldots, n, \quad (30)$$

where $\tilde{x}_i$ is the filtered design variable (density) of the i-th element, H_{ij} is the weight factor and n is the number of elements [35–37]. Every equation containing the design variables has to be adjusted to allow filtering (including partial derivations)

$$\widehat{\frac{\partial c}{\partial x_e}} = \frac{\sum_{j=1}^{n} H_{ij} \frac{\partial c}{\partial x_e}}{\sum_{j=1}^{n} H_{ij}}, \quad \widehat{\frac{\partial v}{\partial x_e}} = \frac{\sum_{j=1}^{n} H_{ij} \frac{\partial v}{\partial x_e}}{\sum_{j=1}^{n} H_{ij}}. \quad (31)$$

In this case, the design variable of the element is averaged over its neighborhood. It ensures the smoothing of stiffness. The filtered density is applied during the construction of the global stiffness matrix before solving the static structural analysis. Generally, the density tends to have a value of 0 or 1 (which could be ideal), but after applying the filter, regions of intermediate (grey density) design variable appear, which are then penalized by the SIMP model. Deformation energy is also averaged and shared with the neighborhood of elements.

Sensitivity Filter

This filter is based on filtering of the sensitivity (i.e., of the first partial derivation of Lagrange function). Experience has proven that filtering sensitivity ensures mesh independence of results and is time-effective [7]. It is also easy to implement and does not increase the complexity of the problem. The filter is purely heuristic but has been proven to yield similar results as the gradient constraint method [5]. The fundamental equation of this filter is

$$\widehat{\frac{\partial c}{\partial x_i}} = \frac{1}{\max(10^{-3}, x_i) \cdot \sum_{j=1}^{n} H_{ij}} \cdot \sum_{j=1}^{n} H_{ij} \cdot x_i \cdot \frac{\partial c(x_i)}{\partial x_e}. \tag{32}$$

Grayscale Filter

The last filter is an addition to the previous filters called the Grayscale filter [38]. Applying this filter should reduce the grey areas. The objective of the filter is also to penalize the volume constraint in the OC algorithm. The parameter q is implemented in the OC scheme and its value can be constant or gradually increasing by multiplication (in our case, coefficient q was multiplied by 1.01 each iteration). The amended OC scheme is described as

$$x_e^{new} = \begin{cases} \max(0, x_e - m), & \text{if } x_e \cdot B_e^{\eta} \leq \max(0, x_e - m) \\ \min(1, x_e + m), & \text{if } x_e \cdot B_e^{\eta} \geq \min(1, x_e + m) \\ (x_e \cdot B_e^{\eta})^q & \text{otherwise.} \end{cases} \tag{33}$$

This filter is activated after 15 iterations of optimization. Usually, it is limited by the maximum value of the coefficient q (in this paper, the maximum was set to 5). If the q coefficient maximum was set to 1, the filter would be deactivated. The idea of the filter is to underestimate the intermediate density leading to a zero value (void density). Underestimation occurs also in the inner iteration. This leads to fewer iterations needed for finding solutions.

2.1.7. Displacement Solver and Termination Criteria

To solve the displacement Equation (4), it is possible to use the direct solver present in MATLAB. However, if the number of degrees of freedom is high (above sixty thousand), the pursuit to provide an accurate result is too ambitious. In such cases, therefore, it is appropriate to switch to the iteration solver. MATLAB includes a solver using the conjugate gradient method with preconditioning [39]. It is recommended to use the simplest preconditioning matrix, the diagonal (Jacobi) preconditioner. The reason is that to ensure the best possible solving stability, assembly of the preconditioner is needed in every iteration and, hence, the simpler is the preconditioner, the faster is the solution. Use of a different (more complex) preconditioner, such as Incomplete Cholesky factorization (with various settings), could reduce the number of iterations but the computing time would be the same or higher due to a slow preconditioner assembly (we tested these in the preliminary stage but the detailed results are not presented here). An alternative approach would be to prepare a "universal" preconditioner from the reference matrix only once and to use in every iteration [40]. In the presented study, however, we used the first approach with assembling the diagonal preconditioner each iteration.

It is important to determine the termination criteria. The maximal number of iterations is the first common criterion. Usually, the value ranges between 200 and 500. It should be noted that a complex task (such as a complex geometry or complex loads) requires more iterations but from our experience, 100 iterations are enough for a simple task. The second criterion is defined as the tolerance of sufficient optimization at which the calculation is terminated. There are two possible options.

The first option is to calculate the change in the values of design variables between the current and previous iteration [7]. This change is compared with the chosen tolerance value and once the change in the design variables is below the tolerance value, the calculation is terminated. Thus, if the tolerance is set to a low value, the number of iterations will increase. This is defined as

$$\max |x_{e_i} - x_{e_{i-1}}| \leq \varepsilon, \tag{34}$$

where x_{e_i} are the design variables of the current iteration, $x_{e_{i-1}}$ are the design variables of the previous iteration and ε is the tolerance value.

The other option is to calculate the ratio of the change of the objective function to the current objective function value. It is assumed that the changes of the objective function near the stationary point are minimal. Compared to the first option, the number of iterations is lower (it may be reduced by as much as half). This option is defined as

$$\left| \frac{c(x_{e_i}) - c(x_{e_{i-1}})}{c(x_{e_i})} \right| \leq \varepsilon, \tag{35}$$

where $c(x_{e_i})$ is the value of the objective function in the i-th iteration, $c(x_{e_{i-1}})$ is the value of the objective function of the previous iteration $(i-1,)$ and ε is the tolerance value. This method, however, comes with a risk of premature termination; for this reason, the first option of terminating criterion was used.

2.2. Key Formulations

It is apparent from the previous sections that the optimization comes with many formulations and parameters. Each parameter can be changed, which might reduce the number of iterations, improve the values of objective functions or of the desired volume fraction; on the other hand, the changes may also lead to solver instability, premature termination, or ineffective shape of the part. In this paper, five key formulations are analyzed and evaluated:

Formulation of the filter radius mentioned in Section 2.1.6 is important because the radius defines the element's neighborhood. If the defined range is too small, the energy is distributed only to a few elements. However, if the neighborhood is too large, the energy is scattered to a point where it is difficult to evaluate the optimum.

Formulation of the filter type was already mentioned in Section 2.1.6 but the theory does not provide an answer to the question of which filter should perform best.

Formulation of the penalization was also mentioned in Section 2.1.3; the theory, however, is able to provide only the lower boundary, not the upper one.

Formulation of the element approximation mentioned in Section 2.1.4 is a necessary step in the initiation of optimization. The element approximations greatly affect the accuracy of the solved displacement and the value of the objective function.

The formulation of the type of the weight factor mentioned in Section 2.1.6 is defined by two functions. However, theory does not provide enough evidence to decide, which function offers the better performance.

These key formulations were tested on several numerical examples including planar and spatial problems (see Figure 6) but to ensure the clarity of this paper, only one example is presented.

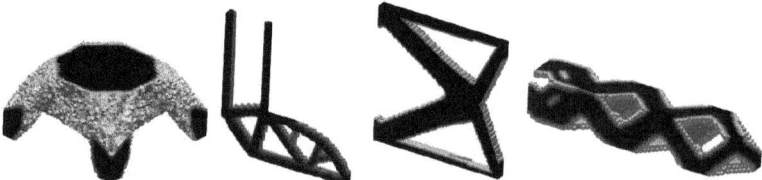

Figure 6. Results of numerical examples of topology optimization; from left to right-four point bending, L problem, two-loadcase cantilever plate, beam with square cross-section subjected to torsion.

2.3. Description of Numerical Test

The sensitivity analysis was performed using a numerical test-cantilever beam. It is a standard problem of mechanics, well-known to every designer. As the optimal shape can be found intuitively, results evaluation is easier.

The boundary conditions were simple. The beam was fixed on one end and the force acted on the other end's edge (bottom edge). The beam had a rectangular cross-section and was made of steel. It was assumed that the forces caused only a small displacement and the original material model was linear isotropic. All finite element meshes were made of solid elements, see Table 1. The authors used linear HEX elements and linear TET elements. In addition, quadratic HEX and TET elements have been calculated and results (mesh statistics) presented in Section 3.4. The geometry and discretization are shown in Figure 7. The figure also contains material and force parameters. The objective of optimization was to minimize the compliance (i.e., to maximize stiffness). In our paper, the constraint was defined as the volume fraction of 30%.

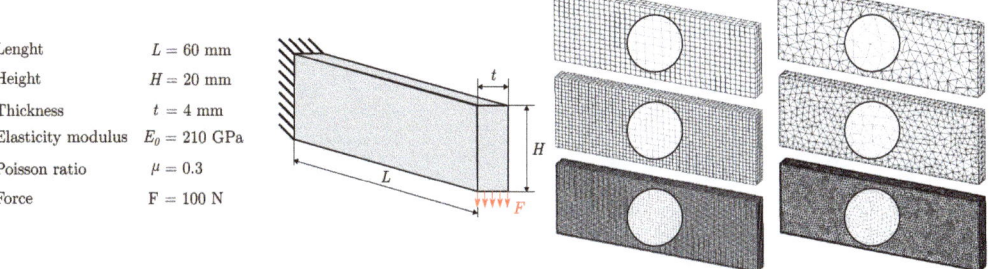

Lenght	$L = 60$ mm
Height	$H = 20$ mm
Thickness	$t = 4$ mm
Elasticity modulus	$E_0 = 210$ GPa
Poisson ratio	$\mu = 0.3$
Force	$F = 100$ N

Figure 7. Geometry, discretization and parameters of the numerical test-cantilever beam.

Table 1. Finite element mesh statistics.

		TET Elements	HEX Elements
Coaser mesh	Number of nodes	686	2460
	Number of elements	2034	1680
	Element size [mm]	1.500	2.000
Normal mesh	Number of nodes	1403	6405
	Number of elements	4458	4800
	Element size [mm]	1.000	1.000
Fine mesh	Number of nodes	10,850	44,650
	Number of elements	38,740	38,400
	Element size [mm]	0.355	0.500

3. Results

Optimization using MATLAB software was performed on multiple meshes. The maximal number of iterations was set to 200. The maximal change of the design variable (i.e., density) was chosen as the terminating criterion. The tolerance was set to 0.01. The cut-off limit of the design variable for element deactivation (white in the Figures) was 0.01 unless stated otherwise. The penalty was set to a constant value of $p = 3$ unless stated otherwise. Unless stated otherwise, the Density filter was used throughout the paper. In tables, two variants of the objective function are shown. The non-normed value indicates the deformation energy. The normed value is the objective function divided by the reference value of the initial objective function (i.e., the objective function describing the original structure before optimization). In other words, the normed value shows how many times the resulting structure is more compliant than the original reference. The normed value in the linear static analysis should be the same (or, at least, similar) regardless of the applied force. Data and shapes presented in this chapter were prepared in Ansys Workbench (AWB).

3.1. Formulation of Filter Radius

The radius of the filter is an important parameter since it defines the element's neighborhood. In this case, the radius is dependent on the element size (ES). For hexahedral elements, multipliers were set to 1.2 ES, 1.5 ES, 2.0 ES, and 3.0 ES, respectively. For clarification, Figure 4 displays the mentioned radiuses. For tetrahedral elements, multipliers were twice as high (i.e., 2.4 ES, 3.0 ES, 4.0 ES and 6.0 ES) to prevent possible inactivation of filters.

Results for the hexahedral and tetrahedral meshes are shown in Figure 8, which is displayed in the dominant (planar) view.

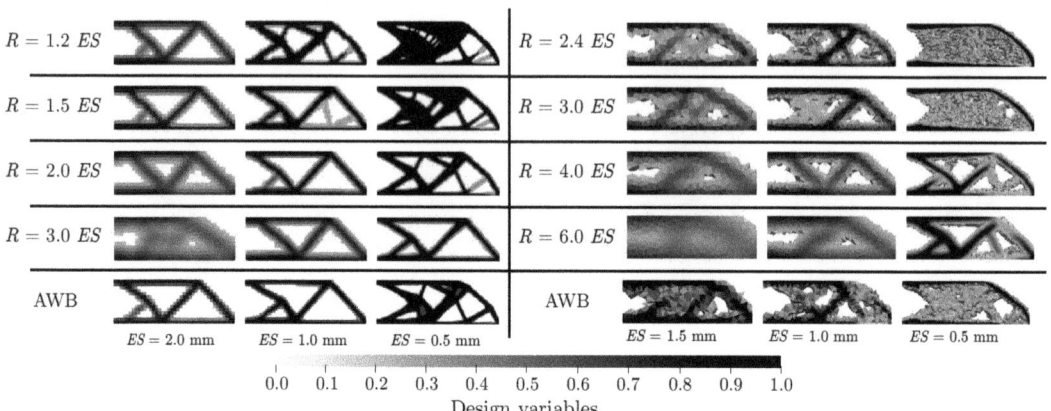

Figure 8. Final shapes for linear HEX elements (**left**) and linear TET elements (**right**) with various radii, Density filter (D).

From Table 2, it was apparent that the radius heavily affected the distortion of design variables over the design region. This means if the radius was increased, the value of the objective function (both deformation energy and normed value) also increased. Besides, the volume fraction increased due to the distortion. A fine mesh with small radii yielded a great stiffness-volume ratio, the optimized structure was approximately 2.5 times more compliant than the original structure but the volume reduction was as high as 53%.

A few notes: Radius should never be smaller than 1.5 ES/2.4 ES. A radius such as 1.2 ES greatly limited the capability of filters. In the case of uniform mesh, the radius should be appropriately chosen from the range between 1.5 ES and 3.0 ES. In the case of the non-uniform (tetrahedral) mesh, the radius should be within the range of 2.4 ES to 4.0 ES. Radii exceeding the upper value of the mentioned ranges make the structure more compliant. In the case of a coarser TET mesh, the radius $R = 6.0\ ES$ caused an over two-fold increase in the deformation energy than radius $R = 2.4\ ES$. Besides, the shapes of the coarser meshes with larger radii were not acceptable due to the high representation of "grey" areas (see Figure 8).

Table 2. Results of optimization with different radius, for meshes with linear HEX elements (first multiplier) or linear TET elements (latter multiplier).

Radius	Value	HEX Elements			TET Elements		
		2.0 mm	1.0 mm	0.5 mm	1.5 mm	1.0 mm	0.355 mm
1.2/2.4 ES	Deformation energy [mJ]	6.1	4.1	3.4	8.0	5.6	3.6
	Normed value [-]	4.2	2.8	2.3	5.8	3.9	2.5
	Iteration	110	96	200	200	200	200
	Volume fraction [%]	63.6	56.0	46.9	75.6	67.4	54.4
1.5/3.0 ES	Deformation energy [mJ]	7.2	4.6	3.7	9.8	6.7	3.9
	Normed value [-]	5.0	3.1	2.5	7.1	4.8	2.7
	Iteration	83	200	200	200	200	200
	Volume fraction [%]	64.6	57.6	50.2	81.1	72.0	59.9
2.0/4.0 ES	Deformation energy [mJ]	9.7	5.2	4.0	12.6	9.0	4.3
	Normed value [-]	6.7	3.5	2.7	9.1	6.3	3.0
	Iteration	167	200	200	200	200	200
	Volume fraction [%]	78.2	55.3	50.2	87.4	75.7	58.5
3.0/6.0 ES	Deformation energy [mJ]	14.5	7.2	4.5	18.4	12.4	5.1
	Normed value [-]	10.1	4.9	3.1	13.3	8.7	3.5
	Iteration	200	200	200	139	200	200
	Volume fraction [%]	89.7	66.1	51.6	98.3	87.0	62.8
AWB	Deformation energy [mJ]	5.7	4.9	3.9	2.7	5.5	4.7
	Normed value [-]	3.9	3.4	2.7	2.0	3.9	3.2
	Iteration	35	42	33	31	57	33
	Volume fraction [%]	51.7	51.3	40.3	78.4	58.6	58.5

3.2. Formulation of Filter Type

The importance of filters was already mentioned in the previous section. The following effects were evaluated: the number of iterations, objective function, and volume fraction. This particular aim of our study was to find the appropriate filter leading to a 0–1 design (and low objective function). Five variants were calculated: No filter (NoF), Density filter (D), Sensitivity filter (S), Grayscale filter with Density filter (G), and Grayscale filter with sensitivity filter (SG). Figure 9 shows each filter.

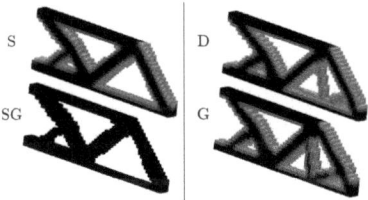

Figure 9. Results of individual filtering algorithms: Density filter (D), Sensitivity filter (S), Grayscale filter (G),Combination of the Sensitivity and Grayscale filters (SG).

At the first look, it should be apparent that the SG combination leads in this case to a purely 0-1 design (with the required volume fraction).

Two variants of the radius were selected: In the first variant, the radius was considered to depend on the element size, namely, it was defined as 1.5 ES. Results of this approach are displayed in Figure 10 and Table 3. The way of how the filters acted when the radius was not constant over different meshes should be noted.

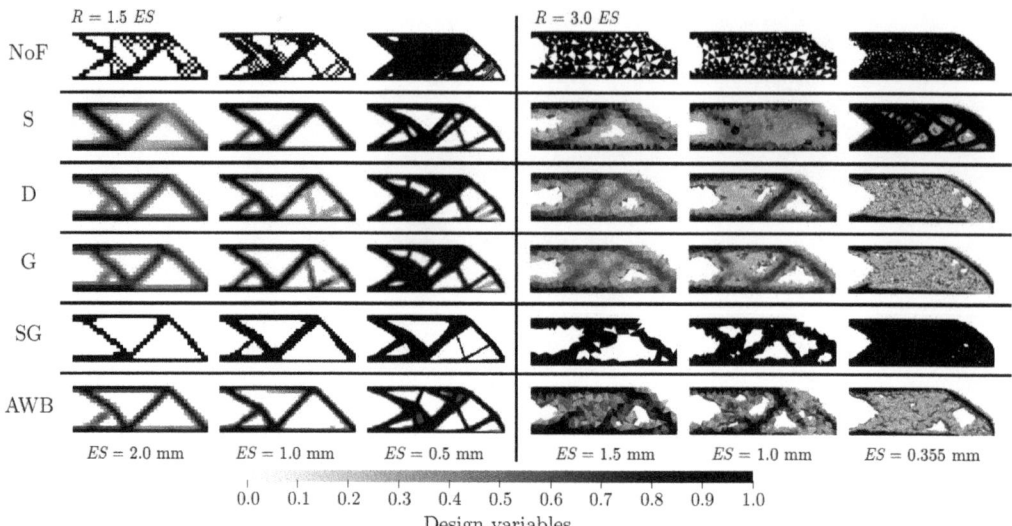

Figure 10. The final shape for linear HEX elements (**left**) and linear TET elements (**right**) with various filters, HEX radius $R = 1.5\ ES$ and TET radius $R = 3.0\ ES$.

Table 3. Results of optimization for various filters for each mesh, dependent radius, filter radius of linear HEX mesh is $R = 1.5\ ES$ an filter radius of linear TET mesh is $R = 3.0\ ES$.

Filter	Value	HEX Elements			TET Elements		
		2.0 mm	1.0 mm	0.5mm	1.5 mm	1.0 mm	0.355 mm
S	Deformation energy [mJ]	6.1	3.9	3.4	7.7	9.3	3.5
	Norm value [-]	4.3	2.7	2.3	5.5	6.6	2.4
	Iteration	117	53	200	80	26	167
	Volume fraction [%]	71.2	54.1	45.5	82.4	93.4	58.8
G	Deformation energy [mJ]	7.6	4.9	3.9	10.6	7.5	4.1
	Norm value [-]	5.3	3.3	2.6	7.6	5.3	2.8
	Iteration	35	34	34	34	36	34
	Volume fraction [%]	63.4	59.5	51.0	82.9	72.6	61.8
D	Deformation energy [mJ]	7.2	4.6	3.7	9.8	6.7	3.9
	Norm value [-]	5.0	3.1	2.5	7.1	4.8	2.7
	Iteration	83	200	200	200	200	200
	Volume fraction [%]	64.6	57.6	50.2	81.1	72.0	59.9
SG	Deformation energy [mJ]	3.8	3.3	3.1	3.7	3.3	3.0
	Norm value [-]	2.6	2.2	2.1	2.7	2.4	2.1
	Iteration	74	45	47	47	51	49
	Volume fraction [%]	30.0	30.0	30.1	30.0	30.0	30.0
AWB	Deformation energy [mJ]	5.7	4.9	3.9	2.7	5.5	4.7
	Norm value [-]	3.9	3.4	2.7	2.0	3.9	3.2
	Iteration	35	42	33	31	57	33
	Volume fraction [%]	51.7	51.3	40.3	78.4	58.6	58.5

In the second variant, the radius was not considered to be dependent on the element size and was assigned a constant value of $R = 4$ mm. This radius was defined as a 1.5 multiple of the element size from the coarse tetrahedral mesh. Results displayed in Figure 11 and Table 4 demonstrate the independence of the results on the mesh (results have a similar shape and value of the objective function). In this variant, TET meshes performed poorly, especially in combination with the Sensitivity filter, which acted unpredictably at best. A coarser TET mesh could result in an acceptable shape (i.e., a shape similar to that derived using the HEX mesh) but finer TET meshes did not lead to volume reduction, but rather to its increase. In addition, the combination of TET mesh with Sensitivity filter often resulted in premature termination of the computation.

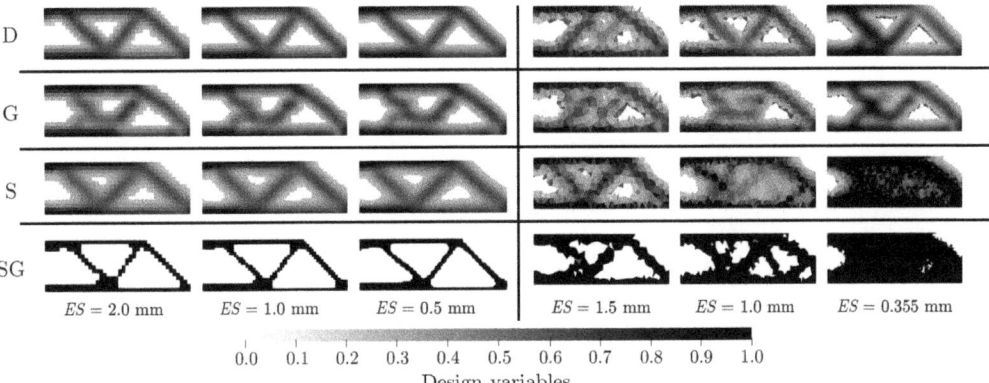

Figure 11. The final shape for linear HEX elements (**left**) and linear TET elements (**right**) with various filters, radius $R = 4$ mm.

Table 4. Results of optimization for various filters for each mesh, independent radius, filter radius $R = 4$ mm.

		HEX Elements			TET Elements		
Filter	Value	2.0 mm	1.0 mm	0.5 mm	1.5 mm	1.0 mm	0.355 mm
S	Deformation energy [mJ]	9.9	9.8	9.6	7.4	8.2	5.1
	Norm value [-]	6.9	6.7	6.5	5.3	5.8	3.5
	Iteration	101	117	135	54	50	24
	Volume fraction [%]	84.9	84.5	84.3	82.8	92.1	92.7
G	Deformation energy [mJ]	11.1	10.9	10.6	9.6	10.4	9.6
	Norm value [-]	7.7	7.4	7.2	6.9	7.4	6.6
	Iteration	41	35	39	36	34	36
	Volume fraction [%]	78.8	79.8	79.6	80.4	80.2	80.3
D	Deformation energy [mJ]	9.7	9.6	9.5	8.7	9.0	9.7
	Norm value [-]	6.7	6.6	6.44	6.3	6.3	6.7
	Iteration	167	158	200	200	200	200
	Volume fraction [%]	78.2	76.9	79.6	77.2	75.7	77.5
SG	Deformation energy [mJ]	3.9	3.8	3.7	3.7	4.1	3.0
	Norm value [-]	2.7	2.6	2.5	2.7	2.9	2.0
	Iteration	47	48	38	34	37	50
	Volume fraction [%]	30	30	29.2	30.0	30.0	30.0

It should be noted that unwanted effects, such as shear locking, can occur while solving the static structural analysis and cause an increase in the relative error of the objective function. In case of the radius of $R = 4$ mm, the deformation energy of the Density filter was slightly higher for HEX elements than for TET elements. Hence, relative errors were small enough to justify acceptation of the results.

Our results indicate that all filters discussed in this paper are suitable for use with a uniform mesh. The recommended filter combines the Density or Sensitivity filter with the

Greyscale filter to ensure a low number of iterations. In the case of a fine mesh, the Density filter without the Greyscale filter reached a maximal number of iterations while with the Greyscale filter, only 34 iterations were needed. The deformation energies of both results were similar. In the case of a non-uniform mesh, only the Density and Greyscale filters can be used effectively.

One could argue that the combination of the Sensitivity and Greyscale filters got us a perfect black and white design with the required volume fraction (volume reduction of 70%) while increasing the compliance of the structure only approximately 2.1 times; however, the shape was likely to be prone to buckling since compared to other results, the parts were thin.

3.3. Formulation of Penalization

Penalization is the parameter of the SIMP material model. Its correct choice is crucial as incorrect penalization would invalidate the model. For the Poisson ratio of $\mu = \frac{1}{3}$, it is recommended to use the penalization of $p = 3$ and higher as mentioned above.

Figure 12 demonstrates that the penalizations $p = 1$ and $p = 2$ should not be used since the shapes are not fully optimized (due to the premature termination of optimization). Literature, however, does not set the upper limit of this inequality. The values of deformation energy grew more or less predictably up to a penalization value of $p > 6$, see the values in Table 5. With high penalization values, such as $p > 6$, it was, nevertheless, clear that the results were becoming mesh-dependent. This could be caused by the convergence of the solution to a local minimum rather than a global minimum.

One could choose the continuation method with a gradual increase in the penalization value. This approach should help in acquisition of a reasonable solution. It should be, however, mentioned that the continuation method might not be capable of yielding a true "black and white" design, as reported by Stolpe and Svanberg [41]. This paper does not fully study this strategy; the risks are that a too fast increase of the penalization could lead to numerical difficulties that only increase the number of iterations.

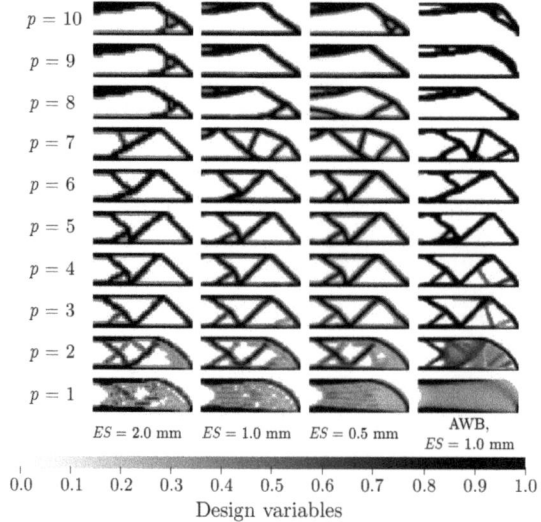

Figure 12. The final shapes for different values of penalization p, Density filter, radius of filter $R = 2$ mm.

Table 5. Values of objection function, deformation energy c [mJ] for each values of penalization.

Penalization	ES = 2.0 mm	ES = 1.0 mm	ES = 0.5 mm	AWB, ES = 1.0 mm
1	3.02	2.90	2.85	2.90
2	4.68	4.78	4.76	4.23
3	5.54	5.68	5.66	4.96
4	6.78	6.88	6.90	5.44
5	8.69	8.97	8.66	6.02
6	9.40	11.02	10.31	6.66
7	10.18	14.16	14.39	7.82
8	24.27	27.67	36.70	11.91
9	30.66	26.97	24.55	21.58
10	34.16	29.08	29.98	62.78

3.4. Formulation of Element Approximation

During the mesh preparation, it is necessary to choose an element approximation (usually linear or quadratic displacement approximation). Literature suggests that problems such as the checkerboard patterns should be less common with quadratic elements [26]. The advantage is that quadratic elements are less stiff than linear ones. However, the need to solve a larger number of unknowns (DOF) is a considerable disadvantage of this approach. Table 6 lists the mesh statistics.

Table 6. Finite element mesh statistics for linear and quadratic elements.

		Linear Elements	Quadratic Elements
HEX	Number of nodes	6405	23,930
	Number of elements	4800	4800
	Element size [mm]	1.0	1.0
TET	Number of nodes	1391	8269
	Number of elements	4351	4351
	Element size [mm]	1.0	1.0

In our case, the optimization settings were slightly altered. Only the Density filter with a radius of $R = 1.5\ ES$ for HEX elements and $R = 3.0\ ES$ for TET. Element size of $ES = 1.0$ mm was used. The maximal number of iterations was set to 100.

Figure 13 shows that the previous statement about the checkerboard pattern being less common with quadratic elements is only partially true. In the case of quadratic HEX elements, a checkerboard pattern was still present in the front part of the structure (although less than when linear elements were used). In the case of TET elements, not even higher order approximations did reduce the checkerboard pattern. Hence, filtering is highly recommended regardless of whether quadratic or linear elements are used. Table 7 obviates that the linear elements are stiffer, i.e., that they provide lower values of the objective function. However, the time needed to solve the problem with quadratic elements is up to thirty times longer than with linear elements.

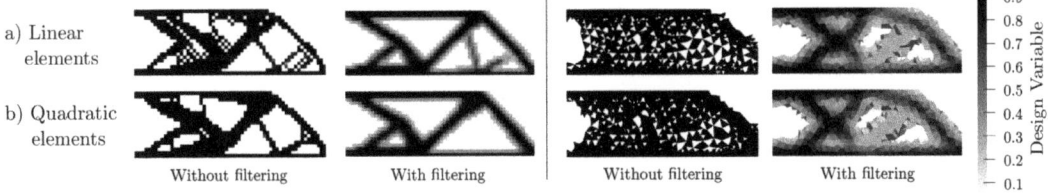

Figure 13. The final shapes for evaluation of element approximations of HEX elements (**left**) and TET elements (**right**); Density filter, radius $R = 1.5\ Es$ for HEX and radius $R = 3.0\ Es$ for TET.

Table 7. Results of optimization for linear elements and quadratic elements.

		Linear Elements		Quadratic Elements	
		With Filter	Without Filter	With Filter	Without Filter
HEX elements	Deformation energy [-]	4.6	3.63	4.79	3.73
	Norm value [-]	3.16	2.49	3.22	2.51
	Iteration	100	33	100	57
	Volume fraction [%]	58.1	30.2	56.4	30.1
	Solving time [s]	140	51.7	4820	3260
TET elements	Deformation energy [-]	6.71	2.83	10.07	4.64
	Norm value [-]	4.74	2.00	4.91	2.27
	Iteration	100	24	100	58
	Volume fraction [%]	73.6	30.1	73.9	30.1
	Solving time [s]	50	17.1	118.9	77.1

3.5. Formulation of Type of Weight Factor

In the theoretical part, two types of weight factors are mentioned—one characterized by a linear function, the other by the Gauss function. Nevertheless, according to the literature, there is no evidence that the Gauss function offers any advantages over the linear one [8].

In the investigation of this formulation, only two meshes were tested. Both meshes used linear elements with element sizes of $ES = 1.0$ mm and $ES = 0.5$ mm. Two radii were chosen as $R = 1.5\ ES$ and $R = 2.0\ ES$, respectively. The maximal number of iterations was 100.

The values of the objective function and volume fraction detailed in Table 8, as well as results shown in Figure 14, indicate a significant resemblance between values acquired using Gaussian and linear weight factors. In the case of a normal mesh ($ES = 1.0$ mm), setting the radius of the linear function to $R = 1.5\ Es$ provided similar results as in the case of the Gauss function with a radius of $R = 2.0\ ES$ (see the highlighted values in Table 8). In the fine mesh ($ES = 0.5$ mm), the similarities were more pronounced than in the normal mesh. This means that the Gauss function does not offer any significant advantage over the linear function as their results are very similar and shapes similar to those produced by the Gauss function can be obtained by simply changing the radius in the linear function/solution. As the preparation of mathematical apparatus is simpler with linear function, we decided to prefer linear solution over the one with the Gauss function.

Table 8. Results of optimization for different weight factors.

		ES = 1.0 mm		ES = 0.5 mm	
		$R = 1.5\ ES$	$R = 2.0\ ES$	$R = 1.5\ ES$	$R = 2.0\ ES$
Linear function	Deformation energy [mJ]	4.61	5.26	3.74	4.02
	Norm value [-]	3.16	3.61	2.53	2.72
	Iteration	100	100	100	100
	Volume fraction [%]	**58.15**	56.2	50.2	50.3
Gauss function	Deformation energy [mJ]	4.11	**4.67**	3.61	3.91
	Norm value [-]	2.82	**3.2**	2.44	2.65
	Iteration	100	100	100	100
	Volume fraction [%]	59.25	**59.15**	50.8	50.0

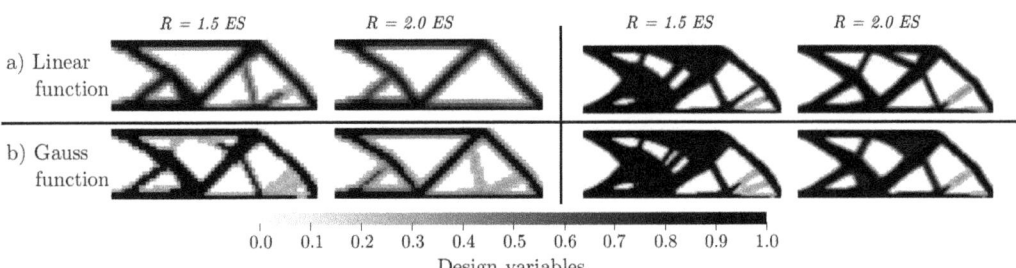

Figure 14. The final shape with different weight functions, mesh with element size $ES = 1.0$ mm (**left**), mesh with element size $ES = 0.5$ mm (**right**), Density filter.

4. Discussion

To fully understand optimization, one should possess advanced experience in math (namely, calculus and finite element method) and computer science (for example, scripting for iteration solvers). One should also be capable of preparing a correct mathematical formulation of the problem, i.e., of determining the objective function of the problem, constraints, and design variables, and of finding equivalent problems, potentially offering an easier solution than the original one.

Five key formulations were analyzed in this paper. The results make it clear that each formulation affects the resulting shape, number of iterations, volume of the part, etc.

The formulation of the filter radius is crucial for determining the element neighborhood. Choosing too small radius might lead to the deactivation of the filter. Choosing a too large radius leads to dissipation of energy in large areas, resulting in too much "grey". Choosing the filter itself can be difficult since the theory does not provide enough knowledge from the application point of view. The SIMP model was used as a simple material model in this study. The penalty value is a key parameter of this model. However, the theory provides only the bottom boundary but does not inform about the upper one. Failure to limit the upper boundary could lead to invalid results that would be highly mesh-dependent. The maximum reasonable penalty value for the steel cantilever in our study was $p = 6$.

The theory also recommends a quadratic approximation of the displacement of the elements. However, it does not provide clear reasons behind this recommendation. Choosing the right approximation might lead to a faster calculation. Lastly, neither the theory nor practice (as demonstrated by our calculations) provide enough evidence or reason for using the Gauss function as the weight function.

It should be noted that there are additional formulations affecting the final shape. For example, various algorithms, described by Zuo et al., can be used in topology optimization [42]. Another approach to this problem could lie in reducing the computing efforts, as reported by Amir et al. [40]. For larger problems, it would be better to prepare a better displacement solver (for example, a parallel displacement solver as suggested, e.g., by Makropoulos et al. [43]. In our paper, however, we did not study other types of structural analysis, such as the heat transfer of flow optimization.

The algorithm used in our scripts, the Optimality Criterion (OC), was designed only for minimizing the compliance and volume constraint. The advantages of the OC algorithm include its simplicity and rapid updating of the design variables. The disadvantage is that it is only capable of solving the minimizing compliance problem (in the current form, it cannot solve the maximizing natural frequency).

Construction of the matrix of the weight factor might be tricky. Open-source scripts usually do not support importing meshes and use a rather simple geometry. Thus, an effective script for importing meshes usually needs to be prepared. The modification of the weight factor calculation allowed us, due to the symmetry of the matrix, to solve only its upper triangle, which halved the calculating time. Another possible modification would lie in splitting the part into sub-regions, which could be calculated in parallel. Of course, the latter approach would come with its own limitations; overlapping would be necessary to be able to combine the individual regions back into the full structure, and, therefore, therefore, already solved regions would have to be calculated again.

The design performance has many criteria such as stiffness, overall solving duration, post-processing, manufacturing, etc. Each formulation has its effect on the design performance. It is difficult to pinpoint settings that are optimal for the design performance in each formulation. However, it is easy to recommend, which settings and parameter values should be avoided. It should be noted that the steps following the optimization (in particular, smoothing) might heavily affect the design performance.

It should be also noted that complex problems might need their own sensitivity analysis. However, this paper should help with the initial estimates. After finding the optimal formulations and their parameters, one could prepare scripts for automated designing of customized structures such as prosthetic aids [14,44] or scripts for automated designing of mountain climbing equipment [45].

5. Conclusions

This paper focuses on the sensitivity analysis of key topology optimization formulations. The novelty of this research comes from the presented results, which might be used in the preparation of custom scripts solving the topology optimization. The solutions were tested on various meshes with various types of elements. The paper contains important theoretical background for the problem of minimizing compliance. To have freedom in choosing such formulations, the authors prepared a MATLAB procedure solving such optimization. The prepared program allows users to import the mesh and boundary conditions. Scripts constructed within this study provided results comparable to the open-source top3d script [9]. The presented paper also includes recommendations on how to choose the parameters of topology optimization.

It is clear that uniform meshes perform generally better in this formulation; this was particularly true during optimization as it allowed the application of multiple filters.

Radius is an important part of the filtering method and correct results depend on the appropriate selection of its value. Too small a radius could possibly lead to difficult manufacturing (even if using additive manufacturing). Using a large radius could produce non-optimized shapes with grey areas. If the mesh is being refined during optimization, it is recommended to use the same or similar radius as in the previous step (coarser mesh).

The combination of the Density and Greyscale filters performed better than the Density filter alone as it yielded similar or even identical values of the objective function as the Density filter in fewer iterations. The combination also performed well on non-uniform meshes. It is obvious that the combined filter still left some grey areas but the success of this filtering was still the best of all tested filters. If the design variable in these grey areas is $x_e < 0.3$, they can be removed when using penalization of $p \geq 3$ since their contribution to stiffness is negligible.

The use of a Gaussian weight factor did not bring any advantage over the linear function as the results calculated using both functions were very similar. As constructing the matrix for the Gaussian weight factor is more difficult, a linear function is more suitable for these purposes.

In this paper, the authors used in most cases linear elements, which led to several conclusions: (i) The usage of linear tetrahedral elements is not recommended in any case. They are too stiff due to the locking effects, which greatly affects the value of the objective function. The only advantage lies in the fast calculation of displacement because the tetrahedral mesh usually has fewer nodes. However, a fine mesh would be needed to get reasonable results, which negates this only advantage. (ii) The uniform mesh provides acceptable results even if the linear approximation is used. For that reason, it is recommended to use the "Cartesian mesher" which provides uniform meshes even for complex geometries. A small disadvantage is represented by the differences in the shape of the resulting structure depending on the method of mesh creation. (iii) Quadratic elements might be less stiff but solving them would be time-consuming. If one would like to use only quadratic elements, it would be recommended to spend time preparing a better solver of linear equation systems (for example, a parallel solver).

The authors recommend performing a sensitivity analysis of the key formulations presented in the paper for each problem, regardless of whether or not the designer has previous experience with similar problems. Without the suggested analysis, more doubts arise and the creation of "false-optimal" shapes is not prevented.

Author Contributions: Conceptualization, M.S. and P.M.; methodology, M.S.; software, M.S.; validation, M.S., D.R. and P.M.; formal analysis, M.S. and D.R. ; investigation, M.S.; resources, P.M.; data curation, M.S.; writing—original draft preparation, M.S.; writing—review and editing, M.S., P.M., M.F. and D.G.; visualization, M.S.; supervision, P.M.; project administration, P.M.; funding acquisition, P.M., M.F. and D.G. All authors have read and agreed to the published version of the manuscript.

Funding: This work was supported by The Ministry of Education, Youth and Sports from the Specific Research Project SP2021/66, by The Technology Agency of the Czech Republic in the frame of the project TN01000024 National Competence Center-Cybernetics and Artificial Intelligence and by Structural Funds of the European Union within the project Innovative and additive manufacturing technology—new technological solutions for 3D printing of metals and composite materials, reg. no. CZ.02.1.01/0.0/0.0/17_049/0008407 and by the European Regional Development Fund under Grant No.CZ.02.1.01/0.0/0.0/15_003/0000493 (Centre of Excellence for Nonlinear Dynamic Behaviour of Advanced Materials in Engineering) with institutional support RVO:61388998.

Institutional Review Board Statement: Not applicable.

Informed Consent Statement: Not applicable.

Data Availability Statement: Data sharing not applicable.

Conflicts of Interest: The authors declare no conflict of interest.

Abbreviations

The following abbreviations are used in this manuscript:

3D	Three Dimensional
2D	Two Dimensional
AWB	ANSYS Workbench
KKT	Karush-Kuhn-Tucker
SIMP	Solid Isotropic Material with Penalty
TET	Tetrahedral
HEX	Hexahedral
DOF	Degrees of Freedom
OC	Optimality Criterion
SAO	Sequential Approximate Optimization
1D	One Dimensional
ES	Element Size
NF	No Filter
D	Density filter
S	Sensitivity filter
G	Gray scale filter
SG	Combination of Sensitivity and Gray scale Filters

References

1. COMSOL Multiphysics Modeling Software. 2020. Available online: https://www.comsol.com/optimization-module (accessed on 3 March 2021).
2. Hecht, F. New development in FreeFem++. *J. Numer. Math.* **2012**, *20*, 251–265. [CrossRef]
3. A FreeFem++ Toolbox for Shape Optimization (Geometry and Topology). 2008. Available online: http://www.cmap.polytechnique.fr/~allaire/freefem_en.html (accessed on 3 March 2021).

4. Hunter, W. ToPy—Topology Optimization with Python. 2017. Available online: https://github.com/williamhunter/topy (accessed on 3 March 2021).
5. Bendsøe, M.P.; Sigmund, O. *Topology Optimization*, 2nd ed.; Corrected Printing ed.; Springer: Berlin, Germany, 2004.
6. Bendsøe, M.P.; Sigmund, O. Material interpolation schemes in topology optimization. *Arch. Appl. Mech. (Ingenieur. Archiv.)* **1999**, *69*, 635–654. [CrossRef]
7. Sigmund, O. A 99 line topology optimization code written in Matlab. *Struct. Multidiscip. Optim.* **2001**, *21*, 120–127. [CrossRef]
8. Sigmund, O. Morphology-based black and white filters for topology optimization. *Struct. Multidiscip. Optim.* **2007**, *33*, 401–424. [CrossRef]
9. Liu, K.; Tovar, A. An efficient 3D topology optimization code written in Matlab. *Struct. Multidiscip. Optim.* **2014**, *50*, 1175–1196. [CrossRef]
10. Hunter, W. Predominantly Solid-Void Three-Dimensional Topology Optimisation Using Open Source Software. Master's Thesis, University of Stellenbosch, Stellenbosch, South Africa, 2009.
11. Michell, A. LVIII. The limits of economy of material in frame-structures. *Lond. Edinb. Dublin Philos. Mag. J. Sci.* **1904**, *8*, 589–597. [CrossRef]
12. Hunar, M.; Jancar, L.; Krzikalla, D.; Kaprinay, D.; Srnicek, D. Comprehensive View on Racing Car Upright Design and Manufacturing. *Symmetry* **2020**, *12*, 1020. [CrossRef]
13. Pagac, M.; Hajnys, J.; Halama, R.; Aldabash, T.; Mesicek, J.; Jancar, L.; Jansa, J. Prediction of Model Distortion by FEM in 3D Printing via the Selective Laser Melting of Stainless Steel AISI 316L. *Appl. Sci.* **2021**, *11*, 1656. [CrossRef]
14. Sotola, M.; Stareczek, D.; Rybansky, D.; Prokop, J.; Marsalek, P. New Design Procedure of Transtibial Prosthesis Bed Stump Using Topological Optimization Method. *Symmetry* **2020**, *12*, 1837. [CrossRef]
15. Sotola, M. Study of Topological Optimization. Master's Thesis, VSB-TUO, Ostrava, Czech Republic, 2020.
16. Hughes, T.J.R. *The Finite Element Method*, 1st ed.; Prentice Hall: Hoboken, NJ, USA, 1987.
17. Ravindran, A.; Ragsdell, K.; Reklaitis, G. *Engineering Optimization: Methods and Applications*, 2nd ed.; John Wiley and Sons: Hoboken, NJ, USA, 2007. [CrossRef]
18. Allaire, G. *Shape Optimization by the Homogenization Method*, 1st ed.; Springer: New York, NY, USA, 2002.
19. Allaire, G.; Bonnetier, E.; Francfort, G.; Jouve, F. Shape optimization by the homogenization method. *Numer. Math.* **1997**, *76*, 27–68. [CrossRef]
20. Bourdin, B.; Chambolle, A. Design-dependent loads in topology optimization. *ESAIM Control. Optim. Calc. Var.* **2003**, *9*, 19–48. [CrossRef]
21. Auricchio, F.; Bonetti, E.; Carraturo, M.; Hömberg, D.; Reali, A.; Rocca, E. A phase-field-based graded-material topology optimization with stress constraint. *Math. Model. Methods Appl. Sci.* **2020**, *30*, 1461–1483. [CrossRef]
22. Burger, M. A framework for the construction of level set methods for shape optimization and reconstruction. *Interfaces Free Boundaries* **2003**, *5*, 301–329. [CrossRef]
23. Dostal, Z. *Optimal Quadratic Programming Algorithms*, 1st ed.; Springer: New York, NY, USA, 2009; [CrossRef]
24. Bendsøe, M.P.; Díaz, A.; Kikuchi, N. Topology and Generalized Layout Optimization of Elastic Structures. *Topol. Des. Struct.* **1993**, *227*, 159–205. [CrossRef]
25. Díaz, A.; Sigmund, O. Checkerboard patterns in layout optimization. *Struct. Optim.* **1995**, *10*, 40–45. [CrossRef]
26. Sigmund, O.; Petersson, J. Numerical instabilities in topology optimization: A survey on procedures dealing with checkerboards, mesh-dependencies and local minima. *Struct. Optim.* **1998**, *16*, 68–75. [CrossRef]
27. Rozvany, G.I.N. On symmetry and non-uniqueness in exact topology optimization. *Struct. Multidiscip. Optim.* **2011**, *43*, 297–317. [CrossRef]
28. Bathe, K.J. *Finite Element Procedures*, 1st ed.; Prentice Hall: Hoboken, NJ, USA, 2006.
29. Bhatti, M.A. *Fundamental Finite Element Analysis and Applications*; John Wiley: Hoboken, NJ, USA, 2005.
30. Bhatti, M.A. *Advanced Topics in Finite Element Analysis of Structures*, 2nd ed.; John Wiley: Hoboken, NJ, USA, 2006.
31. Ansys Workbench Help. 2020. Available online: https://ansyshelp.ansys.com (accessed on 10 January 2021).
32. Munro, D.P. A Direct Approach to Structural Topology Optimization. Ph.D. Thesis, University of Stellenbosch, Stellenbosch, South Africa, 2017.
33. Wang, M.Y.; Wang, S. Bilateral filtering for structural topology optimization. *Int. J. Numer. Methods Eng.* **2005**, *63*, 1911–1938. [CrossRef]
34. Bruns, T.E.; Tortorelli, D.A. An element removal and reintroduction strategy for the topology optimization of structures and compliant mechanisms. *Int. J. Numer. Methods Eng.* **2003**, *57*, 1413–1430. [CrossRef]
35. Andreassen, E.; Clausen, A.; Schevenels, M.; Lazarov, B.S.; Sigmund, O. Efficient topology optimization in MATLAB using 88 lines of code. *Struct. Multidiscip. Optim.* **2011**, *43*, 1–16. [CrossRef]
36. Bourdin, B. Filters in topology optimization. *Int. J. Numer. Methods Eng.* **2001**, *50*, 2143–2158. [CrossRef]
37. Bruns, T.E.; Tortorelli, D.A. Topology optimization of non-linear elastic structures and compliant mechanisms. *Comput. Methods Appl. Mech. Eng.* **2001**, *190*, 3443–3459. [CrossRef]
38. Groenwold, A.A.; Etman, L.F.P. A simple heuristic for gray-scale suppression in optimality criterion-based topology optimization. *Struct. Multidiscip. Optim.* **2009**, *39*, 217–225. [CrossRef]
39. Golub, G.H.; Loan, C.F.V. *Matrix Computations*, 3rd ed.; Johns Hopkins University Press: Baltimore, MD, USA, 1996.

40. Amir, O.; Sigmund, O. On reducing computational effort in topology optimization. *Struct. Multidiscip. Optim.* **2011**, *44*, 25–29. [CrossRef]
41. Stolpe, M.; Svanberg, K. On the trajectories of penalization methods for topology optimization. *Struct. Multidiscip. Optim.* **2001**, *21*, 128–139. [CrossRef]
42. Zuo, K.T.; Chen, L.P.; Zhang, Y.Q.; Yang, J. Study of key algorithms in topology optimization. *Int. J. Adv. Manuf. Technol.* **2007**, *32*, 787–796. [CrossRef]
43. Markopoulos, A.; Hapla, V.; Cermak, M.; Fusek, M. Massively parallel solution of elastoplasticity problems with tens of millions of unknowns using PermonCube and FLLOP packages. *Appl. Math. Comput.* **2015**, *267*, 698–710. [CrossRef]
44. Marsalek, P.; Sotola, M.; Rybansky, D.; Repa, V.; Halama, R.; Fusek, M.; Prokop, J. Modeling and Testing of Flexible Structures with Selected Planar Patterns Used in Biomedical Applications. *Materials* **2021**, *14*, 140. [CrossRef]
45. Rybansky, D.; Sotola, M.; Marsalek, P.; Poruba, Z.; Fusek, M. Study of Optimal Cam Design of Dual-Axle Spring-Loaded Camming Device. *Materials* **2021**, *14*, 1940. [CrossRef]

MDPI
St. Alban-Anlage 66
4052 Basel
Switzerland
Tel. +41 61 683 77 34
Fax +41 61 302 89 18
www.mdpi.com

Symmetry Editorial Office
E-mail: symmetry@mdpi.com
www.mdpi.com/journal/symmetry

www.ingramcontent.com/pod-product-compliance
Lightning Source LLC
LaVergne TN
LVHW070451100526
838202LV00014B/1705